FINITE MATHEMATICS
FROM SETS TO GAME THEORY

ADELBERT F. HACKERT
SACRAMENTO CITY COLLEGE

FINITE MATHEMATICS

FROM SETS TO GAME THEORY

D. C. HEATH AND COMPANY
Lexington, Massachusetts Toronto London

Published simultaneously in Canada.

Printed in the United States of America.

International Standard Book Number: 0-669-81125-4

Library of Congress Catalog Card Number: 73-6084

PREFACE

Much of what is termed "modern mathematics" is not really very new. What is new is a change of emphasis in the way mathematics is being taught and the intrusion of mathematical ideas into the social and behavioral sciences, fields which only fifty years ago were thought to be beyond mathematical analysis.

This book was written with two purposes in mind, to bring the reader up to date on some of the modern pedagogical ideas and to broaden his appreciation of the role of mathematics in today's world.

The book can be used in a one-semester course by either liberal arts students or prospective teachers of elementary mathematics. Because such students have varied mathematical backgrounds, little mathematics beyond arithmetic skills is expected. Certain of the topics of Chapters 5 through 10 do require an acquaintance with negative numbers, and a few require the ability to solve an equation in one variable. For this reason Appendix A offers a quick review of these algebraic concepts.

The number of topics included in this book is too great to be covered in a one-semester course. This was done intentionally so that the instructor may choose from the topics those which best suit the special needs of his students.

It is recommended that Chapter 1 be covered leisurely and in a most informal manner, in keeping with the theme of the chapter, "Mathematics can be enjoyable." It should be emphasized that the problems in this chapter are merely puzzles and that obtaining their solutions is not a life-or-death proposition. In fact, no insistence on correct solutions should be made at this time. To this end, the answers to these problems are not included in the answer section at the end of the book. (They are included, for the convenience of the instructor only, in the Instructor's Manual.)

Chapters 1, 2, and 3 are a common core of subject material for the book. Most of the remaining chapters are independent of each other except that Sections 8.3 to 8.5 are clearer if Chapter 7 is covered first, and some topics in Chapter 10 depend upon ideas discussed in Chapters 6 and 7.

Starred sections and problems require the algebraic background discussed above. These are optional and may be omitted without loss of continuity.

The following outlines suggest ways in which the topics may be selected to serve different needs.

Suggested Course for Prospective Teachers:

Chapters 1 to 3
Chapter 4, Sections 4.1 to 4.3
Chapter 5
Chapter 7, Sections 7.1 and 7.2
Chapter 8
and
The remaining sections of Chapter 4 or the remaining sections of Chapter 7 or one or more of Chapters 6, 9, and 10 as time permits

Suggested Course for Liberal Arts Students:

Chapters 1 to 3
Chapter 4, Sections 4.1 to 4.3
and

Chapter 6, Sections 6.1 to 6.4 Chapter 7 Chapter 10, Sections 10.1 to 10.4	**or**	Chapter 6 Chapter 7 Chapter 9	**or**	Any combination of Chapters 5 to 10 keeping in mind that Chapters 6 and 7 are prerequisites for Chapter 10

CONTENTS

1 THE LIGHTER SIDE OF MATHEMATICS

About the time that the writer was considering how best to introduce this chapter, he was confronted by a student who said, "I'm going to be in your class next semester and I think I ought to warn you, I hate mathematics!"

Such declarations are not uncommon; many people fear and dislike mathematics for a variety of reasons. One reason that is often given is that "mathematics is dull." Some subjects, literature for instance, provide drama, suspense, and even romance. Others, such as biology and chemistry, get right down to the nitty gritty of life itself. If there is any romance in mathematics, it is not to be found in adding a collection of figures. There is very little increase in the flow of adrenalin caused by the imminence of the answer to a long division problem. What is the social or ecological significance of the product of two fractions? Yes, as it is traditionally presented in school, mathematics has little to offer in the way of excitement. A certain number of facts must be learned and there are a few skills which must be mastered. Both textbook and teacher set out to drum in the facts and drill in the skills in a no nonsense, "you will find out where this can be used when the time is right" manner. This is unfortunate, for mathematics is really not at all dull. One does not even have to be particularly good at it to derive enjoyment from it.

In this chapter, for the benefit of those who have not yet discovered it, we open the door to this other, lighter, side of mathematics. We do this by means of a number of mathematical games and puzzles. Those in the first section are strictly for fun. The first five are dedicated to the proposition that figures seldom lie but . . .

1.1 LIARS SOMETIMES FIGURE

EXERCISE 1.1

As I was going to St. Ives
I met a man with seven wives;
Every wife had seven sacks;
Every sack had seven cats;
Every cat had seven kits.
Kits, cats, sacks, and wives,
How many were going to St. Ives?

Of course you have heard that one before, but did you know that one version of it was found on an Egyptian papyrus dating from about 1600 B.C.?

Here is another classic of a more recent vintage.

EXERCISE 1.2 Three men go to a hotel and rent a room for \$30. After they are taken to their room, the hotel clerk discovers that he has overcharged them; the room rents for only \$25. He thereupon sends a bellhop upstairs with the \$5 change. The bellhop decides to return only \$3 and keep the other \$2 for his trouble. Now the room originally cost \$30, but the men had \$3 returned to them. This means that they paid only \$27 for the room. The bellhop kept \$2. \$27 + \$2 = \$29. What happened to the extra dollar?

EXERCISE 1.3 I know two men, Mr. Smith and Mr. Jones, who are not related but both have the same sister. How can that be?

EXERCISE 1.4 Many problems involve the arrangement of numbers. For example, the digits 0, 1, 2, 3, 4, 5, 6, 7, 8, 9 are arranged from smallest to largest. What is the rule used in arranging the sequence

8, 5, 4, 9, 1, 7, 6, 3, 2, 0

EXERCISE 1.5 There are 12 dollars in a dozen. How many dimes are there in a dozen?

EXERCISE 1.6 A boy has a strip of paper 1 inch wide and 24 inches long. It is marked at 1-inch intervals so that he can cut off a series of squares. If the boy takes 1 second for each cut, how long will it take to cut off the 24 squares?

EXERCISE 1.7 Two men play a card game, and the stake is \$1 a game. At the end one has won 5 games and the other has won \$5. How many games did they play?

EXERCISE 1.8 In my town there is a barber who shaves everyone who does not shave himself. Who shaves the barber?

EXERCISE 1.9 Can you connect these nine dots with exactly four straight lines without lifting your pencil from the paper once you start? Two lines may cross and a dot may be on more than one line.

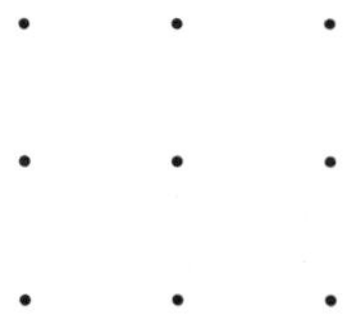

EXERCISE 1.10 I know a man who insists that $5 + 7 = 14$. Could he be right?

Some problems involve a bit of arithmetic. You should find these examples challenging.

EXERCISE 1.11 Write the numbers from 1 to 10 using four 4's for each. For example,

$$1 = \frac{44}{44} \qquad 2 = \frac{4 \times 4}{4 + 4} \qquad 3 = 4 - \left(\frac{4}{4}\right)^4$$

EXERCISE 1.12 A college student sent this telegram to his father:

SEND + MORE = MONEY

Can you replace each of the letters with a different digit to make the message "add up"?

Exercise 1.13 Consider the four cards

1	9
3	11
7	13
5	15

2	10
3	11
6	14
7	15

4	12
5	13
6	14
7	15

8	12
9	13
10	14
11	15

Ask someone to pick a number from 1 to 15 and tell you on which of the cards that number appears. The sum of the numbers on the top left corners of the cards so identified gives you the number he selected.

EXERCISE 1.14 Three thieves have a monkey for a mascot. One day they steal some bananas and take them to their hideout. They decide to get a good night's sleep before dividing their loot. In the middle of the night one thief awakes and thinks, "I do not trust these two fellows, I'd better make sure I get my share." He divides the bananas into three equal parts (there is one banana left over which he throws to the monkey), hides one share, and goes back to sleep. Shortly after this a second thief awakes and has the same thought as the first one. He divides the remaining bananas into three parts (there is one banana left over which he throws to the monkey), hides one of the parts, and gets back into bed. Still later the third thief wakes up and divides the bananas into three parts (there is one banana left over which he throws to the monkey), hides one part, and returns to his bed. In the morning the thieves wake up and divide the remaining bananas into three parts (there is one left over which they throw to the monkey). Each thief takes one part and each goes his own way. What is the least number of bananas they could have stolen?

EXERCISE 1.15 The first astronaut to land on Mars returned with this strange tale. "There are two types of people there. The A's always lie and the B's always tell the truth. I met three inhabitants one day as I was exploring. I asked one of them 'Are you an A or a B?' The Martian answered but there was a burst of static on my communicator which drowned out his reply, and so I asked again, 'Are you an A or a B?' This time a different Martian answered, 'He said that he was a B. He is and so am I.' The third Martian then spoke up, 'That is not true. The first man is an A and I am a B.' From what I had heard I was able to tell who was lying and who was telling the truth." Can you do the same?

1.2 WHAT SHOULD YOU DO?

The exercises of the preceding section were strictly for fun. Those in this section are no less entertaining, but they represent the type of mathematical thinking that can be used to solve practical problems. Do not be discouraged if you cannot solve many of them at this time. All of them are based on mathematical concepts to be discussed at length in the remainder of this book; give them an honest try now, then return to them whenever you think that you have learned something new which might apply. The numbers in the parentheses following each problem number indicate the section of the book in which clues to the solutions may be found.

EXERCISE 1.16 (Section 2.3) Eight teams of the National League are scheduled to play next Saturday. They are Chicago, Cincinnati, Los Angeles, New York, Philadelphia, Pittsburgh, St. Louis, and San Francisco. Your boss calls you into his office and requests that you find how these teams will be paired off. He implies that he would like to have this information before he leaves the office 10 minutes from now. You race back to your desk, pull your transistor radio out of its hiding place, and turn it on just in time to hear sportscaster A predict that the winners will be San Francisco, Los Angeles, St. Louis, and Chicago. You change stations and hear sportscaster B say that the winners will be New York, St. Louis, Pittsburgh, and Los Angeles. In a panic now, you tune in sportscaster C but you are too late; all he tells you is that he thinks Chicago, Cincinnati, St. Louis, and Pittsburgh will be the victors. What can you tell your boss?

EXERCISE 1.17 (Section 2.5) You wish to buy a partnership in an automobile agency in a certain city. You may do so in either the Ford, the Chevrolet, or the Volkswagen agencies. You hire a man to do a market research. He makes a survey of 500 automobile owners in that city and finds that 285 people own Fords, 241 own Chevrolets, 233 own Volkswagens, 93 own both Fords and Chevrolets, 71 own both Chevrolets and Volkswagens, 105 own both Fords and Volkswagens, and 23 own all three makes. You should

(a) Buy into the Ford agency.
(b) Buy into the Chevrolet agency.
(c) Buy into the Volkswagen agency.
(d) Fire the man who made the survey.

EXERCISE 1.18 (Adapted from Lewis Carroll) (Section 3.7) What conclusion must you draw assuming these "facts"?

1. Nobody who really appreciates Beethoven fails to keep silence while the "Moonlight Sonata" is being played.
2. Guinea pigs are hopelessly ignorant of music.
3. No one who is hopelessly ignorant of music ever keeps silence while the "Moonlight Sonata" is being played.

EXERCISE 1.19 (Section 4.6) The board of governors of a certain corporation consists of a chairman, a vice chairman, and two lay members. The charter of the corporation requires that the board members' votes be weighted, with the chairman's vote counting three times that of a lay member and the vice chairman's vote counting twice as much as that of a lay member. Each board member must vote on all measures, and a measure passes provided it receives an affirmative vote count of four or more. You, an electrician, are hired to wire push buttons under the edge of the boardroom table near each member's position in such a way that a light will indicate the passage of a measure whenever a proper combination of buttons are depressed. How will the circuit look?

EXERCISE 1.20 (Section 5.6) You are studying in the library when you hear the fellow next to you mumbling "2 times 3 is 1, 3 times 3 is 4, but 2 plus 3 is 0." Should you correct him?

EXERCISE 1.21 (Section 6.3) You need a raise but there is no use asking for one. Your boss is a tight-fisted, penny-pinching skinflint who runs the office with an iron hand. Probably he would be a decent sort at work if he was not so hen-pecked at home. His wife treats him the same way he treats his office staff. He has a daughter who gets the same treatment from her mother that her father does. Last year the daughter ran off to Europe and created quite a stir when she returned home married to an impoverished Count. It must be true love though because the daughter leads her husband around by the nose. The mother-in-law is thrilled at having royalty in the family, and she thinks the Count's word is the last on any subject. The boss made his new son-in-law a vice president, but he cannot stand the man and will not listen to a word he has to say. He only puts up with the Count because he has never been able to say "no" to anything his daughter asked. Fortunately you are on good terms with everyone in the family except the boss. Which *one* of them should you speak to about your raise in order that the most pressure will be brought to bear on your boss?

EXERCISE 1.22 (Section 7.4) You have a ticket, the very last one available, to the last performance of the hit play of the year and are standing in line waiting for the theater to open when your best friend comes along. He says, "I have been trying to get a ticket for this show for weeks. I've offered to pay $20 for one but no one wants to sell me his. How about a deal? If you can find two people in the line ahead of you who have a birthday on the same day, you sell me your ticket; otherwise I'll give you the $20 and you keep your ticket." There are 45 people in line ahead of you and you know that your friend knows none of them. Your ticket cost you $7.50. You should:

(a) Take the bet. Anyone as dumb as your friend deserves to lose his money.

(b) Take the bet. You can give the money back to your friend after you win. Maybe it will teach him to be more careful how he bets after this.

(c) Get a new friend.

EXERCISE 1.23 (Section 8.3)

1

1 1

1 2 1

1 3 3 1

1 4 6 4 1

1 5 10 10 5 1

1 6 15 20 15 6 1

— — — — — — — —

What numerals should be placed in the blanks to maintain the pattern?

EXERCISE 1.24 (Section 9.4) You are a dietician. You must prepare a mixture of two foods *A* and *B*. Each unit of food *A* contains 10 grams of protein and 20 grams of carbohydrate and costs 20¢. Each unit of food *B* contains 15 grams of protein and 10 grams of carbohydrate and costs 25¢. What is the cheapest mixture that can be obtained if it must contain at least 400 grams of protein and 500 grams of carbohydrate?

EXERCISE 1.25 (Section 10.1) Two companies must build their next branch at one of three cities *A*, *B*, and *C*, which are located at the distances indicated in Figure 1–1. If both companies build in the same city they will split the business evenly. If, however, they build in different cities the company that is closer to a given city will get all that city's business. The potential market is the same in each city. You are hired by one of the companies to determine in which city it should build. You must make your decision without any knowledge of the location the other company will choose. What is your decision?

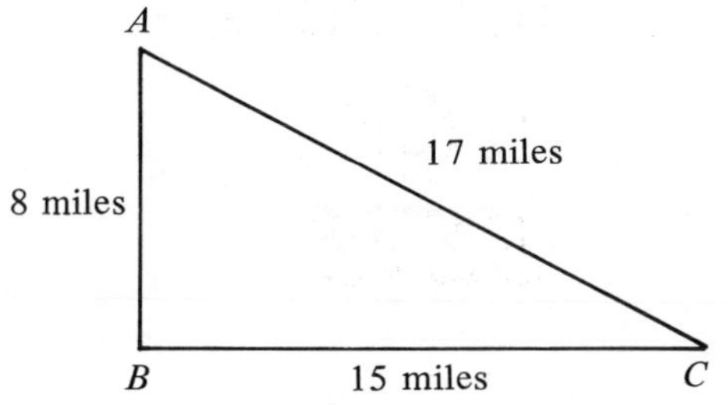

FIGURE 1–1

As you study the topics in the following chapters try to approach them in the same way as you did the problems in this chapter. If you can think of the problems as puzzles to be solved rather than as obstacles to climb over, half your battle will be won before you start.

2 THE ALGEBRA OF SETS

In this chapter we introduce set language and some ideas about sets. These will be used in following chapters to make reasonably precise presentations of linear algebra, probability, and other topics. An investigation of set ideas should help you gain a better understanding of some of the mathematics you already know, and it should also simplify your study of new mathematical ideas.

The use of the word *set*, meaning a collection of objects all with some common characteristic, is not confined to mathematics. We speak of sets of dishes, sets of encyclopedias, sets of tires, etc. Although mathematicians have used set properties since the dawn of civilization, the first formal study of the subject was made barely a century ago by the German mathematician Georg Cantor. Cantor's interest was in certain advanced topics, and to him set theory was merely a tool to use in his investigations. Within 20 years of Cantor's disclosure of his findings, other mathematicians had discovered that set theory is basic to nearly every branch of mathematics. Today sets and set theory are used at every level in which mathematics is taught, even in the primary school where sets are used to explain the principles of counting, addition, and subtraction.

2.1 SETS AND SET NOTATION

2.1.1 Set Notation

Since we must start somewhere, we assume that the meanings of *collection*, *object*, and *belong to* are the same for everyone.

SET A **set** is a collection of objects called **elements** of the set.

One necessary property of a set is that it must be well defined. That is, it must be possible to tell whether or not a given object is an element of the set.

EXAMPLE 2.1 These collections are well defined and so they are sets:

(a) Names of the days of the week
(b) Counting numbers larger than one
(c) Letters of the English alphabet which are neither vowels nor consonants

These collections are not sets:

(d) Older people (*Older than whom?*)
(e) Successful men (*Successful by whose standards?*)

A set is defined by stating which objects are elements of it. There are two ways of doing this. One is by rule as in Example 2.1 and the other is by listing the elements. In either case, the description of the set is usually included within braces, { }.

EXAMPLE 2.2 Use the listing method to specify:

(a) The set of counting numbers between 5 and 10
(b) The set of months of the year whose names start with J
(c) The set of counting numbers greater than 10
(d) The set of counting numbers between 1 and 100

Solutions

(a) {7, 8, 6, 9}
(b) {January, June, July}
(c) {11, 12, 13, 14, . . .}
(d) {2, 3, 4, . . . , 99}

When the listing method is used, it is customary to include each element of the set only once in the list. Sometimes, however, an element is listed more than once. If this is done, the repetitions are ignored when the list is read. Thus {1, 1, 1, 2, 2, 3, 3, 3, 3, 3} defines a set which contains only three elements, the numerals 1, 2, and 3.

Did you notice that the order in which the elements are listed is of no consequence? Perhaps you noticed also that whenever it is impossible or inconvenient to list all of the elements of a set, as was the case in parts (c) and (d) above, then a sufficient number of elements to establish a pattern are listed and three dots, . . ., are used to say "and so on."

We shall denote or name a set with an uppercase letter. When letters are used as elements of a set, they are lowercase letters unless the elements of the set are sets themselves. The symbol $\in$ means "is an element of," and the same symbol with a line through it, $\notin$, means "is not an element of."

EXAMPLE 2.3 Consider set A defined as $\{a, x, *, 8, r\}$. Then $a \in A$, $8 \in A$, but $R \notin A$.

2.1.2 Special Sets

EMPTY AND UNIVERSAL SETS There are two special sets which we need to consider. The first of these is suggested by part (c) of Example 2.1 and is a set which has no members. We call this the **empty** set and denote it $\varnothing$. The other special set is the **universal** set U, a set which contains all the elements under consideration at a particular time.

EXAMPLE 2.4 Suppose we are interested in these sets defined as:

$$A \quad \{1, 2, 3, 4, 5\}$$
$$B \quad \{3, 9, 15, 21, 27, \ldots\}$$
$$C \quad \{1, 8\}$$

For U we could use {counting numbers} here. However, we know that the elements of A, B, or C are precisely $\{1, 2, 3, 4, 5, 8, 9, 15, 21, 27, 33, \ldots\}$, and so this latter set is another U. On the other hand, suppose we are to consider the teams which have won the World Series. Unless one is a baseball buff, one is not likely to know exactly which teams comprise this set. However, nearly everyone knows that only baseball teams play in the World Series; thus a good choice for the universal set U is {baseball teams}.

Exercise Set 2.1

Which of the following descriptions of collections are too ambiguous to define a set?

1. Great books
2. States of the United States before 1839
3. Large cities
4. Hard metals
5. Coins minted before 1900

Describe each of the following sets by the listing method.

6. The set whose members are the first three letters of the English alphabet
7. The set of all counting numbers greater than 3 but less than 5
8. {Counting numbers}
9. {Even counting numbers}
10. {Days of the week whose names start with S}
11. {Months of the year whose names start with C}

Give a rule which describes each of these sets.

12. {Alaska, California, Hawaii, Oregon, Washington}
13. $\{w, x, y, z\}$
14. {1, 2, 3, 4, 5}
15. {3, 6, 9,. . .}

Which of the symbols, $\in$ or $\notin$, should replace the "?" in each expression to make that expression true?

16. 3 ? {1, 2, 3, 4}
17. 50 ? {5, 10, 15,. . .}
18. a ? $\{A, B, C, D\}$
19. Sacramento ? {state capitals}

Consider each of the following collections of sets. What could be a universal set in each case?

20. $\{a, b, c, d\}$, $\{d, c\}$, and $\{b, e, f\}$
21. {U.S. Senators} and {U.S. Representatives}
22. {Senior boys at Podunk U.}, {senior girls at Podunk U.}, and {juniors at Podunk U.}

2.2 SET RELATIONS AND OPERATIONS

2.2.1 Subsets

SUBSET Set A is a **subset** of set B if every element of A is an element of B. The symbol for "is a subset of" is $\subseteq$; the symbol for "is not a subset of" is $\nsubseteq$. If $A \subseteq B$ but B contains at least one element which is not an element of A, then A is a **proper subset** of B. We write $A \subset B$. Students sometimes become confused over which of the symbols $\subseteq$ and $\subset$ to use in a particular problem. A good way to avoid such confusion is to remember that $\subseteq$ can be used in all cases but that $\subset$ can be used only in the cases involving proper subsets. Therefore, when we write $R \subset S$, for example, we mean that S contains one or more elements which are not in R. If we write $R \subseteq S$, we mean that we are not saying whether or not S contains elements besides those which are in R.

EXAMPLE 2.5 What symbol can replace the "?" in each of these expressions:

(a) {Algebra textbooks on my bookshelf} ? {textbooks on my bookshelf}
(b) {3, 5, 8} ? {1, 2, 3, . . ., 10}
(c) $\{a, b, e\}$? {vowels of the English alphabet}
(d) 3 ? {1, 2, 3}

Solution

(a) $\subseteq$ (There is no way of knowing if I have another kind of book on my bookshelf besides algebra texts.)
(b) $\subset$ (We could use either $\subseteq$ or $\subset$, but there are elements in the second set which are not in the first one; thus $\subset$ is a more accurate description of how the sets are related.)
(c) $\nsubseteq$ or $\not\subset$ (b is not a vowel.)
(d) $\in$ (3 is an element, not a set.)

By definition, the empty set is a subset of every set.

EXAMPLE 2.6 Given S: $\{a, b, c\}$, list all the subsets of S.

Solution Every element of S is an element of S; thus $S \subseteq S$. The empty set is a subset. There are three more subsets each of which contains two elements and another three subsets each of which contains one element. The complete listing is:

$\{a, b\}$	$\{a\}$
$\{a, c\}$	$\{b\}$
$\{b, c\}$	$\{c\}$
S	$\varnothing$

In general a set having n elements has 2^n subsets.†

2.2.2 Venn Diagrams

Frequently a **Venn** diagram such as the one in Figure 2–1 is used to illustrate relations between sets.‡ In a Venn diagram the set of points inside some rectangle is taken as the universal set. Subsets of the universal set are

† 2^2 means 2×2, 2^3 means $2 \times 2 \times 2$, 2^n means the product of n 2's.
‡ Named after the English logician, John Venn (1834–1923).

the points in the interiors of circles or other closed curves drawn within the rectangle. Shading indicates particular regions of interest.

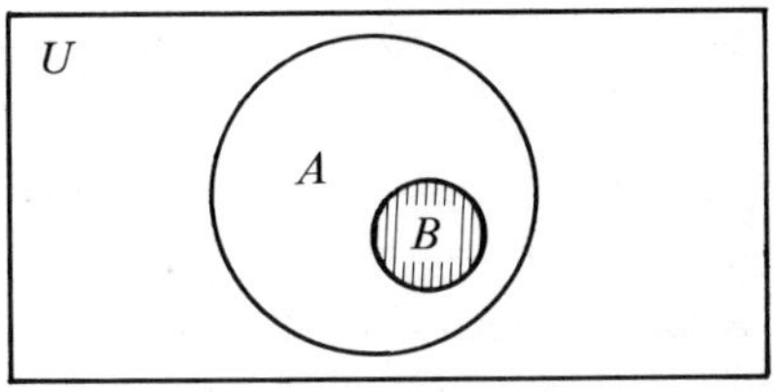

FIGURE 2–1 A Venn diagram showing $B \subset A$

In Figure 2–1, the universal set is pictured as the set of points in the interior of the rectangle. The points in the interior of the larger circle represent elements of set A, and the points in the interior of the smaller circle represent elements of set B. Because every point in the interior of the smaller circle is also in the interior of the larger circle and there are points in the interior of the larger circle which are not in the interior of the smaller circle, this diagram expresses the relation $B \subset A$.

2.2.3 Equal Sets

EQUAL SETS Two sets A and B are **equal** if each set contains exactly the same elements as the other. We write $A = B$.

EXAMPLE 2.7

(a) $\{1, 3, 5, 7, \ldots\} = \{\text{odd counting numbers}\}$
(b) $\{a, b, c\} = \{c, a, b\}$
(c) $\{2, 2, 4, 4, 4, 7,\} = \{2, 4, 7\}$ (*Listing an element more than once does not change the membership.*)
(d) $\{1, 2, 7\} \neq \{1, 2, 5\}$

2.2.4 Set Complementation

COMPLEMENT If $A \subseteq U$, then the complement of A with respect to U is the set which contains every member of U which is not also an element of A. The complement of A is written $\tilde{A}$.

EXAMPLE 2.8 Given $U = \{1, 2, 3, \ldots, 10\}$ and $A = \{2, 4, 6, 8\}$, describe $\tilde{A}$ by listing its members and also by means of a Venn diagram.

Solution

$$\tilde{A} = \{1, 3, 5, 7, 9, 10\}$$

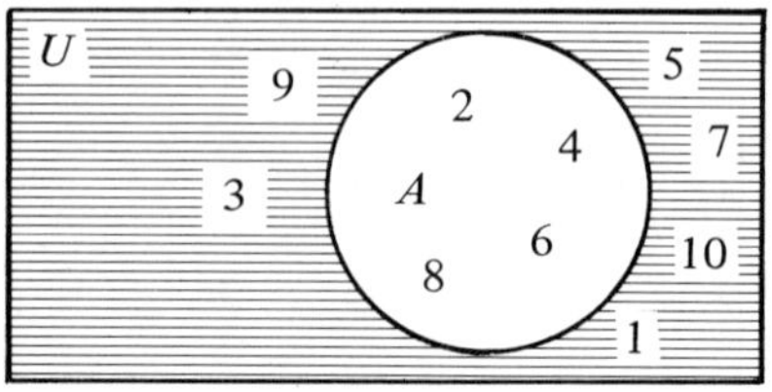

FIGURE 2–2 The shaded region represents $\tilde{A}$

In the Venn diagram (Figure 2–2), the numerals 2, 4, 6, and 8 may be placed anywhere inside the circle which represents set *A*. The remaining numerals may have any location outside *A* but inside the rectangle which represents *U*.

2.2.5 Union and Intersection

UNION The **union** of two sets *A* and *B* is the set whose elements belong to *A* or *B* or to both *A* and *B*. The symbol for union is ∪.

Figure 2–3 shows $A \cup B$ for three different cases: *A* and *B* do not intersect; *A* and *B* have some elements in common; *A* is a proper subset of *B*.

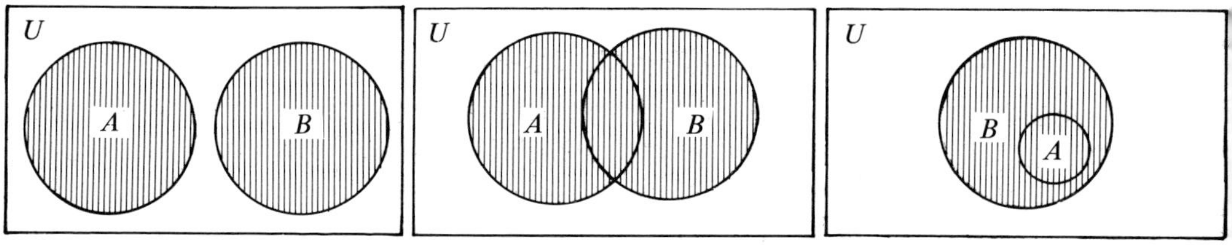

FIGURE 2–3 The shaded region in each figure represents $A \cup B$

INTERSECTION The **intersection** of two sets *A* and *B* is the set whose elements are common to both *A* and *B*. We write $A \cap B$. See Figure 2–4.

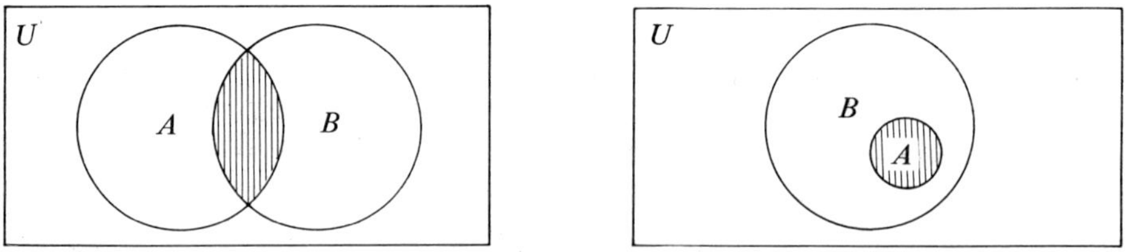

FIGURE 2–4 The shaded region in each figure represents $A \cap B$

If two sets have no common elements, they are said to be **disjoint** sets. When sets are disjoint, their intersection is the empty set. The Venn diagram in part (a) of Figure 2–3 depicts two disjoint sets.

EXAMPLE 2.9 Given that $U = \{a, b, c, d, e\}$, $A = \{a, b\}$, $B = \{b, c, d\}$, and $C = \{c, d, e\}$, list the sets

(a) $A \cup B$ (d) $A \cap C$
(b) $A \cup C$ (e) $\tilde{B}$
(c) $A \cap B$ (f) $A \cup \tilde{B}$

Solution

(a) $A \cup B = \{a, b, c, d\}$
(b) $A \cup C = \{a, b, c, d, e\}$
(c) $A \cap B = \{b\}$
(d) $A \cap C = \varnothing$
(e) $\tilde{B} = \{a, e\}$
(f) $A \cup \tilde{B} = \{a, b, e\}$

Exercise Set 2.2

1. Union and intersection are called **binary** operations because they combine two sets. Complementation is an **unary** operation because it is an operation on a single set. What operations of arithmetic are **binary**? Can you name any which are unary?
2. Form all possible subsets of $G = \{\$, \&\}$.
3. How many subsets does Q have when it contains:

(a) 4 elements? (c) 1 element?
(b) 7 elements? (d) 0 elements?

Given $U = \{1, 2, 3, 4, 5, 6, 7, 8, 9, 10\}$, $A = \{1, 5, 7, 10\}$, $B = \{5, 6, 8, 10\}$, and $C = \{4, 6, 8\}$, list the elements in the sets in Problems 4–14.

4. $A \cap B$
5. $B \cap C$
6. $A \cap (B \cap C)$. Do the operation within the parentheses first.
7. $(A \cup B) \cup C$
8. $(A \cap B) \cup (B \cap C)$
9. $\tilde{A}$
10. $\tilde{B}$
11. $\tilde{A} \cap \tilde{B}$

12. $\widetilde{A \cap B}$. This is the complement of the set in Problem 4.

13. $\widetilde{A \cap (B \cap C)}$

14. $\widetilde{\tilde{A} \cap \tilde{B}}$. This is the complement of the set in Problem 11.

15–25. Figure 2–5 is a Venn diagram for the sets used in Problems 4 to 14. By means of shading, identify the region of the diagram described in each of Problems 4 to 14.

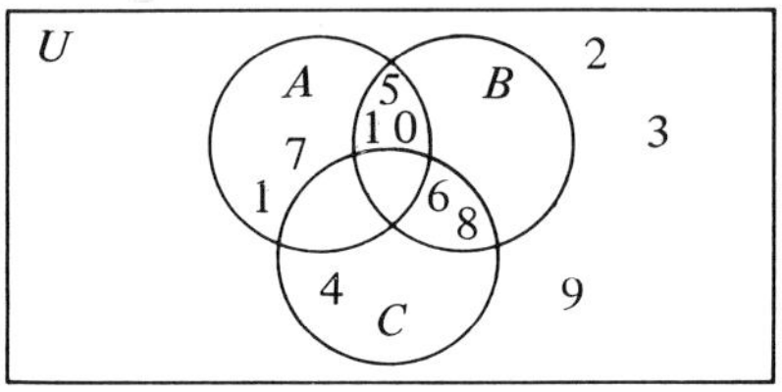

FIGURE 2–5

26. Describe $\tilde{U}$ and $\tilde{\varnothing}$.

2.3 MORE ON VENN DIAGRAMS

In making Venn diagrams, unless specific information is given to the contrary, the drawing always shows all possible intersections. Thus the typical diagram for two sets has two overlapping circles; that for three sets has three overlapping circles, as in Figure 2–5. A Venn diagram for more than three sets is complicated, as can be seen in Figure 2–6.

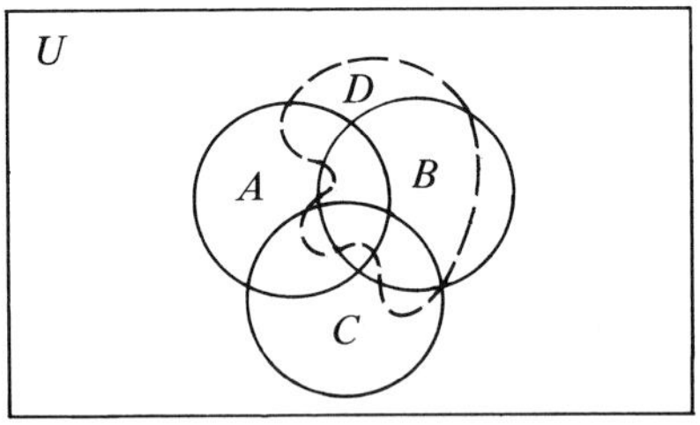

FIGURE 2–6 A Venn diagram showing four sets.

Once a diagram showing all possible intersections has been made, all that remains is to identify and shade the appropriate region or regions.

EXAMPLE 2.10 By a Venn diagram, show $(A \cap B) \cap \tilde{C}$.

Solution Draw a rectangle to represent the universal set, and inside it draw three overlapping circles to represent sets A, B, and C. The problem is to identify that region which is inside both A and B but outside C. This is the shaded region of Figure 2–7.

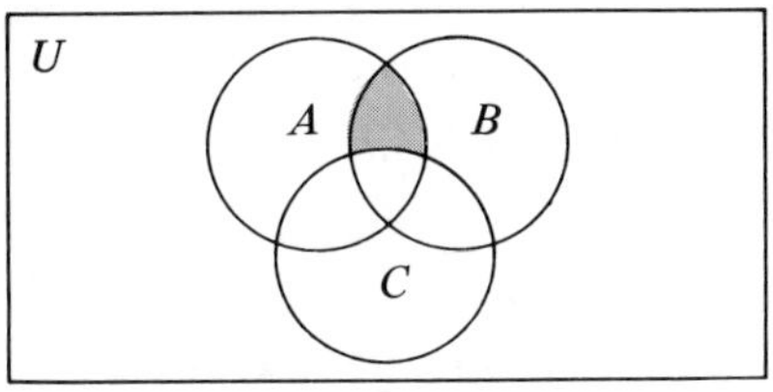

FIGURE 2–7 $(A \cap B) \cap \tilde{C}$.

EXAMPLE 2.11 Use a Venn diagram to show the most general set $A \cup (B \cap C)$.

Solution A problem such as this one, involving both union and intersection, should be solved by parts. Recall that parentheses enclosing an operation indicate that that operation is to be completed first. Accordingly we first locate and shade in the region $B \cap C$ in Figure 2–8 using vertical rules. Next we shade all of A. The total area shaded and/or ruled is $A \cup (B \cap C)$.

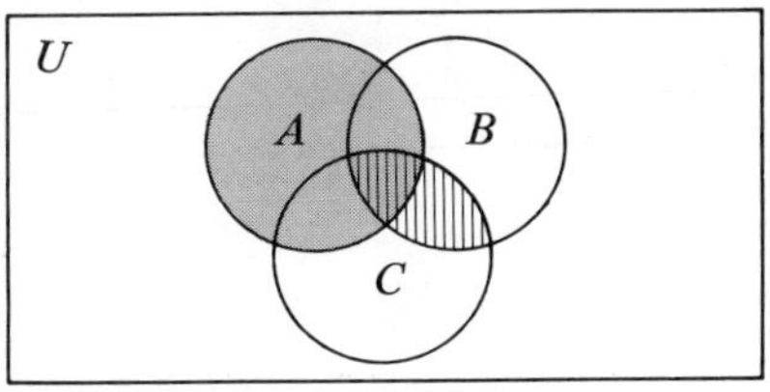

FIGURE 2–8 $A \cup (B \cap C)$.

Notice that in the Venn diagram of Example 2.11 two different types of shading (ruling and ordinary shading) are used in order to keep track of the different parts of the problem. Sometimes it is convenient to use several methods of shading.

EXAMPLE 2.12 On a Venn diagram show $\widetilde{(A \cup B) \cap (B \cup C)}$.

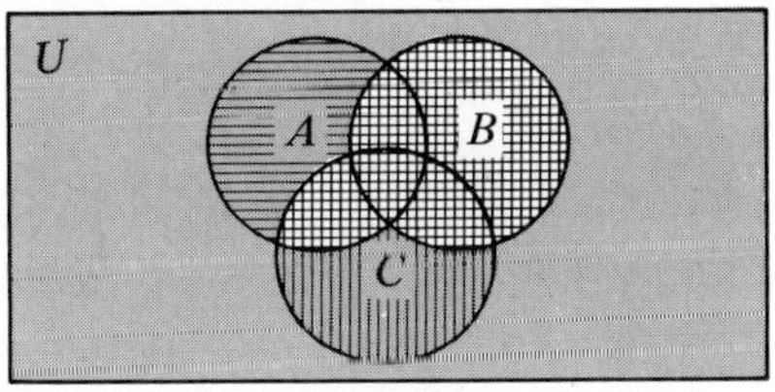

FIGURE 2–9 $\widetilde{(A \cup B) \cap (B \cup C)}$.

Solution To diagram the complement of a set it is a good idea to make a preliminary diagram of the set and then shade its complement. In the present problem this means we first identify $(A \cup B) \cap (B \cup C)$. That set is itself the intersection of two sets, $(A \cup B)$ and $(B \cup C)$. The entire problem should be done in four steps:

1. Rule $A \cup B$ (horizontal shading in Figure 2–9).
2. Rule $B \cup C$ (vertical shading in Figure 2–9).
3. Identify $(A \cup C) \cap (B \cup C)$ (the regions which are ruled both vertically and horizontally in Figure 2–9).
4. Shade everything inside U except $(A \cup C) \cap (B \cup C)$ (the shaded regions of Figure 2–9).

Exercise Set 2.3

Show by a Venn diagram the most general set:

1. $A \cup B$
2. $A \cap \tilde{B}$
3. $\tilde{A} \cap \tilde{B}$ (Shade $\tilde{A}$ and $\tilde{B}$ first, and then shade their intersection using a different type of shading.)
4. $\tilde{A} \cup \tilde{B}$
5. $\widetilde{A \cap B}$
6. $\widetilde{A \cup B}$
7. $A \cup (B \cup C)$
8. $(\tilde{A} \cap \tilde{B}) \cap C$
9. $A \cap (B \cap \tilde{C})$
10. $(\tilde{A} \cup \tilde{B}) \cup C$
11. $\widetilde{A \cup B} \cup C$ (Find $\widetilde{A \cup B}$ first.)
12. $\widetilde{A \cup B} \cup \tilde{C}$
13. $\widetilde{A \cap \tilde{B}}$ (Find $A \cap \tilde{B}$ first.)
14. $(\widetilde{A \cap B}) \cup \tilde{C}$

Draw a Venn diagram to show each set:

15. $A \cap B$ if $B \subset A$
16. $(A \cup B) \cup C$, if $A \subset B$ and $B \subset C$
17. $A \cup B$, if $A \cap B = \varnothing$ (Since $A \cap B = \varnothing$, A and B should be shown as two circles that have no points in common.)
18. $(A \cap B) \cup C$, if $A \cap C = \varnothing$, $B \cap C = \varnothing$, and $A \cap B \neq \varnothing$

On a Venn diagram similar to the one in Figure 2–6 locate each of the following:

19. $(A \cap \tilde{B}) \cap (C \cap \tilde{D})$
20. $(A \cap B) \cup (\tilde{C} \cap D)$

2.4 VENN DIAGRAMS AND EQUAL SETS

Venn diagrams may be used to show that two sets are equal. Recall that two sets are equal if they contain exactly the same elements. Therefore, if two expressions have identical diagrams, they must be equal.

EXAMPLE 2.13 Show that $A \cup (\tilde{A} \cap B) = A \cup B$.

Solution Let us diagram $A \cup (\tilde{A} \cap B)$. This is a two-step problem. First rule $\tilde{A} \cap B$, which consists of all points inside B but outside A (the vertical shading in Figure 2–10). Next shade all of A. All the regions shaded represent $A \cup (\tilde{A} \cap B)$. But these are precisely the same regions that would be shaded if $A \cup B$ were diagrammed [see Figure 2–3(b)]. We conclude that $A \cup (\tilde{A} \cap B)$ and $A \cup B$ are the same set.

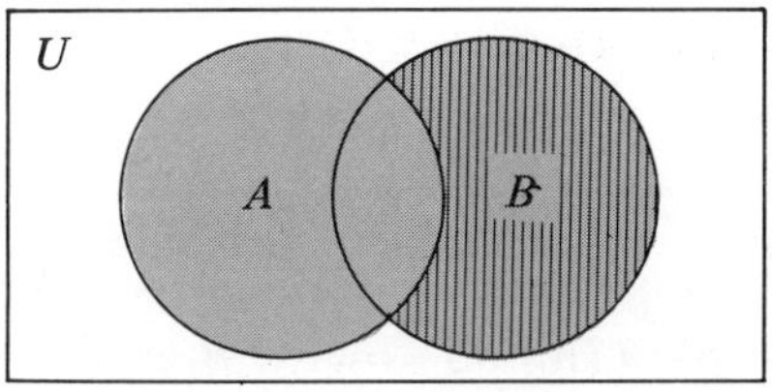

FIGURE 2–10 $A \cup (\tilde{A} \cap B)$.

In a more complicated problem it is usually necessary to make a separate diagram of each of the sets that are to be shown equal.

EXAMPLE 2.14 Show that $(A \cup B) \cap (A \cup C) = A \cup (B \cap C)$.

Solution $A \cup B$ is shaded vertically in Figure 2–11(a), and $A \cup C$ is shaded horizontally. Their intersection is the cross-hatched region. In Figure 2–11(b), $B \cap C$ is the region with vertical shading, and the union of A with $(B \cap C)$ is the entire shaded region. Since the same region is described in each diagram, we conclude that $(A \cup B) \cap (A \cup C) = A \cup (B \cap C)$.

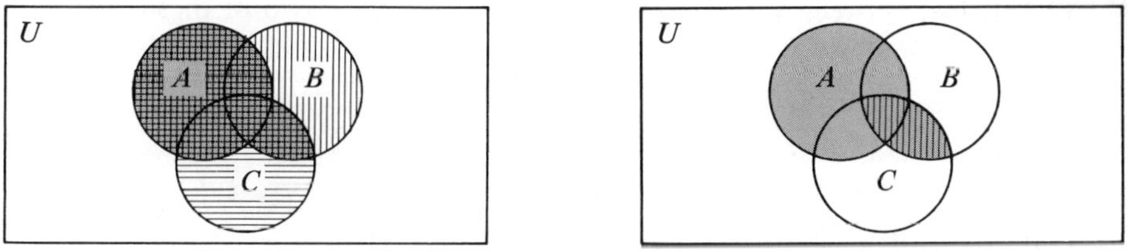

(a) $(A \cup B) \cap (A \cup C)$ (b) $A \cup (B \cap C)$

FIGURE 2–11

Exercise Set 2.4

Use Venn diagrams to show:

$$\left.\begin{array}{l} \text{1. } A \cup B = B \cup A \\ \text{2. } A \cap B = B \cap A \end{array}\right\} \textit{Commutative Laws}$$

3. $A \cup \tilde{A} = U$

4. $A \cup (A \cap B) = A$

5. $A \cup (\tilde{A} \cap B) = A \cup B$

$$\left.\begin{array}{l} \text{6. } (A \cup B) \cup C = A \cup (B \cup C) \\ \text{7. } (A \cap B) \cap C = A \cap (B \cap C) \end{array}\right\} \textit{Associative Laws}$$

$$\left.\begin{array}{l} \text{8. } A \cap (B \cup C) = (A \cap B) \cup (A \cap C) \\ \text{9. } A \cup (B \cap C) = (A \cup B) \cap (A \cup C) \end{array}\right\} \textit{Distributive Laws}$$

$$\left.\begin{array}{l} \text{10. } \widetilde{A \cup B} = \tilde{A} \cap \tilde{B} \\ \text{11. } \widetilde{A \cap B} = \tilde{A} \cup \tilde{B} \end{array}\right\} \textit{DeMorgan's Laws}\dagger$$

12. $(A \cap B) \cup (\tilde{A} \cap B) \cup (A \cap \tilde{B}) = A \cup B$

2.5 CARDINAL NUMBERS

2.5.1 Equivalent Sets

A **one-to-one** correspondence between two sets A and B is a pairing of the elements of the two sets in such a way that each element of A is paired with exactly one element of B and each element of B is paired with exactly one element of A.

EXAMPLE 2.15 There are six ways in which a one-to-one correspondence can be established between the sets $\{a, b, c\}$ and $\{\#, *, \%\}$. Two of these are:

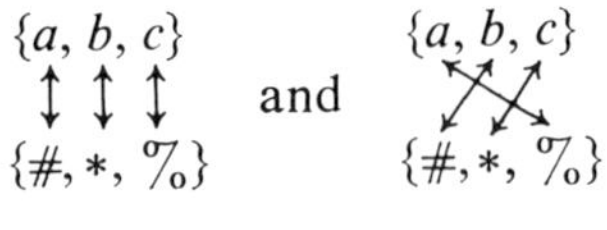

Can you find the other four?

† Named after Augustus DeMorgan (1806–1871). DeMorgan was one of the foremost logicians of his time.

Two sets are **equivalent** if their elements can be placed into a one-to-one correspondence. If A is equivalent to B, we write $A \sim B$.

Notice that there is a fundamental difference between **equality** and **equivalence**. The two sets of Example 2.15 are equivalent, but they are not equal because their elements are not the same. On the other hand, $\{3, a, b, *\}$ and $\{a, b, 3, *\}$ are not only equivalent but also equal.

You have probably observed that two sets which are said to be equivalent each have the same number of elements. Why then do we not count the elements of two sets to determine if they are equivalent? There are two good reasons why this is not done. First, counting is itself the establishing of a one-to-one correspondence between two sets, as shall be shown. The other reason is illustrated by the following.

EXAMPLE 2.16 Are $A = \{1, 2, 3, \ldots\}$ and $B = \{2, 4, 6, \ldots\}$ equivalent?

Solution Let us attempt to answer this question by counting the elements in each set. How many elements are there in A? How many in B? Counting elements sometimes has its shortcomings.

The sets are equivalent, however, as we can show by establishing a one-to-one correspondence between them. Notice that each number in B is twice as large as some number in A and that every number in A is half as large as some number in B. We can use this to make this correspondence:

$$\begin{array}{rl} A = \{ & 1, \; 2, \; 3, \; 4, \ldots, \; n, \ldots\} \\ & \updownarrow \;\; \updownarrow \;\; \updownarrow \;\; \updownarrow \qquad \;\; \updownarrow \\ B = \{ & 2, \; 4, \; 6, \; 8, \ldots, 2n, \ldots\} \end{array}$$

Let us take a look at the process we call counting. Consider the set $A = \{3, 8, 2, *, x\}$. Count the elements. The chances are that you looked at 3 and said "one" then looked at the 8 and said "two," etc., until finally you looked at the x and said "five." Do you see that what you actually did was to establish a one-to-one correspondence between A and $\{1, 2, 3, 4, 5\}$?

Now examine $B = \{1, 2, 3, 4, 5, 6, 7\}$. How many elements has it? Do you have to count the elements to find how many there are? Of course not. The numerals in B are arranged in order from 1 to 7 so that all you have to do is to look at the last numeral in the list and that tells how many elements there are in the set.

2.5.2 Ordered Sets and Standard Sets

ORDERED SET When it is desirable to consider the order of the listing of the elements of a particular set, the braces { } are replaced by parentheses. Thus the ordered set B, above, is written

$$B = (1, 2, 3, 4, 5, 6, 7)$$

There are a number of different reasons why we might want to consider ordered sets. Some of these will turn up in later sections. At the present time we must consider ordered sets with these properties:

1. The elements are counting numbers.
2. The elements are ordered from the smallest to the largest.
3. If a set contains the counting number k, it contains every counting number less than k.

STANDARD SETS Ordered sets with the above properties are called **standard sets.**

EXAMPLE 2.17 The sets in this sequence are standard sets:

$$(1)$$
$$(1, 2)$$
$$(1, 2, 3)$$
$$(1, 2, 3, 4)$$
$$\cdots\cdots\cdots$$
$$(1, 2, 3, 4, \ldots, k)$$
$$\cdots\cdots\cdots\cdots\cdots$$

It is an easy matter to determine the number of elements in a standard set; we merely inspect the last element in the listing of the set.

2.5.3 Cardinal Numbers

The number of elements in any set A is the same as the number of elements in the equivalent standard set. We call this number the **cardinal** number of the set and denote it $N(A)$. To make things work out nicely, the cardinal number of the empty set is zero, that is $N(\varnothing) = 0$.

EXAMPLE 2.18 Given $U = \{1, 2, 3, 4, 5\}$, $A = \{2, 3, 5\}$, and $B = \{4\}$, find:

(a) $N(A)$
(b) $N(B)$
(c) $N(\tilde{A})$
(d) $N(A \cap B)$

Solution

(a) $N(A) = 3$ since $A \sim (1, 2, 3)$.
(b) $N(B) = 1$ since $B \sim (1)$.
(c) $N(\tilde{A}) = 2$ since $\tilde{A} = \{1, 4\} \sim (1, 2)$.
(d) $N(A \cap B) = 0$ since $A \cap B = \varnothing$.

EXAMPLE 2.19 Podunk University offers only three foreign languages, French, German and English. A survey of 80 students revealed that 42 were enrolled in English classes, 29 in French, and 21 in German. Five were taking both French and German, seven were taking both French and English, thirteen were taking both German and English, and three were enrolled in all three subjects. How many students were studying exactly one language?

Solution In this type problem a Venn diagram can be a great help. Make the standard diagram with three overlapping circles, labeled E for English, G for German, and F for French. (See Figure 2–12.) Now all the students who study all three languages belong in $F \cap G \cap E$. Place a "3" in that region. Five students study both French and German, but three of the five also study English. Therefore, there are two students in $F \cap G \cap \tilde{E}$; so we place a "2" in that region. Similarly a "4" goes in $F \cap E \cap \tilde{G}$, and a "10" goes in $G \cap E \cap \tilde{F}$. Next consider those students who study one of the languages but not the other two. For example, 29 students study French: we have a total of nine already accounted for in the diagram, so there should be a "20" in the region $F \cap \tilde{G} \cap \tilde{E}$. The remaining areas can be numbered in a similar fashion. We then have the diagram in Figure 2–12.

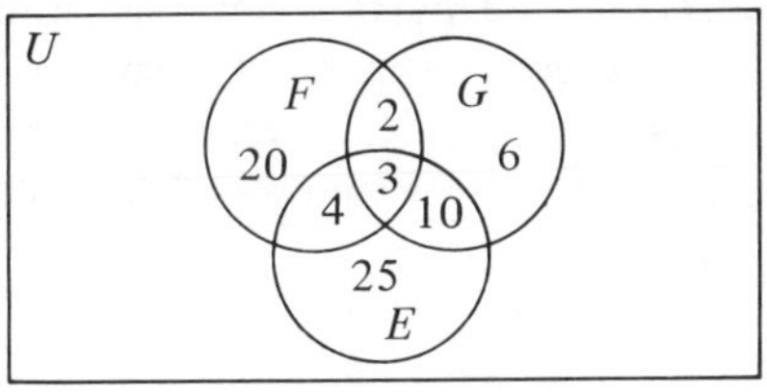

FIGURE 2–12

We observe that there are 20 students who study only French, 6 who study only German, and 25 who study only English. Therefore the number who study exactly one language is $20 + 6 + 25 = 51$.

Exercise Set 2.5

Show all possible one-to-one correspondences between the two sets in each exercise.

1. $\{a, b\}$ and $\{x, y\}$
2. $\{2, 4, 6, 8\}$ and $\{1, 2, 3, 4\}$
3. Give a set which is equivalent but not equal to $\{a, b, c, d\}$.
4. Give a set which is equal to $A = \{\%, ¢, \$\}$. Is the set you have listed equivalent to A?

What is the cardinality of each of the following?

5. $\{x, r, m, n\}$
6. $\{2, 6, 10, 14, 18\}$
7. $\{1, 3, 5, \ldots, 21\}$

Given $U = \{a, b, c, d, e, f\}$, $A = \{b, d, e\}$, $B = \{a, b, d, f\}$, and $C = \{c, e\}$, find:

8. $N(A \cup B)$
9. $N(A \cap B)$
10. $N(A \cap B \cap C)$
11. $N(A \cap \tilde{B} \cap C)$
12. $N(\widetilde{A \cup B})$
13. $N(\widetilde{A \cup B \cup C})$

14. Among the members of a certain fraternity, 70 take English, 40 take math, 40 take chemistry, 15 take math and English, 20 take English and chemistry, 25 take math and chemistry, and 5 take math, English, and chemistry. If all fraternity members take at least one of the three, how many fraternity members are there?

15. The noisemakers at a children's party were horns, poppers, and rachets. There were 15 children at the party. Twelve children received rachets, 10 received horns, 7 received poppers, 7 received horns and rachets, 5 received horns and poppers, 6 received rachets and poppers, and 4 grabbed horns, poppers, and rachets. How many children had:
    (a) Exactly 2 noisemakers?
    (b) Only poppers?
    (c) Only rachets?

Set up a one-to-one correspondence between:

16. $\{1, 2, 3, \ldots\}$ and $\{2, 4, 6, \ldots\}$
17. $\{1, 2, 3, 4, \ldots\}$ and $\{0.5, 1, 1.5, 2, \ldots\}$
18. $\{1, 2, 3, \ldots\}$ and $\{0.1, 0.2, 0.3, \ldots\}$
19. $\{1, 3, 5, 7, \ldots\}$ and $\{2, 4, 6, 8, \ldots\}$

2.6 SETS AND ARITHMETIC

2.6.1 Sets and Addition

In this section we see how some set concepts are related to the basic operations of arithmetic. If you think back to the time when you first received instruction in addition, you will probably remember that to find 2 + 3 you went through the following procedure. Your teacher gave you a piece of paper on which there was a figure similar to Figure 2–13. Each pupil was instructed to count the elements inside the shaded oval and write the number in the blank.

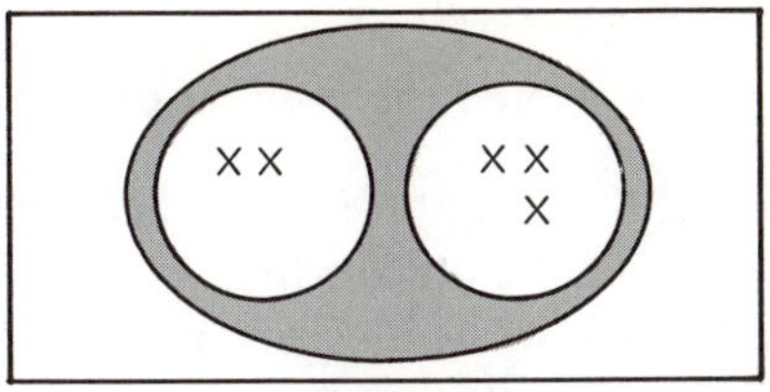

2 + 3 = ____

FIGURE 2–13

The teacher was using (without telling you and perhaps without being aware of it herself) this rule defining addition for the counting numbers:

If A and B are disjoint sets, $N(A) + N(B) = N(A \cup B)$.

A more general definition of addition of counting numbers, which you may easily verify, is:

If A and B are any two sets, $N(A) + N(B) = N(A \cup B) + N(A \cap B)$.

2.6.2 Cartesian Products and Multiplication

ORDERED PAIR An **ordered pair** is an ordered set which contains two elements. The element at the left in the pair is called the **first element,** and the element at the right is its **second element.**

CARTESIAN PRODUCT If A and B are two sets, the **Cartesian product set** $A \times B$ is the set of all ordered pairs which have an element from A as a first element and an element from B as a second element.

EXAMPLE 2.20 If $A = \{x, y\}$ and $B = \{4, 2, 3\}$, then

$$A \times B = \{(x, 4), (x, 2), (x, 3), (y, 4), (y, 2), (y, 3)\}$$

and

$$B \times A = \{(4, x), (4, y), (2, x), (2, y), (3, x), (3, y)\}.$$

We may use the Cartesian product of two sets to define multiplication of counting numbers:

If A and B are any two sets, $N(A)$ multiplied by $N(B) = N(A \times B)$.

EXAMPLE 2.21 Explain how to find the product of 2 and 3.

Solution In Example 2.20, $N(A) = 2$, $N(B) = 3$, and $N(A \times B) = 6$. Thus by the rule that $N(A)$ multiplied by $N(B) = N(A \times B)$, we have $2 \times 3 = 6$.

EXAMPLE 2.22 Explain how to find the product of 3 multiplied by 0.

Solution Suppose $A = \{a, b, c\}$ and $B = \varnothing$. Then $N(A) = 3$ and $N(B) = 0$, and so we need $N(A \times B)$. Let us try to form $A \times B$. We take any element of A, for example a, and pair it with any element of B. But B has no elements. We can form no ordered pairs which have a first element a, b, or c. Thus the set $A \times B$ has no ordered pairs in it. It is the empty set! Thus $N(A \times B) = 0$, and we have 3 multiplied by $0 = 0$.

Exercise Set 2.6

Given $A = \{a, b, c\}$, $B = \{b, c, d\}$, and $C = \{x\}$, list the elements in the following sets:

1. $A \times B$
2. $B \times A$
3. $A \times C$
4. $C \times A$
5. $A \times A$
6. $C \times C$

Using the sets listed for Problems 1 to 6, what is:

7. $N(A) + N(B)$
8. $N(A) + N(C)$
9. $N(A)$ multiplied by $N(B)$
10. $N(A)$ multiplied by $N(C)$
11. Is $N(A) + N(B) = N(B) + N(A)$? Explain using set concepts.
12. Is $A \times B = B \times A$? Is $N(A \times B) = N(B \times A)$? Is $N(A)$ multiplied by $N(B) = N(B)$ multiplied by $N(A)$? Explain each answer.
13. If $B \subseteq A$, then the **complement of A with respect to B**, A/B, is the set of elements of A which are not elements of B. That is, $A/B = A \cap \tilde{B}$. Use $A = \{1, 2, 3, \ldots, 10\}$ and $B = \{2, 4, 7\}$ to illustrate this definition of subtraction: $N(A) - N(B) = N(A/B)$. Show that if $B \nsubseteq A$, this definition does not work.

2.7 INFINITE SETS AND TRANSFINITE NUMBERS

2.7.1 Infinite Sets

Consider the sets $A = \{1, 2, 3, 4, \ldots\}$ and $B = \{3, 6, 9, 12, \ldots\}$. Set B is a proper subset of A; that is, every element of B can be found in A but there are elements in A which are not in B. Clearly A has more elements than B. Or has it? Note that when an element of A is multiplied by 3, the product is an element of B. Note also that the quotient of any element of B divided by 3 is an element of A. These facts suggest this one-to-one correspondence:

$$\begin{array}{l} A = \{1, 2, 3, 4, \ldots, n, \ldots\} \\ \quad\;\; \updownarrow \; \updownarrow \; \updownarrow \; \updownarrow \qquad\; \updownarrow \\ B = \{3, 6, 9, 12, \ldots, 3n, \ldots\}. \end{array}$$

Clearly $A \sim B$ and so $N(A) = N(B)$! How can this be? How can a set have the same number of elements as one of its proper subsets? This question plagued mathematicians for centuries. No one managed to give a satisfactory answer until Cantor offered the one we shall consider next.

A set such as $\{1, 2, 3, 4, \ldots\}$ that is equivalent to one of its proper subsets is said to be **infinite**. A **finite set** is one which is not equivalent to any of its proper subsets.

EXAMPLE 2.23 The sets

{1, 2, 3}
{States in the United States}
{People with blue eyes}

are all finite. The set $C = \{3, 4, 5, \ldots\}$ is a proper subset of $D = \{2, 3, 4, 5, \ldots\}$, but C and D are equivalent since the following one-to-one correspondence can be made:

$$\begin{array}{l} 2, 3, 4, \ldots, k, \quad \ldots \\ \updownarrow \; \updownarrow \; \updownarrow \qquad \updownarrow \\ 3, 4, 5, \ldots, k - 1, \ldots \end{array}$$

Therefore D is an infinite set.

Similarly, the set $\{5, 10, 15, 20, \ldots, 5k, \ldots\}$ is infinite because it can be put into a one-to-one correspondence with its proper subset $\{10, 20, 30, 40, \ldots, \ldots\}$ in the following way:

$$\begin{array}{l} 5, \; 10, 15, 20, \ldots, \; 5k, \ldots \\ \updownarrow \quad \updownarrow \quad \updownarrow \quad \updownarrow \qquad\;\; \updownarrow \\ 10, 20, 30, 40, \ldots, 10k, \ldots \end{array}$$

2.7.2 Transfinite Numbers

We can show that the cardinality of any finite set is one of the counting numbers by putting it into a one-to-one correspondence with one of the standard sets (that is, by counting the elements).

Consider $A = \{1, 2, 3, 4, \ldots\}$. Clearly $N(A)$ is not a counting number because A is an ordered set of all the counting numbers and there is no largest counting number. (Add 1 to any counting number, and the result is another counting number.) Cantor gave this set a special cardinal number, $\aleph_0$ (aleph-null). Any set which is equivalent to the set of counting numbers he called **denumerably infinite**, or more simply, **denumerable**. Thus the sets $\{3, 6, 9, 12, \ldots\}$ and $\{2, 4, 6, 8, \ldots\}$ are denumerable. The cardinality of any denumerable set is $\aleph_0$.

Cantor's proof that the set of all positive rational numbers (numbers which can be written a/b where a and b are counting numbers) is denumerable is a clever one.

EXAMPLE 2.24 The following array, if expanded infinitely, contains all the positive rational numbers.

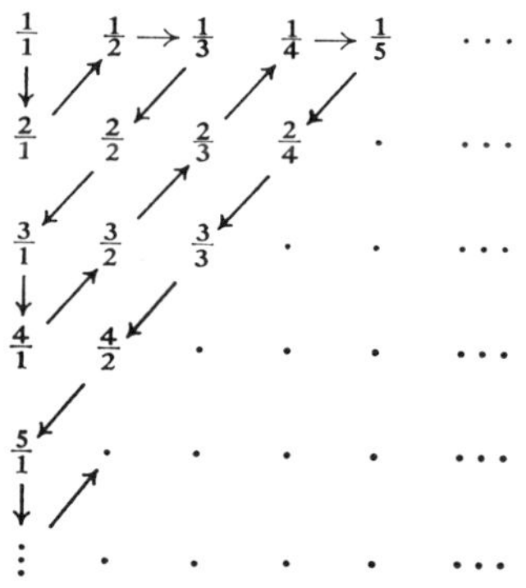

The line goes through every element in the array. Follow the line from $\frac{1}{1}$, and you can make this one-to-one correspondence:

$$\begin{array}{cccccccccccccc} \frac{1}{1} & \frac{2}{1} & \frac{1}{2} & \frac{1}{3} & \frac{2}{2} & \frac{3}{1} & \frac{4}{1} & \frac{3}{2} & \frac{2}{3} & \frac{1}{4} & \frac{1}{5} & \frac{2}{4} & \cdots \\ \updownarrow & \updownarrow & \updownarrow & \updownarrow & \updownarrow & \updownarrow & \updownarrow & \updownarrow & \updownarrow & \updownarrow & \updownarrow & \updownarrow & \\ 1 & 2 & 3 & 4 & 5 & 6 & 7 & 8 & 9 & 10 & 11 & 12 & \cdots \end{array}$$

Therefore the cardinality of the positive rational numbers is $\aleph_0$.

At this point one might suppose that every infinite set is equivalent to the set of counting numbers. Such a supposition would be wrong however. There are more numbers between 0 and 1 than there are counting numbers. This is shown in the following example.

EXAMPLE 2.25 We note that the set of numbers between 0 and 1 contains $T = \{\frac{1}{10}, \frac{1}{100}, \frac{1}{1000}, \ldots\}$ as a subset. Since T can be put into a one-to-one correspondence with the set of counting numbers

$$\begin{array}{ccccccc} \frac{1}{10}, & \frac{1}{100}, & \frac{1}{1000}, & \ldots, & \frac{1}{10^k}, & \ldots \\ \updownarrow & \updownarrow & \updownarrow & & \updownarrow & \\ 1, & 2, & 3, & \ldots, & k, & \ldots \end{array}$$

it follows that there are at least $\aleph_0$ numbers between 0 and 1.

Every number between 0 and 1 can be expressed as an infinite decimal; for example, $\frac{1}{4} = 0.25000\cdots$, $\frac{8}{11} = 0.727272\cdots$, and $\sqrt{0.6} = 0.774597\cdots$ (correct to six digits).

Suppose that all the numbers between 0 and 1 have been written in their decimal form. Suppose further that there are at least $\aleph_0$ of these numbers. Now we make a one-to-one correspondence between $\aleph_0$ of these decimal numbers and the set of counting numbers. One possible correspondence is:

$$\begin{array}{l} 1 \leftrightarrow 0.13742\cdots \\ 2 \leftrightarrow 0.22731\cdots \\ 3 \leftrightarrow 0.40819\cdots \\ 4 \leftrightarrow 0.23121\cdots \\ 5 \leftrightarrow 0.11223\cdots \\ \cdots\cdots\cdots\cdots\cdots \end{array}$$

Follow a diagonal line through the decimals from the top left to the lower right corners. Each time the line goes through a digit write down any digit except that one. Write these digits horizontally with a decimal point before the first one. For the listing above we might write $0.39105\cdots$.

The number generated in this way is not any of the $\aleph_0$ decimals in the list because it is different from the first one in the first digit, different from the second one in the second digit, different from the third one in the third digit, and so on. The new number is between 0 and 1, however, and this proves there are more decimal numbers between 0 and 1 than there are counting numbers.

Since the set of real numbers contains, as a subset, the set of decimal numbers between 0 and 1, it follows that there are more real numbers than there are counting numbers. We shall denote the cardinality of the set of real numbers by c.

Are there other transfinite numbers besides $\aleph_0$ and c? It is easy to show that for any nonempty finite set S the number of subsets of S is greater than the number of elements in S. For example, if $S = \{a, b, c\}$, then $N(S) = 3$, but S has 8 subsets and 8 is greater than 3. Cantor was able to show that the same principle applies to infinite sets; if A is an infinite set, then A has more subsets than it has elements. Thus the number of subsets of the set of counting numbers is a larger number than $\aleph_0$, and the number of subsets of the subsets of the set of counting numbers is a still larger transfinite number, etc.

Cantor also showed that for any transfinite number there is a next larger transfinite number. It can be shown that the smallest transfinite number is $\aleph_0$. Cantor called the next larger one $\aleph_1$ and the next one $\aleph_2$, and so on to $\aleph_{\aleph_0}$. But following $\aleph_{\aleph_0}$ there is a next larger cardinal number, and after that one there is still another, and so on. Among all these alephs is the number c. We have shown that c is not $\aleph_0$, but which of the others is it? Cantor thought it might be $\aleph_1$ but he could not prove it, and *no mathematician knows the answer to this day!*

Exercise Set 2.7

Classify each of the following as finite or infinite:

1. $\{8, 10, 12\}$
2. $\{1, \frac{1}{2}, \frac{1}{3}, \frac{1}{4}, \ldots\}$
3. {Electrons in universe}
4. {Rocks on the moon}
5. {Transfinite numbers}

By establishing one-to-one correspondences with $\{1, 2, 3, 4, \ldots\}$, show that the cardinality of each of the following sets is $\aleph_0$.

6. {Counting numbers greater than 100}
7. $\{\ldots, -3, -2, -1, 0, 1, 2, 3, \ldots\}$
8. $\aleph_0, \aleph_1, \aleph_2, \ldots$

Use what you have learned about infinite sets of counting numbers to answer each of these problems.

9. $\aleph_0 + 1$
10. $\aleph_0 + \aleph_0$
11. $\aleph_0 \times \aleph_0$

12. For finite sets A and B with $B \subseteq A$, subtraction is defined $N(A) - N(B) = N(A \cap \tilde{B})$. If the same definition is used for infinite sets, find $N(A) - N(B)$ when:

    (a) $A = \{1, 2, 3, \ldots\}$ and $B = \{2, 3, 4, 5, \ldots\}$

    (b) $A = \{1, 2, 3, \ldots\}$ and $B = \{2, 4, 6, \ldots\}$

    (c) $A = \{1, 2, 3, \ldots\}$ and $B = \{1, 2, 3, \ldots\}$

    What does the above suggest about the problem $\aleph_0 - \aleph_0$?

SUMMARY

Terms you should know:

1. Set
2. Element
3. Empty set
4. Universal set
5. Subset
6. Ordered set
7. Cardinal number
8. Transfinite number
9. Proper subset
10. Equal sets
11. Complement
12. Union
13. Intersection
14. Ordered pair
15. Standard set
16. Denumerable
17. Disjoint sets
18. Binary operation
19. Venn diagrams
20. One-to-one correspondence
21. Equivalent sets
22. Cartesian product
23. Infinite set

REVIEW EXERCISES

1. Which of the following set descriptions are ambiguous?

    (a) The United States

    (b) The best families

    (c) The inhabitants of Antarctica

2. Which of these are true statements?

    (a) $a \in \{a, b, c\}$

    (b) $a \subset \{a, b, c\}$

    (c) $\{x, y, z\} \notin \{a, b, c\}$

    (d) $x \notin$ consonants

    (e) $\{a, e\} \subseteq$ vowels

    (f) $\{p, q, r\} = \{r, q, p\}$

    (g) $(m, n) \neq (n, m)$

    (h) $\varnothing \in \{a, b, c\}$

    (i) $\varnothing \subset \{\varnothing\}$

    (j) $\varnothing \subset \{\varnothing\}$

3. If $A = \{a, b, c\}$, insert the symbol $\subseteq$ or $\nsubseteq$ which makes each statement true.

    (a) $\{a\}$ ____ A

    (b) $\{a, b, c\}$ ____ A

4. If $A = \{a, b, c\}$, insert the symbol $\subset$ or $\not\subset$ which makes each statement true.

   (a) $\{a\}$ ____ A

   (b) $\{a, b, c\}$ ____ A

5. If $A = \{a, b, c, d\}$:

   (a) What are the elements of A?

   (b) What are the subsets of A?

   (c) Which subsets of A are not proper subsets?

6. Let U be the set of living creatures, A the set of sea creatures, B the set of horses, and C the set of green creatures. Describe the members of:

   (a) $A \cap B \cap C$ (c) $\tilde{A} \cap B$

   (b) $A \cup B \cup C$ (d) $\widetilde{A \cap B}$

7. Shade a Venn diagram to represent each of these sets.

   (a) $A \cup B$ (c) $\widetilde{A \cap B} \cap C$

   (b) $A \cup B \cup \tilde{C}$ (d) $A \cap (B \cup \tilde{C})$

8. Shade the indicated sets of each diagram of Figure 2–14.

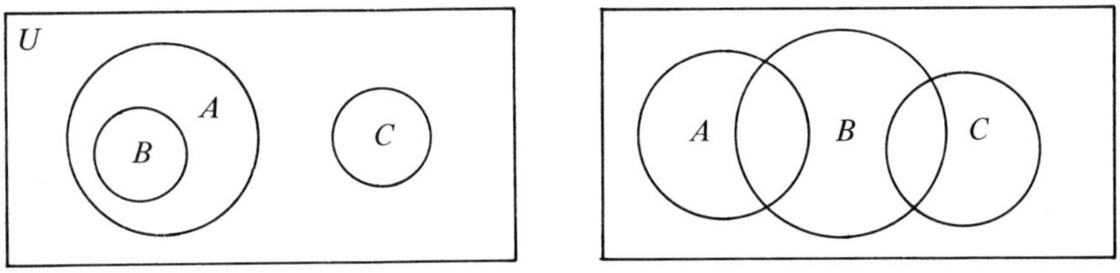

(a) $(A \cap B) \cup C$ (b) $(A \cap B) \cup (B \cap C)$

FIGURE 2–14

9. Which of the following are one-to-one correspondences?

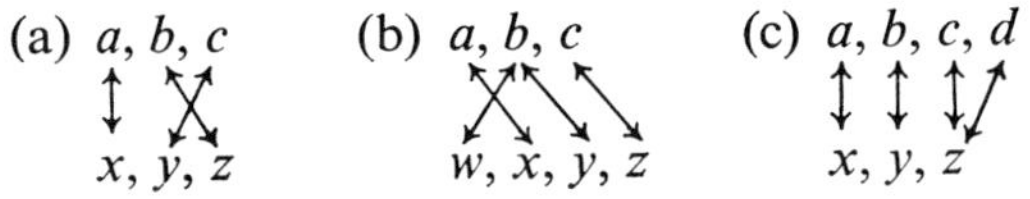

10. Find $N(A)$ when:

    (a) $A = \{2, 4, 6, 8, 10\}$

    (b) $A = \{\varnothing, 1, 2\}$

    (c) $A = \{1, 2, 3, 4, \ldots, m\}$

    (d) $A = \{2, 4, 6, 8, \ldots\}$

11. Given $A = \{a, b\}$, $B = \{x, y\}$, and $C = \{(a, x), (b, x), (x, b), (b, y)\}$:
   (a) Which of the elements of C is not an element of $A \times B$?
   (b) Which elements of $A \times B$ are not elements of C?

3
SYMBOLIC LOGIC

The basic problem of logic is the analysis of methods of reasoning. Reasoning is that special kind of thinking, called inference, in which conclusions are drawn from premises. As an example consider:

All rabbits have four legs.
Charlie is a rabbit.
Therefore, Charlie has four legs.

The premises of this argument are *All rabbits have four legs* and *Charlie is a rabbit*. The conclusion is *Charlie has four legs*. The logician seeks the answer to this question: "If the premises are true, is one justified in saying that the conclusion is also true?" If so, the reasoning that led to the conclusion is valid (logical); if not, the reasoning is invalid (illogical).

The systematic formalization and cataloging of valid methods of reasoning is one of the main tasks of the logician. If his work uses abstract symbols, then it is called **symbolic logic**.

One should not infer, from the above, that the only way one can distinguish between valid and invalid arguments is through the study of formal logic.

Nothing could be further from the truth. A study of logic will help one make the distinction, however, because it will give the student certain techniques for testing the validity of arguments. This knowledge can be used not only with the arguments of others but also in formulation of one's own arguments since when it is known what kinds of mistakes are likely to be made, these mistakes are more easily avoided.

3.1 PROPOSITIONS

3.1.1 Propositions and Their Truth Values

Far too often a debate between well-meaning persons flounders and goes nowhere merely because each party attaches a different meaning to the same term. Consider the different meanings of the word *clean* to the 10-year-old and to his mother who has ordered "clean your room."

To avoid such ambiguities, certain technical terms should be defined in advance of any discussion using them. In this section some of the specialized terms of logic will be defined.

PROPOSITION A **proposition** is a declarative sentence which is true or false but not both.

TRUTH VALUE The **truth value** of a proposition is *true* if the proposition is true, and it is *false* if the proposition is false.

EXAMPLE 3.1 These sentences are propositions:

(a) Sacramento is the capital city of California.
(b) France is one of the United States.
(c) $3 + 2 = 5$.
(d) George is tall.

Assertion (a) is true since it corresponds to reality. Assertion (b) is false since it does not correspond to reality. Assertion (c) is true because by general agreement it is the symbolic statement that the sum of the numbers 3 and 2 is the number 5. Some logicians do not consider a sentence such as (d) to be a proposition because there are many individuals named *George* and because *tall* is a relative term. Here we shall consider that words such as *George*, *I*, *we*, and *they* refer to specific persons and that agreement on standards for words such as *tall*, *hot*, and *old* have been reached. If these assumptions are made, (d) is a proposition whose truth value depends upon the particular George referred to and upon the standard of tallness agreed upon.

3.1.2 Kinds of Propositions

A proposition is **simple** if it does not contain any other proposition as a component part. It is a **compound proposition** if it is composed of two or more propositions.

EXAMPLE 3.2 These are compound propositions:

(a) $1 + 1 = 2$, and 4 is less than 7.
(b) Owls are nocturnal or cats are playful.
(c) If it rains, then I will not attend the ball game.

CONNECTIVES The words *and, or, not, if* ··· *then*, and *if and only if* are the basic words used to connect propositions. They are called **logical connectives**.

The dots in *if* ··· *then* indicate missing words. When *if* ··· *then* is used to connect two propositions, one of the propositions is written in place of the dots and the other one follows the *then* as in part (c) of Example 3.2. Often a compound is an *if* ··· *then* proposition even though one or both of the words *if* and *then* are missing.

EXAMPLE 3.3 These are *if* ··· *then* propositions:

(a) If the day is hot, I will be happy.
(b) It is cold when the wind blows.
(c) A geometric figure that is a square is also a rectangle.

Proposition (a) means: If the day is hot, *then* I will be happy. Proposition (b) does not have an *if* or a *then*, but *when* is a substitute for *if*. The proposition means: If the wind blows, then it is cold. Proposition (c) means: If a geometric figure is a square, then it is a rectangle.

3.1.3 Use of Symbols

The power of symbolic logic lies in its use of symbols to represent propositions and connectives. After an argument has been reduced to symbols, its validity can be judged on the basis of its form rather than on its content. We shall represent propositions with lowercase letters such as $p, q, r, \ldots$. Usually, each letter will stand for a simple proposition, and a compound will be formed by linking two letters together by means of one of the logical connectives. The following list shows the symbol which is used for each connective and also gives the name by which the compound is known.

Connective	Symbol	Name of Compound
not	$\sim$	negation
and	$\wedge$	conjunction
or	$\vee$	disjunction
if $\cdots$ then	$\rightarrow$	conditional
if and only if	$\leftrightarrow$	biconditional

The connective *not* is the only one which can be used with a single simple proposition; all the others must link together two propositions. The following example illustrates the use of the connectives.

EXAMPLE 3.4 Given that p represents "The baby is crying," q represents "The birds are singing," and r represents "The dog is barking," represent the following propositions symbolically:

(a) The dog is not barking
(b) The baby is crying and the birds are singing.
(c) The dog is barking or the birds are singing.
(d) The dog is barking and the birds are not singing.
(e) If the dog is barking and the birds are not singing, then the baby is crying.
(f) The baby cries if and only if the dog barks.

Solution

(a) $\sim r$. This proposition is the negation of the proposition r.
(b) $p \wedge q$. This is the conjunction of p and q.
(c) $r \vee q$. This is the disjunction of r and q.
(d) This proposition could be written $r \wedge (\sim q)$, but whenever $\sim$ is used with one or more of the other connectives it is always assumed that the negation takes place first; therefore there is no ambiguity if we write

$$r \wedge \sim q.$$

(e) $(r \wedge \sim q) \rightarrow p$. The parentheses are necessary because without them the proposition $r \wedge \sim q \rightarrow p$ might be interpreted "The dog is barking and, if the birds are not singing, the baby is crying."
(f) $p \leftrightarrow r$.

EXAMPLE 3.5 Let s and t stand for

s: Sally is smart.
t: Tom is tall.

What is the English sentence that has the same meaning as

(a) $\sim s$
(b) $s \wedge t$
(c) $(t \vee s) \rightarrow t$

Solution

(a) Sally is not smart.
(b) Sally is smart and Tom is tall.
(c) If Tom is tall or Sally is smart, then Tom is tall.

Exercise Set 3.1

1. Which of these sentences are propositions? If a sentence is a proposition is it simple or compound?
   (a) The dog is barking.
   (b) Snakes are mammals.
   (c) Shut the door.
   (d) Iron is hard and gold is soft.
   (e) Oh, what a beautiful morning!
   (f) It will rain or snow today.
   (g) Harry is happy and Susan is sad.
   (h) If I go, you may go with me.
2. Let p, q, and r stand for

   p: Birds fly.
   q: The sky is blue.
   r: The grass is green.

   Write the English sentence with the same meaning as:

   (a) $\sim p$
   (b) $p \wedge q$
   (c) $p \vee q$
   (d) $\sim p \vee q$
   (e) $p \wedge q \wedge r$
   (f) $p \wedge \sim q \wedge \sim r$
   (g) $\sim p \vee (\sim q)$
   (h) $(p \wedge q) \rightarrow r$
   (i) $p \rightarrow (q \vee r)$
   (j) $p \leftrightarrow q$
   (k) $\sim p \leftrightarrow (q \wedge r)$
3. Using the propositions p, q, and r of Problem 2, write the symbolic representation of:
   (a) The sky is blue and the grass is green.
   (b) Birds fly or the sky is blue.

(c) Birds do not fly and the sky is not blue.
(d) It is not the case that the sky is blue and the grass is green.
(e) If the grass is green and the sky is not blue, then the birds do not fly.
(f) The grass is green but the birds do not fly if the sky is blue.

4. Write each of the following in *if· · ·then* form.
   (a) I will attend the ball game if it does not rain.
   (b) All doctors are intelligent.
   (c) When it rains it pours.
   (d) Ice begins to melt when the temperature rises above 0° Centigrade.

3.2 TRUTH VALUES AND TRUTH TABLES

3.2.1 Negation

In the preceding section, the five basic logical connectives were introduced. In this section we consider the form of the compounds and the rules for assigning truth values to each of them.

NEGATION The negation $\sim p$ is defined to be true if p is false and false if p is true.

It is customary to use the symbol T to denote the truth value of a true proposition and F to denote the truth value of a proposition which is false. Using these symbols the relations between the truth values of p and $\sim p$ are shown in the following **truth table**:

TABLE 3–1 Truth table for the negation of p.

p	$\sim p$
T	F
F	T

To use the table, read across a row. The top row, for instance, tells us that when p is a true proposition, $\sim p$ is a false proposition. The second row of the table says that when p is false, $\sim p$ is true.

3.2.2 Conjunction

The word *and* is defined to have the same meaning in logic as in common usage.

CONJUNCTION The proposition $p \wedge q$ is true if p and q are both true, and false in all other cases.

To make a truth table for $p \wedge q$, we note that it is possible that both p and q are true, that p is true and q is false, that p is false and q is true, or that both p and q are false. Thus there are four possible truth-value arrangements for $p \wedge q$. The truth table for the conjunction must have one row for each of these arrangements. Our convention for arranging the truth values in the first column of the truth table is TTFF, and for those in the second column TFTF. (See Table 3–2.)

TABLE 3–2 Standard way of showing all possible arrangements of truth values for the two propositions p and q.

p	q
T	T
T	F
F	T
F	F

From the definition of conjunction we know that $p \wedge q$ is true when p and q are both true and false when p and q are not both true. Therefore the truth table for $p \wedge q$ is completed by adding to Table 3–2 a third column headed $p \wedge q$ and entering the values TFFF under that heading. The completed table is Table 3–3.

TABLE 3–3 Truth table for the conjunction p ∧ q.

p	q	$p \wedge q$
T	T	T
T	F	F
F	T	F
F	F	F

When specific propositions p and q are combined by means of $\wedge$, it is possible to determine their truth values and then the corresponding row of Table 3–3 is used to determine the truth value of the conjunction $p \wedge q$.

EXAMPLE 3.6 What is the truth value of each of the following?

(a) The sum of 1 and 1 is 2 and the 25th of December is Christmas Day.
(b) The number 4 is even and no man has walked on the moon.

Solution (a) T. "The sum of 1 and 1 is 2" and "The 25th of December is Christmas Day" are both true propositions. Row 1 of Table 3–3 tells us that whenever two propositions are true, their conjunction is true.

(b) F. "The number 4 is even" is a true proposition, but "No man has walked on the moon" is false. Row 2 of Table 3–3 tells us that the conjunction of these two propositions must be false.

3.2.3 Disjunction

In common usage, the word *or* may be used in two different ways. A menu which states that a dinner includes "soup or salad" uses *or* in its exclusive sense, for it means that one who orders the dinner may have a choice between soup and salad but he may not have both soup and salad. On the other hand, a weather forecast of "rain or snow tomorrow" uses *or* in its inclusive sense because it provides for the possibility that tomorrow there may be both rain and snow. Here we shall always use the inclusive case of *or*.

DISJUNCTION The proposition $p \vee q$ is defined to be false if p and q are both false and true in all other cases.

TABLE 3–4 Truth table for the disjunction $p \vee q$.

p	q	$p \vee q$
T	T	T
T	F	T
F	T	T
F	F	F

EXAMPLE 3.7 Give the truth value of each proposition.

(a) New York City is in New York State or Washington, D.C. is the capital city of the United States.
(b) $p \vee \sim p$
(c) H_2O is a formula for rubber or a diamond will not scratch glass.

Solution (a) T. "New York City is in New York State" and "Washington, D.C. is the capital city of the United States" are both true (row 1, Table 3–4).

(b) T. The propositions p and $\sim p$ have opposite truth values; thus one of them must be true (row 2 or row 3, Table 3–4).

(c) F. H_2O is a formula for water, not rubber, and diamonds are so hard that they scratch glass. Both components are false, and row 4 of Table 3–4 tells us that their disjunction is also false.

3.2.4 Conditional Propositions

CONDITIONAL The proposition $p \rightarrow q$ is defined to be false if p is true and q is false, and true in all other cases.

TABLE 3–5 *Truth table for the conditional $p \rightarrow q$.*

p	q	$p \rightarrow q$
T	T	T
T	F	F
F	T	T
F	F	T

To say that the proposition "if p then q" is true when p is false and q is true or when both p and q are false may seem a bit strange at first; the following hypothetical situation, however, will lend plausibility to the definition. Suppose a politician promises, "If I am elected, taxes will not be raised." Let p and q represent "I am elected" and "Taxes will not be raised," respectively. Then the politicians promise is $p \rightarrow q$. Now suppose p and q are both true, that is, the politician is elected and taxes do not go up. Then surely the promise has been kept and $p \rightarrow q$ is true (row 1 of Table 3–5). If the politician is elected (p is true) and taxes rise (q is false), then the promise has not been kept ($p \rightarrow q$ is false), and this illustrates the entry in row 2 of Table 3–5. Next let us consider the cases in which the politician is not elected (p is false). His promise makes no mention of that possibility; surely if he is not elected, no one will hold him accountable whether taxes go up or not. Therefore we consider that his promise has been kept in this case also and enter T's in the last two rows of the third column of the truth table.

3.2.5 Biconditional Propositions

BICONDITIONAL The proposition $p \leftrightarrow q$ is defined to be true when p and q have the same truth values and false if they have different truth values.

In a later section of this chapter, it will be shown that because $\leftrightarrow$ represents *if and only if*, the definition above gives the proper truth values.

TABLE 3–6 Truth table for the biconditional $p \leftrightarrow q$.

p	q	$p \leftrightarrow q$
T	T	T
T	F	F
F	T	F
F	F	T

EXAMPLE 3.8 What is the truth value of each of the following?

(a) $(1 + 3 = 2) \rightarrow$ (2 is greater than 1)

(b) If Ohio is a state, then the Rocky Mountains are east of the Mississippi River.

(c) $(1 + 1 = 2) \leftrightarrow$ (4 is greater than 2)

(d) Dallas is the capital of France if and only if Texas is a country in Europe.

Solution (a) T. The truth value of "$1 + 3 = 2$" is F and that of "2 is greater than 1" is T. From row 3 of Table 3–5, we see that in this case the conditional is true.

(b) F. "Ohio is a state" is true, and "The Rocky Mountains are east of the Mississippi River" is false. (See row 2 of Table 3–5.)

(c) T. See row 1 of Table 3–6.

(d) T. See row 4 of Table 3–6.

3.2.6 Constructing Truth Tables

The information shown in Tables 3–1 through 3–6 can be used in constructing truth tables for more complex propositions. The following examples will illustrate how this is done.

EXAMPLE 3.9 Construct a truth table for the proposition $p \wedge \sim q$.

Solution We start the table with the standard four rows and two columns for a compound involving the two propositions p and q. The number of rows is determined by the number of simple statements and not by the logical connectives used.

p	q
T	T
T	F
F	T
F	F

The proposition in question, $p \wedge \sim q$, is the conjunction of the two simpler propositions p and $\sim q$. The table already has the truth values of p; to find corresponding values for $\sim q$, we add a third column to the table and head it $\sim q$.

p	q	$\sim q$
T	T	F
T	F	T
F	T	F
F	F	T

Table 3–1 tells us that $\sim q$ is false whenever q is true. In the table we are constructing, q is true in row 1 and in row 3. Therefore we place F's under $\sim q$ in both of those rows. Similarly, because $\sim q$ is true whenever q is false, we place T's under the $\sim q$ in rows 2 and 4 of our table. Then we add a fourth column headed $p \wedge \sim q$ (see Table 3–7). In row 1, p is true and $\sim q$ is false; thus from row 2 of Table 3–3 we know that we should enter an F in the first row of our table under the $p \wedge \sim q$. In row 2 of the table we are constructing, p is true and $\sim q$ is true; therefore $p \wedge \sim q$ is true for this case. We put a T under the heading $p \wedge \sim q$ in row 2. The entries in row 3 and row 4 of Table 3–7 were computed in a similar manner using the entries under p and $\sim q$ and the rules given by Table 3–3.

TABLE 3–7 Truth table for $p \wedge \sim q$.

p	q	$\sim q$	$p \wedge \sim q$
T	T	F	F
T	F	T	T
F	T	F	F
F	F	T	F

EXAMPLE 3.10 Construct a truth table for $\sim (p \vee q)$.

Solution The truth table is Table 3–8. It was constructed in three steps. First the standard entries in the first and second columns were made. Next the heading $p \vee q$ was added and the entries in it computed using the definition of disjunction and Table 3–4 as a guide. Finally the fourth column heading was added, and the entries under it were computed from those in the third column using the information in Table 3–1 as a guide.

TABLE 3–8 Truth table for $\sim(p \vee q)$.

p	q	$p \vee q$	$\sim(p \vee q)$
T	T	T	F
T	F	T	F
F	T	T	F
F	F	F	T

EXAMPLE 3.11 Construct a truth table for $(p \vee q) \rightarrow p$.

Solution

TABLE 3–9 Truth table for $(p \vee q) \rightarrow p$.

p	q	$p \vee q$	$(p \vee q) \rightarrow p$
T	T	T	T
T	F	T	T
F	T	T	F
F	F	F	T

According to Table 3–5 the proposition $(p \vee q) \rightarrow p$ is false only when $p \vee q$ is true and p is false. Those conditions hold, in Table 3–9, in the third row but in none of the others. Therefore the third entry under $(p \vee q) \rightarrow p$ is an F, and all the others are T's.

Exercise Set 3.2

1. Find the truth value of each of the following propositions.
   (a) The moon is a planet.
   (b) George Washington was the first President of the United States.
   (c) $(2 + 3 = 5)$ and $(1 - 1 = 0)$
   (d) 3 is an odd number and 5 is greater than 2.
   (e) New York is on the eastern seaboard and Alabama is a northern state.
   (f) $p \wedge \sim p$
   (g) $(2 + 2 = 4)$ or $(3 + 1 = 4)$
   (h) $(5 - 2 = 3) \vee (4 + 1 = 3)$
   (i) If Racine is in Wisconsin, then $1 + 1 = 3$.
   (j) If $1 + 1 = 3$, then Racine is in Wisconsin.

(k) $(1 + 1 = 3) \rightarrow$ (2 is an odd number).

(l) A is the first letter of the alphabet if Mexico is in South America.

(m) A is the first letter of the alphabet if and only if Mexico is in South America.

(n) $(1 + 1 = 3) \leftrightarrow (3 + 4 = 5)$

2. Construct truth tables for each of the following:

(a) $\sim p \wedge q$
(b) $\sim (p \wedge q)$
(c) $(\sim p \wedge \sim q)$
(d) $q \rightarrow p$
(e) $p \rightarrow \sim q$
(f) $\sim p \rightarrow p$
(g) $\sim (p \rightarrow q)$
(h) $(p \rightarrow q) \rightarrow p$
(i) $p \rightarrow (q \rightarrow p)$
(j) $p \leftrightarrow (p \rightarrow q)$
(k) $(p \wedge q) \leftrightarrow p$
(l) $p \leftrightarrow (q \rightarrow p)$

3. The symbol $\veebar$ is sometimes used for the *exclusive or*. That is, $p \veebar q$ is true if p is true or if q is true but not if both p and q are true. Make truth tables for:

(a) $p \veebar q$ (b) $(p \veebar q) \rightarrow (p \vee q)$

3.3 TRUTH TABLES FOR COMPLEX PROPOSITIONS

3.3.1 Constructing Truth Tables for Complex Propositions

In Section 3.2 we discussed the problem of making truth tables to show the logical possibilities of some compound propositions. The same methods can be used when the propositions involved are more complex.

EXAMPLE 3.12 Make a truth table for $[p \rightarrow (p \wedge q)] \vee \sim p$.

Solution The truth table is Table 3–10.

TABLE 3–10 Truth table for $[p \rightarrow (p \wedge q)] \vee \sim p$.

p	q	$p \wedge q$	$\sim p$	$p \rightarrow (p \wedge q)$	$[p \rightarrow (p \wedge q)] \vee \sim p$
T	T	T	F	T	T
T	F	F	F	F	F
F	T	F	T	T	T
F	F	F	T	T	T
①	②	③	④	⑤	⑥

The steps in the construction of the table are:

1. Make the standard headings and entries for two propositions p and q (columns ① and ②).
2. Since there are several connectives in the compound proposition, the next step is to break the proposition down into simpler propositions none of which contains more than one connective. For $[p \rightarrow (p \wedge q)] \vee \sim p$, these are p, $p \wedge q$, and $\sim p$. The table should have columns headed by each of these; it already has one headed with p; we add the other headings in columns ③ and ④.
3. The propositions p, $p \wedge q$, and $\sim p$ are used to reconstruct the original proposition. Each time two pieces are connected, a new column heading is added. Thus column ⑤ is headed $p \rightarrow (p \wedge q)$, and column ⑥ is headed with the reconstructed proposition $[p \rightarrow (p \wedge q)] \vee \sim p$.
4. Using the values in columns ① and ②, and the rules given by Table 3–3, the corresponding truth values are entered in column ③.
5. Using the values in column ①, and the rules given by Table 3–1, the corresponding values are entered in column ④.
6. Using the values in columns ① and ③, and the rules of Table 3–5, truth values are computed for column ⑤.
7. The values in column ⑥ are computed from those in columns ⑤ and ④ using the rules given by Table 3–4.

EXAMPLE 3.13 Construct a truth table for $(p \wedge r) \wedge (q \vee \sim r)$.

Solution There are three simple propositions, p, q, and r, in this compound. Proposition p can be either true or false; for each truth value of p, q has two

TABLE 3–11 $(p \wedge r) \wedge (q \vee \sim r)$.

p	q	r	$p \wedge r$	$\sim r$	$q \vee \sim r$	$(p \wedge r) \wedge (q \vee \sim r)$
T	T	T	T	F	T	T
T	T	F	F	T	T	F
T	F	T	T	F	F	F
T	F	F	F	T	T	F
F	T	T	F	F	T	F
F	T	F	F	T	T	F
F	F	T	F	F	F	F
F	F	F	F	T	T	F
①	②	③	④	⑤	⑥	⑦

possible truth values, T and F, and for each truth value of q, r can be either true or false. Therefore there are $2 \times 2 \times 2 = 8$ possible arrangements of truth values for p, q, and r, and there must be eight rows in the truth table—one for each possible arrangement of truth values. The standard way to place the truth values of three simple propositions in a table is to head the columns p, q, and r, respectively, and then put TTTTFFFF in the first column, TTFFTTFF in the second column, and TFTFTFTF in the third column. The completed truth table for $(p \wedge r) \wedge (q \vee \sim r)$ is Table 3–11. The circled numbers are not part of the table but are intended to indicate the order in which the steps in the construction of the table were performed.

Columns ①, ②, and ③ are the standard columns for three simple propositions. Column ④ is filled in with reference to columns ① and ③; column ⑤ is filled in with reference to column ③; column ⑥ is filled in with reference to columns ② and ⑤; and finally, column ⑦ is filled in with reference to columns ④ and ⑥.

EXAMPLE 3.14 Construct a truth table for $[(p \wedge q) \wedge r] \rightarrow (p \vee q)$.

Solution

TABLE 3–12 $[(p \wedge q) \wedge r] \rightarrow (p \vee q)$.

p	q	r	$p \wedge q$	$p \vee q$	$(p \wedge q) \wedge r$	$[(p \wedge q) \wedge r] \rightarrow (p \vee q)$
T	T	T	T	T	T	T
T	T	F	T	T	F	T
T	F	T	F	T	F	T
T	F	F	F	T	F	T
F	T	T	F	T	F	T
F	T	F	F	T	F	T
F	F	T	F	F	F	T
F	F	F	F	F	F	T
①	②	③	④	⑤	⑥	⑦

The truth table is Table 3–12. Column ④ is filled in with reference to columns ① and ②; column ⑤ is filled in with reference to columns ① and ②; column ⑥ is filled in with reference to columns ④ and ③; and column ⑦ is filled in with reference to columns ⑥ and ⑤.

3.3.2 Truth-Table Analysis

Once a truth table of a proposition has been completed, a truth-table analysis of the proposition can be made. An analysis of Table 3–11 shows that the proposition $(p \wedge r) \wedge (q \vee \sim r)$ is true when p, q, and r are all true and false in all other cases.

EXAMPLE 3.15 Under what conditions is the proposition "You are young and married, and you are happy or not married" true?

Solution We are asked to find what the truth values must be for the simple propositions in the compound if the compound is true. The simple propositions may be represented as:

p: You are young.
q: You are happy.
r: You are married.

The compound, in symbols, is:

$$(p \wedge r) \wedge (q \vee \sim r).$$

In Example 3.13 we showed that this proposition is true when p, q, and r are true and it is false in all other cases. Therefore the proposition "You are young and married, and you are happy or not married" is true only when you are young, happy, and married.

With a little care, patience, and time the methods of this section may be used to prepare a table to analyze even very complicated propositions.

Exercise Set 3.3

1. Construct truth tables for each proposition.
   (a) $(p \wedge q) \rightarrow (p \vee q)$
   (b) $\sim (p \vee \sim q)$
   (c) $(p \rightarrow q) \wedge \sim q$
   (d) $(p \vee q) \wedge (\sim p \vee \sim q)$
   (e) $(p \rightarrow q) \leftrightarrow (\sim p \vee \sim q)$
   (f) $(p \wedge q) \wedge \sim r$
   (g) $p \wedge \sim (q \vee r)$
   (h) $(p \rightarrow q) \rightarrow (p \vee r)$
   (i) $[(p \wedge q) \rightarrow r] \leftrightarrow [(p \wedge q) \wedge \sim r]$
   (j) $[(p \rightarrow q) \wedge (q \rightarrow r)] \rightarrow (p \rightarrow r)$
2. Under what conditions is each of these propositions true?
   (a) If you don't drink your milk, you will not have strong bones.
   (b) You drink milk but you do not have strong bones.
   (c) You drink milk if and only if you have strong bones.
   (d) It is not true that if you drink milk you will have strong bones.

(e) The proposition "You drink milk or you do not have strong bones" is false.

(f) You will have strong bones if you drink milk or take calcium tablets.

(g) You do not drink milk, you do take calcium tablets, and your bones are weak.

3. Suppose a proposition involves the four simple propositions p, q, r, and s. How many rows would be necessary in a truth table to show all possible arrangements of truth values?

3.4 LOGICALLY TRUE PROPOSITIONS

A proposition is **logically true** if it is true for every possible arrangement of truth values of its component propositions, and it is **logically false** if it is false for every possible arrangement of truth values of its parts. Logically true propositions are often called **tautologies**.

EXAMPLE 3.16 Consider the propositions:

(a) $p \vee \sim p$

(b) $p \rightarrow (p \vee q)$

(c) $p \wedge \sim p$

(d) $(p \rightarrow q) \wedge (p \wedge \sim q)$

Proposition (a) is a tautology because p and $\sim p$ have opposite truth values; thus one of them must be true in any given case, and so their disjunction is always true. In order for (b) to be false, p must be true and $p \vee q$ false, but this is impossible since $p \vee q$ is true whenever p is true. Thus $p \rightarrow (p \vee q)$ is logically true. Proposition (c) is logically false since p and $\sim p$ cannot both be true at the same time. Proposition (d) is not so easily analyzed. However, its truth table, Table 3–13, shows it to be logically false.

TABLE 3–13 Truth table for $(p \rightarrow q) \wedge (p \wedge \sim q)$.

p	q	$\sim q$	$p \rightarrow q$	$p \wedge \sim q$	$(p \rightarrow q) \wedge (p \wedge \sim q)$
T	T	F	T	F	F
T	F	T	F	T	F
F	T	F	T	F	F
F	F	T	T	F	F

EQUIVALENT Two propositions p and q are **equivalent** if the biconditional $p \leftrightarrow q$ is logically true. If $p \leftrightarrow q$ is logically false, p and q are **contradictories.**

EXAMPLE 3.17 Show that $\sim p \vee q$ is equivalent to $p \rightarrow q$.

Solution From Table 3–14 it can be seen that $(\sim p \vee q) \leftrightarrow (p \rightarrow q)$ is logically true. Thus $\sim p \vee q$ and $p \rightarrow q$ are equivalent, by the definition of equivalent.

TABLE 3–14 Truth table for $(\sim p \vee q) \leftrightarrow (p \rightarrow q)$.

p	q	$\sim p$	$\sim p \vee q$	$p \rightarrow q$	$(\sim p \vee q) \leftrightarrow (p \rightarrow q)$
T	T	F	T	T	T
T	F	F	F	F	T
F	T	T	T	T	T
F	F	T	T	T	T

If two propositions, involving the same basic propositions, are equivalent then both have the same truth values and one can be substituted for the other.

EXAMPLE 3.18 Consider the proposition:

The book is on my desk or it is not in my house.

If "The book is in my house" is represented by p and "The book is on my desk" is represented by q, then the compound is represented by

$$q \vee \sim p$$

and this is equivalent to $\sim p \vee q$. From Example 3.17, we know that $\sim p \vee q$ can be replaced by its equivalent $p \rightarrow q$. Therefore we may rewrite the original proposition as:

If the book is in my house, then it is on my desk.

The two important ideas of this section are these:

1. A proposition that is always true is said to be logically true. Later we shall see that logically true propositions are essential in making a valid argument.
2. If the biconditional $p \leftrightarrow q$ is logically true, p and q have the same truth values and, in an argument, either one of them may be replaced by the other.

Exercise Set 3.4

1. Classify each of the following as logically true, logically false, or neither logically true nor logically false.

   (a) $(p \wedge q) \rightarrow p$

   (b) $p \vee \sim (p \wedge q)$

   (c) $(p \wedge q) \wedge \sim (q \rightarrow p)$

   (d) $[p \vee (p \wedge q)] \rightarrow p$

   (e) $(\sim p \wedge q) \wedge (q \rightarrow p)$

   (f) $\sim [(p \wedge q) \wedge \sim q] \rightarrow \sim p$

2. Some equivalences are so important that names have been given to them for reference purposes. Some of these are:

   (a) $p \vee q; q \vee p$
   (b) $p \wedge q; q \wedge p$
   } *Commutative Properties*

   (c) $p \vee (q \wedge r); (p \vee q) \wedge (p \vee r)$
   (d) $p \wedge (q \vee r); (p \wedge q) \vee (p \wedge r)$
   } *Distributive Properties*

   (e) $p \wedge (q \wedge r); (p \wedge q) \wedge r$
   (f) $p \vee (q \vee r); (p \vee q) \vee r$
   } *Associative Properties*

   (g) $\sim (p \vee q); \sim p \wedge \sim q$
   (h) $\sim (p \wedge q); \sim p \vee \sim q$
   } *DeMorgan's Laws*

   Show that each of the pairs are equivalent.

3. Test each of these pairs for equivalence. Use the properties of Problem 2 and the equivalence shown in Example 3.17 when possible.

   (a) $\sim (\sim p); p$

   (b) $p \vee \sim q; q \rightarrow p$

   (c) $p \wedge \sim q; \sim (q \vee \sim p)$

   (d) $p \rightarrow \sim q; \sim (p \wedge q)$

   (e) $p \vee (\sim p \wedge q); p \vee q$

   (f) $\sim [(p \wedge q) \wedge r]; (p \wedge q) \rightarrow \sim r$

4. Using the properties of DeMorgan from Problem 2, express each of the following in a different form.

   (a) It is not true that the sun is shining and it is raining.

   (b) The game will not be played on Friday or Saturday.

   (c) The sentence "That geometric figure is a square with area 10 square inches" is false.

   (d) The proposition "This metal is not solid or not heavy or it is not gold" is not true.

5. Which of the following propositions have the same meaning as "It is raining or it is snowing"?
   (a) If it is raining, then it is snowing.
   (b) It is raining if it is not snowing.
   (c) It is snowing or it is raining.
   (d) Either it is snowing and it is raining, or it is not raining.
   (e) It is not true that it is not raining and it is not snowing.
6. Compare the commutative, associative, distributive, and DeMorgan's properties of Problem 2 with the corresponding set laws of Section 2.4. Are there any significant differences in the forms between properties which are named the same?

3.5 VARIATIONS OF THE CONDITIONAL

Three propositions closely related to the conditional are its **converse**, its **inverse**, and its **contrapositive**. These are defined as follows:

If $p \rightarrow q$ is a given conditional, then

$q \rightarrow p$ is its **converse**,
$\sim p \rightarrow \sim q$ is its **inverse**,

and

$\sim q \rightarrow \sim p$ is its **contrapositive**.

EXAMPLE 3.19 Consider the proposition:

If it is spring, then the grass is green.

Its converse is:

If the grass is green, then it is spring.

Its inverse is:

If it is not spring, then the grass is not green.

Its contrapositive is:

If the grass is not green, then it is not spring.

The truth tables for $p \rightarrow q$, $q \rightarrow p$, $\sim p \rightarrow \sim q$, and $\sim q \rightarrow \sim p$ are compared in Table 3–15. Observe that a conditional and its contrapositive are equivalent since they have the same truth values, but that a conditional is not equivalent to either its converse or to its inverse.

TABLE 3–15 Truth table comparing truth values of $p \rightarrow q$, its converse, its inverse, and its contrapositive.

p	q	$p \rightarrow q$	$q \rightarrow p$	$\sim p \rightarrow \sim q$	$\sim q \rightarrow \sim p$
T	T	T	T	T	T
T	F	F	T	T	F
F	T	T	F	F	T
F	F	T	T	T	T

There are many propositions which are conditionals although they are not stated in *if · · · then* form. We shall conclude this section with a consideration of a few of the most common of these.

A politician promises "Taxes will fall only if I am elected." Should this promise be interpreted as "If I am elected, then taxes will fall"? Let p represent "Taxes will fall" and q represent "I am elected." The politician's promise is "p is true only if q is true"; and from Table 3–16 it is clear that this is not the same as $q \rightarrow p$ since when the latter proposition is true (rows 1, 2, and 4), it is possible for p to be true (row 2) even though q is false.

TABLE 3–16 Truth table for $q \rightarrow p$.

p	q	$q \rightarrow p$
T	T	T
T	F	T
F	T	F
F	F	T

On the other hand, when $p \rightarrow q$ is true (rows 1, 3, and 4 of Table 3–17), *p is* true *only when q* is true.

We must conclude that the politician's promise is "If taxes fall, it will be because I am elected." He has not promised to lower taxes (although he may hope that the voters think he has), but he has claimed that no one else can lower them.

Suppose that the politician claims "If taxes are to fall, it is necessary that I be elected." He means "I must be elected if taxes are to fall" or "If taxes fall, it will be because I am elected." Thus, using the symbols p and q

TABLE 3–17 Truth table for $p \to q$.

p	q	$p \to q$
T	T	T
T	F	F
F	T	T
F	F	T

as defined above, we must conclude that "q is necessary for p" means "if p then q."

Finally, suppose our friend says "The lowering of taxes is a sufficient reason for my being elected." If p is "the lowering of taxes" and q is "my being elected," his promise is "p is sufficient for q." This is another way of saying "The truth of p is enough to ensure the truth of q." A comparison of Tables 3–16 and 3–17 shows that this condition is satisfied by Table 3–17 but not by Table 3–16. Thus "p is sufficient for q" and "$p \to q$" are equivalent.

In summary, each of the following have the same meaning:

$p \to q$
p only if q
q is necessary for p
p is sufficient for q

EXAMPLE 3.20 Write each of the following as a conditional.

(a) One can pass this course only if one studies.
(b) A necessary condition for a person to vote is that that person be registered.
(c) A sufficient condition for water to be salty is that it be taken from the Pacific Ocean.

Solution

(a) If one passes this course it will be because he has studied.
(b) If one votes, then he is a registered voter.
(c) If water is taken from the Pacific Ocean, then it is salty.

In an earlier section, the promise was made that later you would be shown why we use the truth values that we do in the truth table for $p \leftrightarrow q$. That promise can be kept at this time. Consider Table 3–18 which shows that

$$[(p \to q) \wedge (q \to p)] \leftrightarrow (p \leftrightarrow q)$$

is a tautology.

TABLE 3–18

p	q	$p \to q$	$q \to p$	$(p \to q) \wedge (q \to p)$	$p \leftrightarrow q$	$[(p \to q) \wedge (q \to p)] \leftrightarrow (p \leftrightarrow q)$
T	T	T	T	T	T	T
T	F	F	T	F	F	T
F	T	T	F	F	F	T
F	F	T	T	T	T	T

Therefore $p \leftrightarrow q$ is equivalent to $(p \to q) \wedge (q \to p)$. But $p \to q$ can be read "p only if q" and $q \to p$ can be read "if q then p," and so the proposition $p \leftrightarrow q$ can be read "p if and only if q." This is what prompted the definition given earlier. Also, since $q \to p$ can be interpreted "p is a necessary condition for q" and $p \to q$ can be interpreted as "p is a sufficient condition for q," the proposition $p \leftrightarrow q$ can be read "p is a necessary and sufficient condition for q."

Exercise Set 3.5

1. Let p be "I will miss the train," q be "I will be late to work," and r be "I will lose my job." Represent the following propositions as conditionals.
   (a) If I miss the train and am late to work, I will lose my job.
   (b) Missing the train is a necessary condition for my being late to work.
   (c) My being late to work is a sufficient reason for my losing my job.
   (d) I shall lose my job only if I am late to work.
   (e) I must not be late to work or I will lose my job.
   (f) I shall be late to work only if I miss my train.
2. Write the converse, inverse, and contrapositive of each of the following propositions.
   (a) If it rains, I shall stay indoors.
   (b) If $A \subset B$ then $B \not\subset A$.
   (c) A tree is a conifer if it is a pine.
   (d) The sun is shining only if it is not raining.
3. Suppose p is true and q is false. What is the truth value of:
   (a) $p \to q$
   (b) the converse of $p \to q$
   (c) the contrapositive of $p \to q$
   (d) the inverse of $p \to q$
   (e) q is sufficient for p

(f) q is necessary for p

(g) q only if p

4. Which of these pairs of propositions are equivalent. Use DeMorgan's Laws where they apply.

(a) $p \rightarrow \sim q$; $\sim q \rightarrow p$

(b) $p \rightarrow \sim q$; $q \rightarrow \sim p$

(c) $\sim p \rightarrow \sim q$; $q \rightarrow p$

(d) $p \rightarrow (q \vee r)$; $(\sim q \wedge \sim r) \rightarrow \sim p$

(e) $\sim p \rightarrow (\sim q \wedge r)$; $p \rightarrow (q \wedge \sim r)$

(f) $(p \wedge q) \rightarrow r$; $\sim r \rightarrow (\sim p \vee \sim q)$

(g) $p \vee (q \wedge r)$; $\sim p$ only if $(q \wedge r)$

5. Each of the following is a true conditional. Which of them has a true converse?

(a) If $1 + 1 = 2$, then 2 is even.

(b) If 1 is greater than 2, 3 is even.

(c) If $x = 3$, then x is less than 5.

(d) If $3 + 4 = 5$, then 2 is less than 5.

3.6 PROOF PATTERNS

3.6.1 Arguments

A proposition is a **logical consequence** of a given set S of propositions if it must be true whenever every proposition in set S is true.

EXAMPLE 3.21 Show that q is a logical consequence of the propositions p and $p \rightarrow q$.

Solution Consider the proposition $p \wedge (p \rightarrow q)$. Its truth table is Table 3–19.

TABLE 3–19

p	q	$p \rightarrow q$	$p \wedge (p \rightarrow q)$
T	T	T	T
T	F	F	F
F	T	T	F
F	F	T	F

Since this table shows that both p and $p \rightarrow q$ are true only in row 1 and, in that row, q is also true, we may conclude that q follows logically from p and $p \rightarrow q$.

An argument consists of a set of propositions called the **premises** and one proposition called the **conclusion**. If the conclusion is a logical consequence of the premises, the argument is said to be **valid**; if the conclusion is not a logical consequence of the premises, the argument is **invalid**.

EXAMPLE 3.22 Show that the argument with premises

If you do not study, you will fail.

and

You do not study.

and conclusion

Therefore you fail.

is valid.

Solution Let p be "You do not study" and q be "You fail." Then the premises of the argument are

$$p \rightarrow q \qquad \text{and} \qquad p$$

and the conclusion is

$$q$$

We have seen, in Example 3.21, that q is a logical consequence of p and $p \rightarrow q$. Therefore this argument is valid.

EXAMPLE 3.23 Show that the argument

If you do not study, you will fail.
You failed.
Therefore you did not study.

is not valid.

Solution Let p and q be the propositions:

p: You do not study.
q: You fail.

In symbolic form the premises of the argument are $p \to q$ and q. The conclusion is p. The truth table for $(p \to q) \wedge q$ is Table 3–20.

TABLE 3–20

p	q	$p \to q$	$(p \to q) \wedge q$
T	T	T	T
T	F	F	F
F	T	T	T
F	F	T	F

From this table, we note that $(p \to q) \wedge q$ is true in rows 1 and 3; and in row 1 the conclusion p is true, but in row 3, p is false. Therefore p is not necessarily true when both $p \to q$ and q are true. Thus we conclude that p does not follow as a logical consequence of $p \to q$ and q, and so the argument is invalid.

Note that when an argument is invalid, we cannot say that the conclusion is false but only that it is *impossible to say whether the conclusion is true or false* if we have only the premises by which to judge.

3.6.2 Testing Validity

If an argument has a limited number of premises, two or three for instance, one can always use a truth-table analysis to check the validity of the argument. On the other hand, when there are several premises, the truth-table approach is cumbersome. A more workable way to test validity is to break the argument into sections and compare each section with certain logical argument forms called **proof patterns**. One of these patterns is the **Law of Detachment**.

$$\begin{array}{ll} & p \\ & p \to q \\ \hline \text{Therefore} & q \end{array}$$

This is the argument form verified in Example 3.22. Another proof pattern is the **Law of Syllogism**.

$$\begin{array}{ll} & p \to q \\ & q \to r \\ \hline \text{Therefore} & p \to r \end{array}$$

The Law of Syllogism can be extended to more than two premises, all of which are conditionals. As an example we have

$$\begin{array}{rr} & p \rightarrow q \\ & q \rightarrow r \\ & r \rightarrow t \\ & t \rightarrow s \\ \hline \text{Therefore} & p \rightarrow s \end{array}$$

Note that one "follows the arrows" from proposition p to proposition s in this proof pattern.

The validity of most arguments can be tested using only the Laws of Detachment and Syllogism. To use only these two patterns, it is often necessary to replace one or more of the premises by an equivalent proposition. The most frequently used equivalences are:

	Proposition	Equivalent Proposition
DeMorgan's Laws	$\sim (p \wedge q)$	$\sim p \vee \sim q$
	$\sim (p \vee q)$	$\sim q \wedge \sim q$
Contraposition	$p \rightarrow q$	$\sim q \rightarrow \sim p$
Disjunction and conditional	$\sim p \vee q$	$p \rightarrow q$

EXAMPLE 3.24 Test the validity of each of the following arguments.

(a) My hat is in the closet or it is on the table.
It is not in the closet.
Therefore it is on the table.

(b) If the sun is shining, then the day is warm.
The day is not warm.
Therefore the sun is not shining.

(c) If the sun is shining, then the day is warm.
I do not drink iced tea if the day is not warm.
Therefore, if the sun is shining, I drink iced tea.

Solution (a) Let p and q be defined

p. My hat is in the closet.
q: My hat is on the table.

Then the argument is

$$\begin{array}{rr} & p \vee q \\ & \sim p \\ \hline \text{Therefore} & q \end{array}$$

If $p \vee q$ is replaced by its equivalent

$$\sim p \rightarrow q$$

the argument is

$$\begin{array}{rl} & \sim p \rightarrow q \\ & \sim p \\ \hline \text{Therefore} & q \end{array}$$

and this is the form of the Law of Detachment and is a valid argument.

(b) Let s and w be the propositions

s: The sun is shining.
w: The day is warm.

The argument becomes symbolically

$$\begin{array}{rl} & s \rightarrow w \\ & \sim w \\ \hline \text{Therefore} & \sim s \end{array}$$

If $s \rightarrow w$ is replaced by its contrapositive

$$\sim w \rightarrow \sim s$$

the argument is

$$\begin{array}{rl} & \sim w \rightarrow \sim s \\ & \sim w \\ \hline \text{Therefore} & \sim s \end{array}$$

and this is the Law of Detachment pattern; so the argument is valid.

(c) Using

s: The sun is shining.
w: The day is warm.
t: I drink iced tea.

the argument is

$$\begin{array}{rl} & s \rightarrow w \\ & \sim t \rightarrow \sim w \\ \hline \text{Therefore} & s \rightarrow t \end{array}$$

When $\sim t \rightarrow \sim w$ is replaced by its contrapositive

$$w \rightarrow t$$

the argument is

$$\begin{array}{rl} & s \rightarrow w \\ & w \rightarrow t \\ \hline \text{Therefore} & s \rightarrow t \end{array}$$

This is the proof-pattern form we have called the Law of Syllogism, and therefore the argument is valid.

EXAMPLE 3.25 Show that these arguments are invalid.

(a)

$$\begin{array}{rr} & \sim p \rightarrow q \\ & p \\ \hline \text{Therefore} & \sim q \end{array}$$

(b) If a man is healthy, he will live to be sixty.
Sam celebrated his sixtieth birthday today.
Therefore Sam is healthy.

Solution (a) To show that an argument is invalid, all that one needs to do is to show that the conclusion *can* be false even though the premises are true.

$$\begin{array}{rr} & \sim p \rightarrow q \\ & p \\ \hline \text{Therefore} & \sim q \end{array}$$

Suppose that p and q are both true. Then $\sim p \rightarrow q$ is true and so is p, but $\sim q$ is false. An example which shows that an argument is invalid is called a **counter-example**.

(b) Let p and q be defined

p: A man is healthy.
q: He will live to be sixty.

The argument is

$$\begin{array}{rl} & p \rightarrow q \\ & q \\ \hline \text{Therefore} & p \end{array}$$

As a counter-example to this argument consider what happens when p is false and q is true. In this case both premises of this argument are true, but the conclusion p obviously is not true. (As far as we know Sam may be alive today only through the miracles of modern medicine.)

Exercise Set 3.6

Test the validity of each argument. When an argument is valid, identify the proof pattern or equivalence(s) you would use to verify the argument. When an argument is invalid, give a counter-example.

1. $$\begin{array}{l} p \\ p \to (q \vee r) \\ \hline q \vee r \end{array}$$

2. $$\begin{array}{l} p \vee \sim r \\ \\ \hline r \to p \end{array}$$

3. $$\begin{array}{r} p \\ \sim p \to q \\ \hline q \end{array}$$

4. $$\begin{array}{c} \sim p \vee q \\ \\ \hline \sim (p \wedge \sim q) \end{array}$$

5. $$\begin{array}{l} p \to q \\ p \to r \\ \hline q \to r \end{array}$$

6. $$\begin{array}{c} p \to \sim q \\ \sim q \to \sim r \\ \hline p \to \sim r \end{array}$$

7. $$\begin{array}{l} r \vee s \\ r \\ \hline s \end{array}$$

8. $$\begin{array}{l} p \to q \\ r \to \sim q \\ \hline p \to r \end{array}$$

9. If today is Wednesday, yesterday was Tuesday.
   Yesterday was Tuesday.
   Therefore today is Wednesday.

10. Today is Tuesday or Wednesday.
    Today is not Wednesday.
    Therefore today is Tuesday.

11. Yesterday was Tuesday only if today is Wednesday.
    Today is Wednesday.
    Therefore yesterday was Tuesday.

12. If today is Tuesday, yesterday was Monday.
    If yesterday was Monday, tomorrow is Wednesday.
    Therefore if today is Tuesday, tomorrow is Wednesday.

3.7 VALID ARGUMENTS—PROOF

An argument is valid if its conclusion must be true when its premises are true. A **proof** of an argument is any demonstration of the validity of the argument. The most primitive proof is the kind used in the last section, the proof by truth-table analysis. A more sophisticated kind of proof is the type commonly used in mathematics. It consists of a sequence of propositions. The first proposition in the sequence is one of the premises of the argument, and each proposition after the first one is either a premise or a conclusion that follows logically from one or more of the propositions that precedes it. The last proposition in the sequence must follow logically from propositions which precede it, and it must also be the proposition which was originally stated as the conclusion of the argument.

The following examples show how such proofs are constructed.

EXAMPLE 3.26 Given that p, $\sim q \rightarrow r$, and $p \rightarrow \sim q$ are true, prove that r is true.

Solution The premises p and $p \rightarrow \sim q$ "fit" the premises of the Law of Detachment. That is,

$$\begin{array}{rl} & p \\ & p \rightarrow \sim q \\ \hline \text{Therefore} & \sim q \end{array}$$

Thus $\sim q$ is a logical consequence of p and $p \rightarrow \sim q$. This means that $\sim q$ is true. But by the Law of Detachment,

$$\begin{array}{rl} & \sim q \\ & \sim q \rightarrow r \\ \hline \text{Therefore} & r \end{array}$$

So when $\sim q$ and $\sim q \rightarrow r$ are true, r is also true. Therefore r is true when p, $\sim q \rightarrow r$, and $p \rightarrow \sim q$ are true. To show this proof in a more systematized form, a two-column presentation may be used. The left-hand column is the sequence of propositions in the proof; and in the right-hand column, opposite each proposition, is a brief explanation of why that proposition is true.

When it is written in the two-column form, the above proof is:

1. p is true	1. Premise
2. $p \rightarrow \sim q$ is true	2. Premise
3. $\sim q$ is true	3. Detachment using premises (1) and (2)
4. $\sim q \rightarrow r$ is true	4. Premise
5. Therefore r is true	5. Detachment using (3) and (4) as premises

EXAMPLE 3.27 Given as premises $p \rightarrow q$, $\sim r \rightarrow \sim q$, and $r \rightarrow t$, prove $p \rightarrow t$ is true.

Solution The only proof pattern suggested by the premises of this argument is the Law of Syllogism since that is the one whose premises are all conditionals. We note that $\sim r \rightarrow \sim q$ is equivalent to $q \rightarrow r$ by contraposition. We have

1. $\sim r \rightarrow \sim q$ is true	1. Premise
2. $q \rightarrow r$ is true	2. Contrapositive of (1)
3. $p \rightarrow q$ is true	3. Premise
4. $r \rightarrow t$ is true	4. Premise
5. Thus $p \rightarrow t$ is true	5. Syllogism—premises are (3), (2), and (4)

EXAMPLE 3.28 Prove the argument:

If John wins the contest, he will travel to Europe.
If he travels to Europe, he will visit Paris.
If he does not visit Venice, he will not visit Paris.
John won the contest.
Therefore he will visit Venice.

Solution Let w, t, p, and v be defined by

w: John wins the contest.
t: He will travel to Europe.
p: He will visit Paris.
v: He will visit Venice.

In symbols the argument is

$$\begin{array}{rc} & w \rightarrow t \\ & t \rightarrow p \\ & \sim v \rightarrow \sim p \\ & w \\ \hline \text{Therefore} & v \end{array}$$

A proof is:

1. $\sim v \rightarrow \sim p$ is true	1. Premise
2. $p \rightarrow v$ is true	2. Contrapositive of (1)
3. $w \rightarrow t$ is true	3. Premise
4. $t \rightarrow p$ is true	4. Premise
5. $w \rightarrow v$ is true	5. Syllogism—(3), (4), and (2)
6. w is true	6. Premise
7. Therefore v is true	7. Detachment—(5) and (6)

EXAMPLE 3.29 Prove:

> Little Miss Muffet was not frightened by a spider or Jack jumped over the candlestick.
> If Jack jumped over the candlestick, the cow jumped over the moon.
> The cow did not jump over the moon.
> Therefore Little Miss Muffet was not frightened by a spider.

Solution

p: Miss Muffet was frightened by a spider.
q: Jack jumped over the candlestick.
r: The cow jumped over the moon.

Then

$$\begin{array}{rl} & \sim p \vee q \\ & q \rightarrow r \\ & \sim r \\ \hline \text{Therefore} & \sim p \end{array}$$

We wish to show that $\sim p$ is true provided the premises are all true; therefore it seems desirable to try replacing one of the premises with (*something*) $\rightarrow \sim p$. The only one of the premises which can be replaced in this way is $\sim p \vee q$, which is equivalent to $\sim q \rightarrow \sim p$. In order to draw the conclusion $\sim p$ from $\sim q \rightarrow \sim p$, we must show that $\sim q$ is true. We can do this, using the premise $\sim r$, if we can first show that $\sim r \rightarrow \sim q$ is true, and this we can do since $q \rightarrow r$ is the contrapositive of $\sim r \rightarrow \sim q$ and $q \rightarrow r$ is one of the given premises. We write:

1. $q \rightarrow r$	1. Premise
2. $\sim r \rightarrow \sim q$	2. Contrapositive of (1)
3. $\sim r$	3. Premise
4. $\sim q$	4. Detachment—(3) and (2)
5. $\sim p \vee q$	5. Premise
6. $\sim q \rightarrow \sim p$	6. Disjunction and conditional (5)
7. $\sim p$	7. Detachment—(4) and (6)

The proofs in the last four examples are merely sample proofs; there are many other sequences of propositions which could have been used in each case. The important thing to remember, when you make proofs, is that you should not expect things to go smoothly at first. Take a couple of premises and try to make them fit either the detachment or the syllogism pattern. If they will not do so, lay one of them aside and try another premise in its place. When you do get two premises to fit one of the proof patterns, write that pattern down and, using its conclusion as a new premise, try to form a second proof pattern with one of the original premises. Keep this up until you have used up all the premises. Don't be afraid to replace a premise with an equivalent expression when necessary, and don't be afraid to scrap the whole thing and start over when it becomes evident that you have wandered down a blind alley. With luck and perseverance you will create proofs like those above. Remember, if all else fails you can always revert to truth tables.

Exercise Set 3.7

1. For each of the following arguments give a proof of the argument if it is valid, or show that it is possible for the conclusion to be false even though the premises are all true if the argument is invalid.

(a) Given: $p \rightarrow q$, $r \rightarrow \sim q$, and r
Prove: $\sim p$

(b) Given: $p \rightarrow q$ and $\sim r \rightarrow \sim q$
Prove: $\sim r \rightarrow \sim p$

(c) Given: $p \vee q$ and $\sim q$
Prove: p

(d) Given: $p \rightarrow s$ and $\sim p \rightarrow q$
Prove: $q \vee s$

(e) Given: $p \rightarrow q$, $q \rightarrow \sim s$, $\sim t \rightarrow s$, and $r \rightarrow \sim t$
Prove: $p \rightarrow \sim r$

(f) Given: $\sim (p \wedge q)$ and p
Prove: $\sim q$

(g) Given: $(p \wedge q) \rightarrow r$ and r
Prove: $p \wedge q$

(h) Given: $p \rightarrow r$, $r \rightarrow \sim s$, and $\sim s$
Prove: p

(i) If wishes were horses, beggars would ride.
Wishes are not horses.
Therefore beggars cannot ride.

(j) If the cup is gold, then it is heavier than water.
John can carry the cup only if it is not heavier than water.
Therefore if the cup is gold, John cannot carry it.

(k) If prices are high, then wages are high.
Prices are high or there are price controls.
If there are price controls, then there is not inflation.
There is inflation.
Therefore wages are high.

(l) When the hour is late, I become sleepy.
My being sleepy is sufficient reason for my taking a nap.
I am not taking a nap.
Therefore the hour is not late.

(m) If this is December, then last month was November.
If last month was November, then six months ago it was June.
If six months ago it was June, then eleven months ago it was January.
If next month will be January, then this is December.
Last month was November.
Therefore, this is December. (Use only the premises given.)

(n) Smith is liable if the contract is legal.
If Smith is liable then he must go to jail.
Smith will lose his job only if he goes to jail.
The contract is legal.
Therefore Smith will lose his job.

2. Find a valid conclusion, if possible, for each of the following sets of premises.

(a) $(\sim p \vee q) \rightarrow r$
$\sim (p \wedge \sim q)$
Therefore...

(b) $\sim r \rightarrow (s \rightarrow \sim t)$
$\sim r \vee w$
$\sim p \rightarrow s$
$\sim w$
Therefore...

(c) $p \vee q$
$p \leftrightarrow r$
$\sim q$
Therefore...

(d) If Jones committed the murder, then he was at the scene of the crime.
Jones did not commit the murder.
Therefore...

(e) If the international situation is tense, the stock market will fall.
The international situation is not tense.
Therefore...

(f) If Smith did not enter into the contract or if he performed the contract, then Jones will not win the lawsuit.
If Smith did not finish construction by March 1, he did not perform the contract.
Smith did enter into the contract but he did finish construction by March 1.
Therefore...

(g) The horse named Leadfoot will win the race if and only if the one named Snail's Pace shows.
If Snail's Pace shows, Slow Lightning will come in last.
Slow Lightning will not come in last or Seven-Year-Itch will tie for first.
If Leadfoot does not win, then Seven-Year-Itch will tie for first.
Seven-Year-Itch will never tie for first.
Therefore...

SUMMARY

Terms you should know:

1. Proposition
2. Truth value
3. Negation
4. Conjunction
5. Disjunction
6. Truth table
7. Conditional
8. Biconditional
9. Logically true
10. Logically false
11. Tautology
12. Equivalence
13. Converse
14. Inverse
15. Contrapositive
16. Validity
17. Proof

Proof patterns and equivalent propositions:

1. Law of detachment:

$$\begin{array}{rl} & p \\ & p \rightarrow q \\ \hline \text{Therefore} & q \end{array}$$

2. Law of syllogism:

$$\begin{array}{rl} & p \rightarrow q \\ & q \rightarrow r \\ \hline \text{Therefore} & p \rightarrow r \end{array}$$

3. Law of contrapositive:

$$p \rightarrow q \text{ is equivalent to } \sim q \rightarrow \sim p$$

4. DeMorgan's laws:

$$\sim (p \wedge q) \text{ is equivalent to } \sim p \vee \sim q$$
$$\sim (p \vee q) \text{ is equivalent to } \sim p \wedge \sim q$$

5. Disjunction and conditional:

$$\sim p \vee q \text{ is equivalent to } p \rightarrow q$$

REVIEW EXERCISES

1. Define p, q, and r:

p: Mathematics is easy.
q: Mathematics is fun.
r: Mathematics is useful.

Write in reasonable English, the proposition represented by:

(a) $p \wedge q$ (e) $(p \wedge q) \wedge r$
(b) $p \vee q$ (f) $\sim p \rightarrow r$
(c) $\sim (p \vee q)$ (g) $(p \wedge q) \rightarrow \sim r$
(d) $\sim (p \vee q)$ (h) $q \leftrightarrow (p \wedge r)$

2. Let p, q, and r be propositions such that p is true, q is false, and r is false. Decide whether each of the following is true or false.

(a) $\sim p$ (e) $\sim p \vee \sim (q \vee r)$
(b) $(p \vee q) \vee r$ (f) $(\sim p \vee \sim q) \wedge [(p \vee \sim r) \wedge (q \vee r)]$
(c) $(p \wedge q) \vee (q \wedge r)$ (g) $p \rightarrow (q \vee r)$
(d) $\sim p \vee (q \vee r)$ (h) $(\sim r \rightarrow q) \rightarrow (\sim p \vee q)$

3. Classify the following propositions as logically true, logically false, or neither.

(a) $p \wedge \sim p$ (d) $[p \vee (q \vee \sim r)] \wedge [\sim p \vee (q \vee r)]$
(b) $p \vee \sim p$ (e) $\sim p \rightarrow (p \rightarrow q)$
(c) $(p \wedge q) \vee (p \wedge \sim q)$ (f) $[p \wedge (p \rightarrow q)] \rightarrow q$

4. Write in reasonable English, the converse, the inverse, and the contrapositive of each proposition.

(a) I will get wet if it rains.
(b) If the wind blows or it rains, I shall stay indoors.

5. Let p be "Candy is sweet" and q be "Ice cream is cold." Translate the following into symbolic form:

(a) Candy is sweet only if ice cream is cold.
(b) A necessary condition for ice cream to be cold is that candy is sweet.

(c) A sufficient condition for candy to be sweet is that ice cream be cold.

(d) A necessary and sufficient condition that candy is sweet is that ice cream is cold.

6. Test the validity of these arguments.

(a) Given: $p \to q$ and $r \to \sim q$
Therefore $p \to \sim r$

(b) Given: p, q, $\sim p \to r$, and $q \to \sim r$
Therefore $\sim r$

(c) Given: $\sim s$, $p \to r$, and $(p \to q) \to s$
Therefore r

(d) Given: $p \to (r \to s)$, $\sim r \to \sim p$, and p
Therefore s

(e) Given: $\sim p \to r$, $p \to \sim q$, and $\sim s \to q$
Therefore $\sim r \to s$

(f) Given: $(p \vee q) \to \sim t$, $w \vee r$, $w \to p$, and $r \to q$
Therefore $\sim t$

(g) Given: $(p \wedge q) \to (s \leftrightarrow t)$, $(w \vee x) \to r$, $(s \leftrightarrow t) \to (w \vee x)$, and $\sim r$
Therefore $\sim p \vee \sim q$

(h) We have a good team or it does not win many games.
Our team does not win many games.
Therefore we do not have a good team.

(i) If John makes a success of himself, he will become famous.
If he becomes famous, he will be rich.
If he is rich, he will not be happy.
Therefore, if he makes a success of himself, he will not be happy.

4 ABSTRACT MATHEMATICAL SYSTEMS

The set algebra of Chapter 2 and the logic of Chapter 3 are examples of mathematical systems. A mathematical system consists of a set of elements, at least one operation, at least one relation (usually equivalence), and some properties (rules) which the elements, operations, and relations must obey.

A third mathematical system, which should be familiar to the reader, is the real number system. Set algebra, logic, and the real number system are called **concrete** systems because the elements and properties are meaningful. If no meaning is given to the constituent parts of a system, then it is said to be **abstract**.

It is a phenomenon of mathematics that frequently a system is developed as an exercise in pure mathematics, and then years later someone finds a practical problem which the system seems to explain. One such system, boolean algebra, will be discussed in this chapter. Before we can talk about any particular system, however, we must become familiar with some of the properties which are common to many different systems.

4.1 AN ABSTRACT SYSTEM

4.1.1 Operations

A mathematical system must have at least one operation. Before we attempt a formal definition of an operation, let us examine one within our common experience, ordinary addition of real numbers. Consider the following problems:

$$\begin{array}{cccc} 2 & 1 & 2.5 & \frac{1}{2} \\ +3 & +5 & +3.4 & +\frac{3}{4} \\ \hline \end{array}$$

What do they all have in common? Simply this, each consists of two numbers to be added (the **addends**). Even when an addition problem consists of three or more addends, we still must add them two at a time. This is because addition is a **binary operation**.

Now, how should we read each of the problems above? We should say "add 3 to 2, add 5 to 1, add 3.4 to 2.5, and add $\frac{3}{4}$ to $\frac{1}{2}$." That is, in each case the addends should be considered to be an ordered pair of numbers. It may be difficult to grasp the significance of this at this time since it is a well known fact of ordinary arithmetic that *3 added to 2* and *2 added to 3* both have the same sum, 5. As we proceed with our study, however, it should become clear that it is quite possible to have legitimate mathematical systems in which $2 + 3$ and $3 + 2$ do not have the same answer. We shall study one of these systems in Chapter 6.

Finally, let us consider the answers to the above problems. These are 5, 6, 5.9, and $1\frac{1}{4}$. In general, the answer to any addition problem involving real numbers is a single real number.

We may summarize the addition of real numbers in this way: Addition of real numbers is a process that assigns a single real number to each ordered pair of real numbers. Similarly, multiplication of real numbers is a process that assigns a real number to each ordered pair of real numbers.

We may now give a formal definition of a binary operation.

BINARY OPERATION A **binary operation** on a set S is a process which assigns an element of set T to each of the ordered pairs of S. If $T \subseteq S$, then S is said to be **closed** relative to the operation.

In ordinary arithmetic, the set of counting numbers is closed relative to addition since the sum of any two counting numbers is a counting number. On the other hand, the set of counting numbers is not closed relative to division, for 3 divided by 4 is $\frac{3}{4}$, and $\frac{3}{4}$ is not a counting number. An example which shows that a proposition is not true for all cases is called a **counter-example**. (Recall that we used counter-examples in Chapter 3 to show that some arguments are invalid.) The preceding example is just one of many counter-examples which show that the set of counting numbers is not closed relative to division.

In the mathematical system called *set algebra*, the elements of the system are sets. Union and intersection are binary operations since $A \cup B$ and $A \cap B$ each assign a unique set to the sets A and B. Complementation is a **unary** operation because $\tilde{A}$ assigns a unique set to the single set A. What then can be said about $A \subset B$? Does $\subset$ represent an operation? If so, the definition of an operation tells us that $A \subset B$ must be the name of some set. But $A \subset B$ is not the name of a set; it is merely a statement about sets A and B which may or may not be true. We call such a statement a **relation.**

RELATION In a mathematical system, a **relation** is a proposition involving pairs of elements of the system.

If A and B are two sets, then $A = B$ is a proposition. Therefore *equality* is another of the relations of set algebra.

In arithmetic in the real number system, $3 + 1$ is a symbol for the number 4; thus addition is an operation. On the other hand, $3 + 1 = 4$ is a proposition, and so equality ($=$) is a relation.

4.1.2 A Mathematical System

We are now ready to give an example of an abstract mathematical system. The set of elements in this system will be

$$G = \{A, B, C, D\}$$

The system will have one relation, equality ($=$), and one binary operation. We will need a symbol and a name for the operation. These are to have no meaning, and so let us call the operation "star" and use $*$ as its symbol. The star operation is defined by Table 4–1.

TABLE 4–1

$*$	A	B	C	D
A	B	A	D	C
B	A	B	C	D
C	D	C	A	B
D	C	D	B	A

To perform the star operation on an ordered pair of elements of G, find the horizontal row of the table which is headed (on the left) by the first element and the vertical column headed (at the top) by the second element. The result of the operation is the element in the intersection of the row and column. For example, the shaded part of the table shows that $B * C = C$.

EXAMPLE 4.1 Using the Table 4–1, compute $B * A$, $C * B$, and $D * C$.

Solution

$$B * A = A$$
$$C * B = C$$
$$D * C = B$$

4.1.3 Properties of Systems

We are now in a position to investigate some of the properties of the star system. We observe that all entries in the body of Table 4–1 are themselves elements of the set G. Because of this the system is closed relative to the operation $*$. (Of course, the star table was deliberately constructed so that the system would exhibit closure; we could as easily have put some element other than A, B, C, or D into the body of the table. In that case the system would not have been closed relative to the operation. Closure it 1ot a necessary property for a mathematical system; sometimes it may not even be desirable.)

Note that the results of $B * C$ and $C * B$ are both C. If one takes the trouble to check, it will be seen that the result of performing the star operation on any two elements of the set is the same without regard to the order in which the two elements are selected. The reason for this is that the star operation table was constructed in such a way that "star" would be a **commutative** operation. In formal language we say:

COMMUTATIVE PROPERTY A binary operation $*$ is **commutative** if for every pair of elements x and y, $x * y = y * x$.

Testing a mathematical system for commutativity with respect to a particular operation would be a formidable task if it were not for one fact. When the elements of a system obey a commutative property relative to an operation, then the operation table for that operation displays a symmetry with respect to a diagonal line from the upper left to the lower right corner of the table. That is, the top row of the table is identical to the left-hand column; the second row from the top of the table is identical to the second column from the left; and so on. As an example, consider Table 4–1. The elements in the first row are B, A, D, C in that order and the elements of the left-hand column are also B, A, D, C in that order. The elements of the second row and second column are A, B, C, D in that order, and so on.

The star operation is a binary operation, that is, it may only be used to combine two elements at a time. When a series of three or more elements are to be operated upon by the same binary operation, combine the first element with the second and then that result with the third element and so on until all the elements have been combined.

EXAMPLE 4.2 Compute $A * C * B * D$.

Solution First operate on A with C, then operate on that result with B, then operate on that result with D; thus

$$\begin{aligned} A * C * B * D &= D * B * D \\ &= D * D \\ &= A \end{aligned}$$

Parentheses are used, as they are in arithmetic and algebra, to set off parts of a problem that are to be treated separately.

EXAMPLE 4.3 Compare the results of $(D * A) * A$ and $D * (A * A)$.

Solution Perform the operation in the parentheses first. Thus

$$\begin{aligned} (D * A) * A &= C * A \\ &= D \end{aligned}$$

and

$$\begin{aligned} D * (A * A) &= D * B \\ &= D \end{aligned}$$

Example 4.3 illustrates a third property of the star system, **associativity**.

ASSOCIATIVE PROPERTY An operation $*$ is **associative** if for all elements x, y, and z of the system, $(x * y) * z = x * (y * z)$.

Example 4.3 does not prove that the star system is associative. To do that we would have to check every problem that could be made by replacing the x, y, and z of the definition by A's, B's, C's, and D's.

Exercise Set 4.1

Using the "star" system compute the following:

1. $A * B$
2. $C * (A * D)$

3. $A * (D * D) * A$

4. $(D * C) * (A * C)$

In Problems 5 to 7 consider the system in which the elements are $S = \{1, 2, 3, 4, 5, 6\}$, the only relation is equality (=), and the operation $\times$ is defined by the table.

$\times$	1	2	3	4	5	6
1	1	2	3	4	5	6
2	2	1	6	5	4	3
3	3	5	1	6	2	4
4	4	6	5	1	3	2
5	5	3	4	2	6	1
6	6	4	2	3	1	5

5. Is the system closed relative to $\times$? Explain.

6. Compute 3×4 and 4×3. Do the results indicate that $\times$ is commutative? Explain.

7. Compute $(1 \times 4) \times 3$ and $1 \times (4 \times 3)$. Do the results show that $\times$ is associative? Explain.

Test the following operation tables for closure and commutativity. Which of the systems appear to be associative?

8.

!	*A*	*B*	*C*	*D*
A	*D*	*C*	*B*	*A*
B	*C*	*D*	*A*	*B*
C	*B*	*A*	*D*	*C*
D	*A*	*B*	*C*	*D*

9.

+	0	1	2	3	4
0	0	1	2	3	4
1	1	2	3	4	5
2	2	3	4	5	6
3	3	4	5	6	7
4	4	5	6	7	8

10.

@	#	$	%	¢
#	#	$	%	¢
$	$	#	%	%
%	%	%	#	$
¢	¢	%	$	#

11.

M	1	2	3	4
1	1	2	3	4
2	2	4	2	4
3	3	2	1	4
4	4	4	4	4

12. Give counter-examples to prove that subtraction and division are not commutative operations in the system of ordinary arithmetic.

13. For the system of logic: (*a*) What are the elements? (*b*) What are the binary operations? (*c*) There is a unary operation—what is it? (*d*) Name a relation of the system.

14. Define a binary operation $\sim$ on the set $\{0, 1, 2, 3, 4, 5\}$ as follows:

    If $a = b$, $a \sim b = 0$.
    If $a \neq b$, $a \sim b =$ the larger of the numbers minus the smaller one.
    For example, $2 \sim 2 = 0$, $5 \sim 3 = 2$, $2 \sim 3 = 1$.

    (*a*) Construct the $\sim$ table.

    (*b*) Show that the set is closed relative to $\sim$.

    (*c*) Show that $\sim$ is commutative.

    (*d*) Show that $\sim$ is not associative.

4.2 OTHER PROPERTIES OF SYSTEMS

4.2.1 Identity and Inverse Elements

Let us reconsider the system of Section 4.1. The table for this system is

$*$	A	B	C	D
A	B	A	D	C
B	A	B	C	D
C	D	C	A	B
D	C	D	B	A

There is one element, B, which leaves every element unchanged with respect to the $*$ operation. We call B the **identity** element for $*$, that is, $B * x = x$ and $x * B = x$ for every $x \in G$.

IDENTITY ELEMENT If there is a single element I of a system with operation $*$ such that for every other element x, $x * I = x$ and $I * x = x$, then I is the **identity element** for the operation $*$.

Note that in general an operation may not be commutative over the entire set of elements, but in order for I to be an identity element the operation must be commutative over every two-element subset, one of whose elements is I.

If there exists an identity element, then there may be pairs of elements called **inverses**. That is:

INVERSES If I is an identity element for the operation $*$ and x and y are two elements, (or possibly the same element) and $x * y = I$ and $y * x = I$, then x and y are said to be $*$ **inverses**.

EXAMPLE 4.4 Consider the set of numbers $\{\ldots, -3, -2, -1, 0, 1, 2, 3, \ldots\}$ with the operation being that of ordinary addition. What is the identity element for addition? Do any elements have inverses under addition?

Solution Since 0 is the only number which does not change a number when added to the number, 0 is the identity for addition. By definition of negative numbers, we know that for any number a, $a + (-a) = 0$ and $(-a) + a = 0$. Therefore 1 and -1 are inverses, 2 and -2 are inverses, 3 and -3 are inverses, and so on. Also 0 is its own inverse.

EXAMPLE 4.5 Which elements of the star system of Section 4.1 are inverses?

Solution The identity element is B; therefore we are looking for pairs of elements which combine to produce B. We see that

$$A * A = B$$
$$B * B = B$$
$$C * D = B \text{ and } D * C = B$$

Therefore A is its own inverse, B is its own inverse, and C and D are inverses of each other.

4.2.2 Distributivity

EXAMPLE 4.6 Consider the system whose elements are $\{a, b, c\}$ in which two operations, $\oplus$ and $\otimes$ are defined by the following tables:

TABLE 4–2

$\oplus$	a	b	c
a	a	a	a
b	a	c	b
c	a	b	c

$\otimes$	a	b	c
a	a	b	c
b	b	c	a
c	c	a	b

Compare the results of

$$b \oplus (a \otimes c) \qquad \text{and} \qquad (b \oplus a) \otimes (b \oplus c)$$

Solution

$$\begin{aligned} b \oplus (a \otimes c) &= b \oplus c \\ &= b \\ (b \oplus a) \otimes (b \oplus c) &= a \otimes b \\ &= b \end{aligned}$$

The results are the same.

Example 4.6 illustrates a property called the **distributive** property of $\oplus$ over $\otimes$. The rule is:

DISTRIBUTIVE PROPERTY If for every three elements x, y, and z of a system with two operations $\oplus$ and $\otimes$

$$x \oplus (y \otimes z) = (x \oplus y) \otimes (x \oplus z)$$

then $\oplus$ is said to **distribute** over $\otimes$.

EXAMPLE 4.7 In ordinary arithmetic $3 \times (4 + 2) = (3 \times 4) + (3 \times 2)$, and in general, the product of any number and the sum of two numbers is the same as the result when both numbers inside the parentheses are multiplied by the number outside and then the products are added. Therefore we say that multiplication distributes over addition.

A system with several operations may have no distributive property, it may have one, as is true in ordinary arithmetic, or it may have a distributive property for more than one operation. As an example of the latter of these possibilities, recall that in set algebra, union distributes over intersection and intersection distributes over union.

In a noncommutative system with operations $+$ and $\times$ it may happen that $x + (y \times z) = (x + y) \times (x + z)$ and $(y \times z) + x = (y + x) \times (z + x)$ but that $x + (y \times z) \neq (y \times z) + x$ for every possible replacement of x, y, and z by elements of the system. Or it may happen that one of $x + (y \times z) = (x + y) \times (x + z)$ or $(y \times z) + x = (y + x) \times (z + x)$ is true but the other is not. In such cases it is necessary to consider both a left distributive property, $x + (y \times z) = (x + y) \times (x + z)$, and a right distributive property, $(y \times z) + x = (y \times x) + (z \times x)$.

Exercise Set 4.2

1–4. Check each of the systems in Problems 8 to 11 of Exercise Set 4.1 to determine:

(a) What element, if any, is the identity?

(b) What elements, if any, have inverses?

5. Consider the system whose operation table is:

#	0	1	2	3	4	5
0	0	1	2	3	4	5
1	1	0	1	2	3	4
2	2	1	0	1	2	3
3	3	2	1	0	1	2
4	4	3	2	1	0	1
5	5	4	3	2	1	0

(a) Is the system closed relative to the operation?

(b) Is the system commutative for #?

(c) Give a counter-example showing that the system is not associative.

(d) What is the identity for #?

(e) What elements are inverses?

6. Given the system whose set is {0, 1} with + and • defined:

+	0	1
0	0	1
1	1	1

•	0	1
0	0	0
1	0	1

(a) Does + distribute over • ?

(b) Does • distribute over + ?

(c) What is the identity for + ? For • ?

(d) Which elements have inverses relative to each operation?

7. In ordinary arithmetic, what is the multiplicative identity? What is the multiplicative inverse of $\frac{3}{4}$?

4.3 SPECIAL SYSTEMS

4.3.1 Clock Arithmetic

Constructing an abstract system is a game anyone can play. All one has to do is select some set of elements, decide upon a symbol for an operation, and fill in the corresponding operation table. Of course the finished system may have few or none of the properties we have been discussing. The game becomes more of a challenge when one tries to build a system having all of the properties or some specific properties but not all of them. Another variation of the game is to pick a pattern for the operation and then when the table has been completed to test it to see what properties it exhibits.

EXAMPLE 4.8 Consider the "four-hour clock" of Figure 4–1. The clock has only one hand, which moves in the usual "clockwise" direction. We shall take

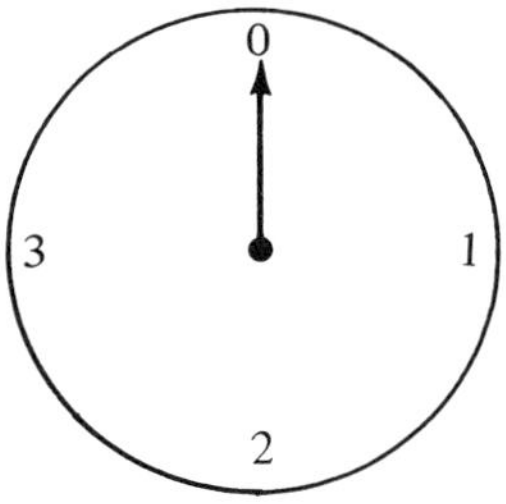

FIGURE 4–1 A four-hour clock

as the elements of a system the numerals {0, 1, 2, 3} from the face of the clock. Addition, +, will be defined in the following way. To find the sum $3 + 2$, let the hand of the clock start at 3 and then move two more divisions in the clockwise direction. The hand will come to rest on 1; thus $3 + 2 = 1$. Similarly, $1 + 3 = 0$ because if the hand moves from the 1 through three divisions it stops at 0. The complete table for this operation is:

TABLE 4–3

+	0	1	2	3
0	0	1	2	3
1	1	2	3	0
2	2	3	0	1
3	3	0	1	2

Is the system closed relative to addition? Is the operation + commutative? Is it associative? What is the identity for +? Which elements have inverses?

We may define multiplication, ×, on the four-hour clock in the following way. The product 2×3 is found by computing the sum $2 + 2 + 2$. To perform this computation, start the hand at 2 and then move it two divisions at a move for two moves. The hand will come to rest on 2; therefore $2 \times 3 = 2$. Similarly 0×2 is computed by starting the clock hand at 0 and then *not moving it* (0 + 0). Thus $0 \times 2 = 0$. Following is the multiplication table for arithmetic on the four-hour clock. Is the system closed relative to $\times$? Is $\times$ commutative? Is $\times$ associative? What is the identity for $\times$? Which elements have inverses under $\times$?

TABLE 4–4

×	0	1	2	3
0	0	0	0	0
1	0	1	2	3
2	0	2	0	2
3	0	3	2	1

Clock arithmetic is an example of a special mathematical system called a **modular arithmetic**. Modular arithmetics are used extensively in the study of number theory. We shall investigate some properties and applications in Chapter 5.

4.3.2 A Permutation System

Another interesting system has as its elements the set of motions that bring an equilateral triangle into coincidence with itself.

EXAMPLE 4.9 Cut an equilateral triangle out of stiff paper and label its vertices *A*, *B*, and *C*. Enter the labels on the reverse side of the paper, also, so that each vertex has the same letter on the back as on the front.

Begin with the triangle placed on a level surface so that one side, *AB* for instance, is horizontal and the opposite vertex *C* lies above *AB* as shown in Figure 4–2.

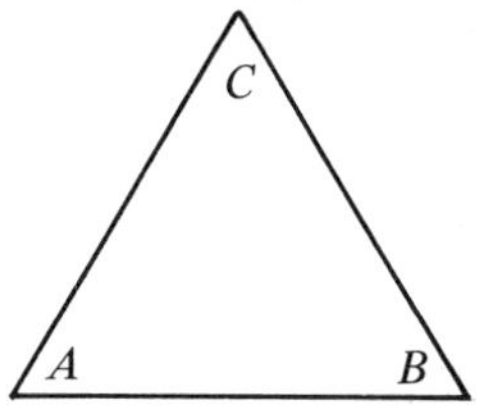

FIGURE 4–2

As we discover motions that bring the triangle into coincidence with itself, we shall assign a name to each one (see Figure 4–3). One motion that will work is a clockwise rotation of 120° about the point *P*. This motion is shown in the second row of Figure 4–3. The result of such a rotation puts *A* in place of *C*, *C* in place of *B*, and *B* in place of *A*. Let us call this rotation 1. Another motion which works is a clockwise rotation of 240° (see row three of Figure 4–3.) It puts *A* in place of *B*, *B* in place of *C*, and *C* in place of *A*. Let us call this rotation 2. Another "motion" that works is no motion at all (row 1 of Figure 4–3.) Let us call this motion 0.

We shall consider any two motions to be the same if they have the same effect. A clockwise rotation of 360° is the same as no motion at all so we will call it 0 also. A counterclockwise rotation of 120° has the same effect as a

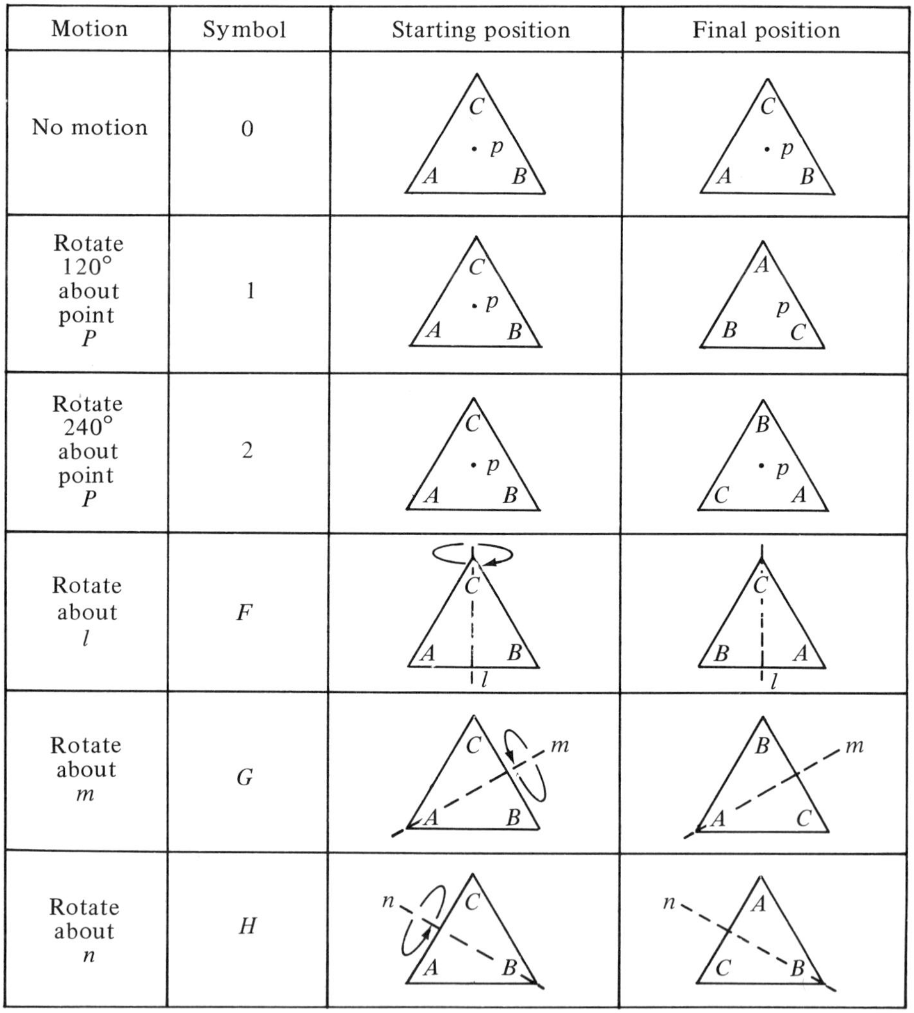

FIGURE 4–3

clockwise rotation of 240°, and so we will call it 2. Thus if we do not lift the triangle from the table, there are only three classes of motions which will place the triangle in coincidence with itself: a 1 motion, a 2 motion, and a 0 motion.

If we allow the triangle to be lifted from the table, there are three additional motions. In one of them we flip the triangle over to bring the back face up while leaving the top vertex where it is. Starting with the triangle in the position shown in Figure 4–2, this motion interchanges A and B and leaves C alone. Let us call this motion F (row 4 of Figure 4–3). A similar motion which keeps the vertex on the left fixed, while the top vertex and the one on the right are interchanged, we will call G (row 5 of Figure 4–3). One which keeps the vertex on the right fixed but interchanges the one on the left and the top one we will call H (row 6 of Figure 4–3).

We now have a system with six elements, 0, 1, 2, F, G, and H. We define an operation & as follows: If x and y are any two of these motions, the composition $x \,\&\, y$ is the motion that results when the two motions are performed one after the other with x performed first and y taking over where x leaves off. For example, $2 \,\&\, F$ means first perform 2 and then F. If we start with the triangle positioned as in Figure 4–2, 2 puts B at the top, C on the left, and A on the right (see Figure 4–3). Motion F applied to the triangle in this position leaves B at the top and puts A on the left and C on the right. The effect is the same as if the single motion G had been used. Thus $2 \,\&\, F = G$. Some other combined movements are:

$$\begin{aligned} 2 \,\&\, 2 &= 1 \\ G \,\&\, 1 &= H \\ G \,\&\, H &= 1 \end{aligned}$$

The construction of the table for operation & is left as an exercise.

4.3.3 Groups and Fields

GROUP A mathematical system with one operation is called a **group** if:

1. It is closed with respect to the operation.
2. It is associative with respect to the operation.
3. It has an identity for the operation.
4. Every element has a unique inverse with respect to the operation.

If, in addition to the above properties, the system is commutative it is said to be a **commutative group**.

A group is the simplest mathematical system which, ordinarily, receives any extensive treatment. There are other systems which are important enough to have names; these are **rings**, **integral domains**, and **fields**. Unfortunately we have neither the time nor the space to treat all of these here. We will describe a field since the mathematical system with which you have the most experience is one of these.

FIELD A **field** is a set with two operations $+$ and $\cdot$ which satisfies the following:

1. The set is closed relative to both operations.
2. Both operations are commutative.
3. Both operations are associative.
4. There is a unique identity 0 for $+$.
5. There is a unique identity 1 for $\cdot$
6. For every element a, there is an element $-a$, such that $a + (-a) = (-a) + a = 0$; a and $(-a)$ are inverses.
7. For every element a, except 0, there is an element a^{-1} such that $a \cdot a^{-1} = a^{-1} \cdot a = 1$.
8. $\cdot$ distributes over $+$.

The choice of $+$ and $\cdot$ as symbols for the operations and 0 and 1 as symbols for the identities is an arbitrary one; any other symbols would serve as well.

Exercise Set 4.3

1. Is the system {0, 1, 2, 3} with operation $+$ as defined in Example 4.8 a group? Explain.
2. Is the system {0, 1, 2, 3} with operation $\times$ as defined in Example 4.8 a group?
3. Following the method of Example 4.8, construct addition and multiplication tables for arithmetic on a five-hour clock. Is the system with operation $+$ a group? Is the system with operation $\times$ a group? If 0 is not considered, is the system a group?
4. Consider the system $\{1, -1, i, -i\}$ with operation $\cdot$ defined by this table:

	1	-1	i	$-i$
1	1	-1	i	$-i$
-1	-1	1	$-i$	i
i	i	$-i$	-1	1
$-i$	$-i$	i	1	-1

Given that $\cdot$ is an associative operation in this system, show that this system is a commutative group.

5. Complete the operation table for Example 4.9, and then answer these questions:

   (a) What is the identity for &? (b) What is the inverse of 2? Of F? (c) Is the system commutative relative to &? (d) Given that the system is associative, is it a group?

6. Consider the system $\{a, b, c\}$ with two operations + and · defined:

+	a	b	c
a	a	b	c
b	b	c	a
c	c	a	b

·	a	b	c
a	a	a	a
b	a	b	c
c	a	c	b

   Does the system satisfy all the field properties?

7. Does the system of Problem 3 satisfy all the field properties?

4.4 BOOLEAN ALGEBRA

In this section, we shall consider a particular abstract system, **boolean algebra**. In the remaining sections of this chapter, we shall see how boolean algebras may be used to solve problems involving simple electrical networks.

BOOLEAN ALGEBRA A system is a **boolean algebra**† if it consists of a set B, closed relative to two operations + and · and satisfies these properties (axioms).

A_1: Each of the operations is commutative.

A_2: There are unique identity elements in B, one for each operation. Call the identity for +, 0, and that for ·, 1. Then for each element a in B, $a + 0 = 0 + a = a$ and $a \cdot 1 = 1 \cdot a = a$.

A_3: For every element a in B, there exists a unique element a' in B such that

$$a + a' = 1 \qquad \text{and} \qquad a \cdot a' = 0$$

Call a' the **complement** of a.

A_4: Each operation distributes over the other. That is,

$$a + (b \cdot c) = (a + b) \cdot (a + c)$$

and

$$a \cdot (b + c) = (a \cdot b) + (a \cdot c)$$

† Named for the British mathematician George Boole (1815–1864) who first studied this particular system from an algebraic viewpoint.

Once again it should be noted that there is no particular reason for the choice of $+$ and $\cdot$ to represent the operations. Any other representations would be acceptable. The same is true for the choice of 0 and 1 for the identities.

EXAMPLE 4.10 Show that the operation tables below define a boolean algebra on the set $B = \{0, 1\}$.

Solution

TABLES 4–5

+	0	1
0	0	1
1	1	1

·	0	1
0	0	0
1	0	1

A casual glance at the tables is enough to verify the closure properties, and it is not much harder to verify properties A_1 and A_2.

To verify property A_3, we must find, for each element a in B, a unique element a' such that $a + a' = 1$ and $a \cdot a' = 0$. There are only two elements in the system, 0 and 1, and surely enough $1 + 0 = 0 + 1 = 1$ and $1 \cdot 0 = 0 \cdot 1 = 0$. Therefore 0 and 1 are complements ($0 = 1'$ and $1 = 0'$).

Verification of A_4 takes a bit longer. To show that addition (+) distributes over multiplication (·), we must verify, by the above tables, that:

$$\begin{aligned}
1 + (1 \cdot 1) &= (1 + 1) \cdot (1 + 1)\\
1 + (1 \cdot 0) &= (1 + 1) \cdot (1 + 0)\\
1 + (0 \cdot 1) &= (1 + 0) \cdot (1 + 1)\\
1 + (0 \cdot 0) &= (1 + 0) \cdot (1 + 0)\\
0 + (1 \cdot 1) &= (0 + 1) \cdot (0 + 1)\\
0 + (1 \cdot 0) &= (0 + 1) \cdot (0 + 0)\\
0 + (0 \cdot 1) &= (0 + 0) \cdot (0 + 1)\\
0 + (0 \cdot 0) &= (0 + 0) \cdot (0 + 0)
\end{aligned}$$

To show that "multiplication" distributes over "addition," we must verify:

$$\begin{aligned}
1 \cdot (1 + 1) &= (1 \cdot 1) + (1 \cdot 1)\\
1 \cdot (1 + 0) &= (1 \cdot 1) + (1 \cdot 0)\\
1 \cdot (0 + 1) &= (1 \cdot 0) + (1 \cdot 1)\\
1 \cdot (0 + 0) &= (1 \cdot 0) + (1 \cdot 0)\\
0 \cdot (1 + 1) &= (0 \cdot 1) + (0 \cdot 1)\\
0 \cdot (1 + 0) &= (0 \cdot 1) + (0 \cdot 0)\\
0 \cdot (0 + 1) &= (0 \cdot 0) + (0 \cdot 1)\\
0 \cdot (0 + 0) &= (0 \cdot 0) + (0 \cdot 0)
\end{aligned}$$

The actual verification is up to you, the student.

4.4.1 Properties of a Boolean Algebra

A boolean algebra obeys a number of properties (theorems) which are logical consequences of A_1 through A_4. The following are some of these:

T_1: (*Principle of duality*) Every statement which can be derived from the axioms remains valid if the operations $+$ and $\cdot$ are interchanged and the elements 0 and 1 are interchanged.

T_2: (*Idempotent laws*) For each element, a, $a + a = a$ and $a \cdot a = a$.

T_3: For every element a,

$$a + 1 = 1 \qquad \text{and} \qquad a \cdot 0 = 0$$

T_4: (*Absorbtion laws*) For each pair of elements a and b,

$$a + (a \cdot b) = a \qquad \text{and} \qquad a \cdot (a + b) = a$$

T_5: (*Associative laws*) For every choice of a, b, and c,

$$(a + b) + c = a + (b + c) \qquad \text{and} \qquad (a \cdot b) \cdot c = a \cdot (b \cdot c)$$

T_6: (*Double complementation law*) For every element a,

$$(a')' = a.$$

T_7: $0' = 1$ and $1' = 0$.

T_8: (*DeMorgan's laws*) For every a and b,

$$(a \cdot b)' = a' + b' \qquad \text{and} \qquad (a + b)' = a' \cdot b'$$

It is simple enough to verify T_1 through T_8 for the boolean algebra of Example 4.10. It is quite another thing to do so for boolean algebras in general. Rest assured that no one expects you to be able to do this. The next example is offered merely to illustrate how proofs may be done.

EXAMPLE 4.11 Prove T_3: $a + 1 = 1$ and $a \cdot 0 = 0$.

Solution Let us first prove $a \cdot 0 = 0$.

$$\begin{aligned}
a \cdot 0 &= 0 + (a \cdot 0) && (\text{by } A_2)\\
&= (a \cdot a') + (a \cdot 0) && (\text{by } A_3)\ (a \cdot a' = 0)\\
&= a \cdot (a' + 0) && (\text{by } A_4)\ (\cdot \text{ distributes over } +)\\
&= a \cdot a' && (\text{by } A_2)\ (a' + 0 = a')\\
&= 0 && (\text{by } A_3)\ (a \cdot a' = 0)
\end{aligned}$$

To prove that $a + 1 = 1$, we use T_1, the principle of duality. In $a \cdot 0 = 0$ (above) replace the $\cdot$ by $+$ and the 0 by 1 and we have:

$$a + 1 = 1.$$

It is a characteristic of mathematics that few proofs are difficult after someone shows how to do them. Look at another one; this type will occur in the exercises.

EXAMPLE 4.12 Prove $a + ([a \cdot b] + [a \cdot c]) = a$.

Solution

$$\begin{aligned} a + [(a \cdot b) + (a \cdot c)] &= [a + (a \cdot b)] + (a \cdot c) && (by\ T_5) \\ &= a + (a \cdot c) && (by\ T_4) \\ &= a && (by\ T_4) \end{aligned}$$

4.4.2 Comparison of Boolean Algebra, Symbolic Logic, and Set Algebra

In your study of symbolic logic and set algebra in the preceding chapters you may have noticed similarities between the two systems. Did it occur to you that they are not really different at all? Both of them are concrete applications of boolean algebra. The following table shows the similarities between the three systems.

Boolean Algebra	Symbolic Logic	Set Algebra
Set $B = \{a, b, c, \ldots\}$	Set of all propositions	Set of all subsets of some universal set U
Binary operation, $+$	Binary operation, $\vee$	Binary operation, $\cup$
Binary operation, $\cdot$	Binary operation, $\wedge$	Binary operation, $\cap$
Unary operation, a'	Unary operation, $\sim p$	Unary operation, $\tilde{A}$
Special element, 1	Any logically true proposition	The universal set U
Special element, 0	Any logically false proposition	The empty set $\varnothing$
Relation $=$	Relation $\leftrightarrow$	Relation $=$

The interpretations of the Axioms A_1 through A_4 are shown below.

Boolean Algebra	Symbolic Logic	Set Algebra
A_1 *Commutativity*		
$a + b = b + a$	$p \vee q \leftrightarrow q \vee p$	$A \cup B = B \cup A$
$a \cdot b = b \cdot a$	$p \wedge q \leftrightarrow q \wedge p$	$A \cap B = B \cap A$
A_2 *Identity Elements*		
0 for +	A logically true proposition for $\vee$	$\varnothing$ for $\cup$
1 for ·	A logically false proposition for $\wedge$	U for $\cap$
A_3 *Complementation*		
$a + a' = 1$	$(p \vee \sim p) \leftrightarrow$ any logically true proposition	$A \cup \tilde{A} = U$
$a \cdot a' = 0$	$(p \wedge \sim p) \leftrightarrow$ any logically false proposition	$A \cap A = \varnothing$
A_4 *Distributivity*		
$a \cdot (b + c) = (a \cdot b) + (a \cdot c)$	$p \wedge (q \vee r) \leftrightarrow (p \wedge q) \vee (p \wedge r)$	$A \cap (B \cup C) = (A \cap B) \cup (A \cap C)$
$a + (b \cdot c) = (a + b) \cdot (a + c)$	$p \vee (q \wedge r) \leftrightarrow (p \vee q) \wedge (p \vee r)$	$A \cup (B \cap C) = (A \cup B) \cap (A \cup C)$

Each of theorems T_1 through T_8 have parallels in logic and in set algebra, and in fact, any theorem which can be proved in one of the systems corresponds to a similar theorem in each of the other systems. In the following sections we shall see how any expression in boolean algebra can be interpreted physically as an electric circuit. It is possible, therefore, to write a logical argument in terms of the symbols of boolean algebra and then build an electrical model of the argument. Conversely, it is possible to describe an electric circuit by means of the symbols of logic. This connection between logic and electric circuitry has been fundamental to the continuing improvement of the modern digital computer.

Exercise Set 4.4

1. Show that the set $\{a, b, c, d\}$ with operations + and · defined in these tables is a boolean algebra. (NOTE: In this system a is the symbol for the additive identity 0, and b is the symbol for the multiplicative identity, 1.)

+	a	b	c	d
a	a	b	c	d
b	b	b	b	b
c	c	b	c	b
d	d	b	b	d

·	a	b	c	d
a	a	a	a	a
b	a	b	c	d
c	a	c	c	a
d	a	d	a	d

2. For the system of Problem 1 show that these four expressions are equal:

   (a) $(a + b) \cdot (a' + c) \cdot (b + c)$

   (b) $(a \cdot c) + [(a') \cdot b] + (b \cdot c)$

   (c) $(a + b) \cdot (a' + c)$

   (d) $[a \cdot c] + [(a') \cdot b]$

3. For the system of Problem 1 show that

$$(a \cdot b \cdot c) + (a' \cdot b' \cdot c') = a$$

4. Show that the set $\{a, b, c, d, e\}$ with $+$ and $\cdot$ defined by these tables is not a boolean algebra. Which axioms are not satisfied?

+	a	b	c	d	e
a	a	b	c	d	e
b	b	b	b	b	b
c	c	b	c	b	b
d	d	b	b	d	b
e	e	b	b	b	e

·	a	b	c	d	e
a	a	a	a	a	a
b	a	b	c	d	e
c	a	c	c	a	a
d	a	d	a	d	a
e	a	e	a	a	e

Supply the proper axiom or theorem numbers to complete these proofs concerning boolean algebras in Problems 5 and 6.

5. Prove: $x \cdot (x' + y) = x \cdot y$

   Proof:

$$\begin{aligned} x \cdot (x' + y) &= (x \cdot x') + (x \cdot y) && (\textit{Why?}) \\ &= 0 + (x \cdot y) && (\textit{Why?}) \\ &= x \cdot y && (\textit{Why?}) \end{aligned}$$

6. Prove: $x' + (x \cdot y') = (x \cdot y)'$

   Proof:

$$\begin{aligned} x' + (x \cdot y') &= (x' + x) \cdot (x' + y') && (\textit{Why?}) \\ &= 1 \cdot (x' + y') && (\textit{Why?}) \\ &= (x' + y') && (\textit{Why?}) \\ &= (x \cdot y)' && (\textit{Why?}) \end{aligned}$$

Using the axioms and theorems of boolean algebra, prove:

7. $x + (x' \cdot y) = x + y$
   (HINT: Use Problem 6)
8. $(x \cdot y) + (x \cdot y') = x$
9. $(x + y + x') = 1$
10. $(x' \cdot y)' = x + y'$

4.5 SWITCHING CIRCUITS

4.5.1 Types of Circuits

A **switching circuit** is an electric circuit involving two-state devices (switches, diodes, transistors, etc.). We shall speak only of switches, although what we say will apply equally well to any of the other devices.

We will denote a switch by a lowercase letter, such as x. Figure 4–4 shows the two basic components of a type of switching circuit which can be described using boolean algebra.

When we say that current flows in one of the circuits of Figure 4–4, we mean that the switches in that circuit are set in such a way that the light at L will be lit.

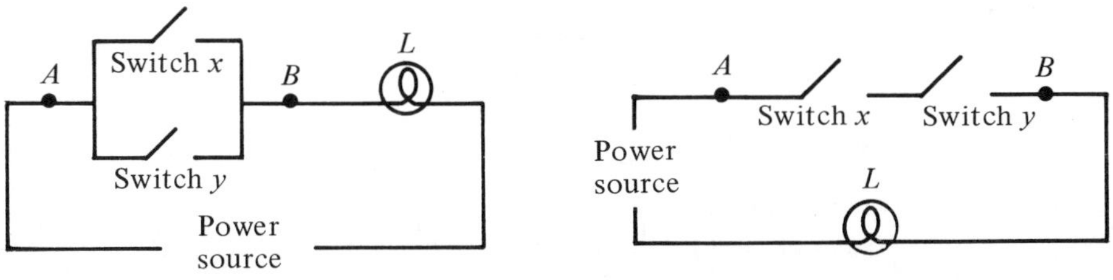

(a) Two switches in parallel. (b) Two switches in series.

FIGURE 4–4

In the circuit of Figure 4–4(a), the light will be on provided that current flows through either switch x or switch y or through both of them. In this circuit the switches are said to be connected in **parallel**, and the circuit is called a **parallel** circuit.

In a **series** circuit, Figure 4–4(b), the switches are connected in such a way that current must flow through both of the switches in order for the light to be on.

Simplified diagrams of the circuits of Figure 4–4 are shown in Figure 4–5.

Note that the simplified diagrams show only the parts of the diagrams in Figure 4–4 which involve the switches and that the light and power source have been omitted. When simplified diagrams are used, it is assumed that a light will be lit (or a bell will ring or some other physical manifestation will occur) whenever the switches are set in such a way that there is current flow between terminals A and B.

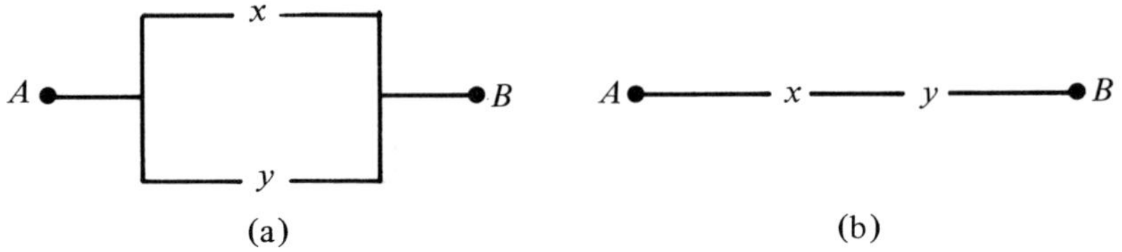

FIGURE 4–5

If current will flow through a switch, we say that it is **closed**; otherwise the switch is **open**.

Let us consider a switch (letter) to be a variable which takes on the value 1 if it is closed and the value 0 if it is open. Let us also designate the state of a light in the circuit as being 1 if the light is lit and 0 if the light is not lit. We shall define $+$ to mean "the switches x and y are in parallel" and $\cdot$ to mean "the switches x and y are in series." We may then construct operation tables showing the state of the light for the various settings of the switches. The operation table for $+$ is:

TABLE 4–6

	$+$	State of switch y: 0	1	
State of switch x	0	0	1	→ State of light
	1	1	1	

The operation table for $\cdot$ is:

TABLE 4–7

		State of y: 0	1
State of x	0	0	0
	1	0	1

These tables are precisely the tables of Example 4.10. Thus the boolean algebra of that system is the one we shall use here.

Using this system, the algebraic expression that describes two switches x and y in parallel is $x + y$, and for the same two switches in series it is $x \cdot y$ (or simply xy).

EXAMPLE 4.13 Give the algebraic expression which describes each of the circuits of Figure 4–6.

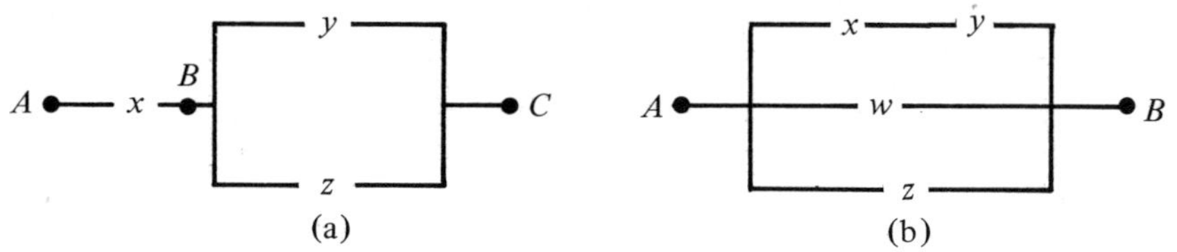

FIGURE 4–6

Solution (a) The circuit consists of two circuits in series: the circuit from A to B and the one from B to C. The first part is represented by x, the second part by $y + z$. The circuit from A to C is represented

$$x(y + z)$$

(b) This one has three circuits in parallel. The representation is

$$xy + w + z$$

If two switches are arranged so that one is open when the other is closed and closed when the other is open, one of the switches is named with the letter x and the other is named x'. If two or more switches are **ganged** so that all are closed or all are open simultaneously, all of them are named with the same letter.

EXAMPLE 4.14 Draw a circuit diagram which represents the expression

$$z(x + y')(x + y)$$

Solution The expression $z(x + y')(x + y)$ is the product of three simpler expressions, z, $x + y'$, and $x + y$. A circuit representing the first letter, z, is a circuit containing a single switch identified as switch z, Figure 4–7(a). The second factor $x + y'$ is represented by two switches in parallel, Figure 4–7(b), and the third factor $x + y$ is represented by switches x and y in parallel, Figure 4–7(c). Since the product of boolean variables has a representation as a

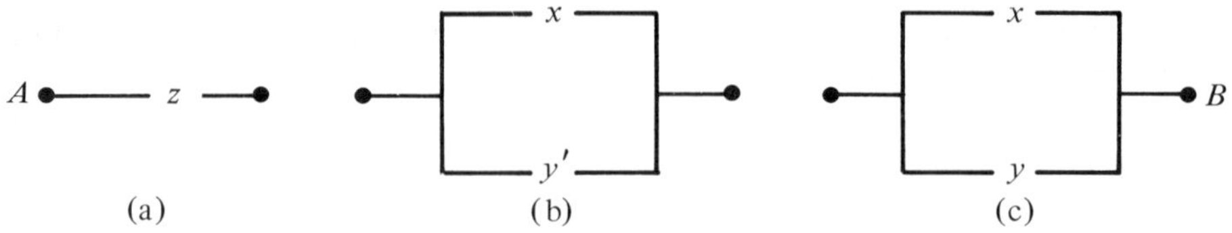

FIGURE 4–7

series circuit, the expression $z(x + y')(x + y)$ can be represented by a circuit consisting of the three circuits of Figure 4–7 connected in series. This circuit is shown in Figure 4–8.

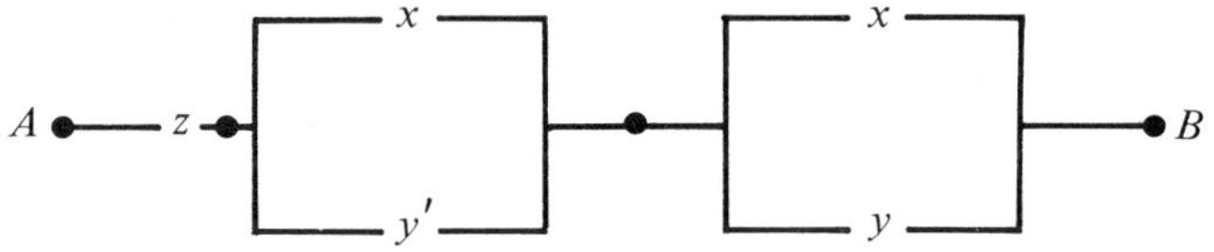

FIGURE 4–8 Circuit diagram representing $z(x + y')(x + y)$.

4.5.2 Closure Properties

We are interested in the **closure properties** of a given circuit, that is, the conditions which determine whether or not current will flow in that circuit (that is, when will the bulb light up?). The easiest way to investigate these properties is with a table of closure properties which is identical to a truth table except that its entries are 0's and 1's.

EXAMPLE 4.15 Construct a closure-property table for the circuit which realizes the expression $f = (x + y + z')(x' + y + z)$.

Solution The completed table is Table 4–8.

TABLE 4–8

x	y	z	f
1	1	1	1
1	1	0	1
1	0	1	1
1	0	0	0
0	1	1	1
0	1	0	1
0	0	1	0
0	0	0	1

There are three variables in the system (there are six switches, but switch x and switch x' are represented by only one variable). Thus there are eight rows in the table. The entries under x, y, and z are 1's and 0's in the same pattern as were the T's and F's in a truth table for a compound of p, q, and r. The entries under f in the table are the states of a light bulb and the circuit, 1 if the light is on and 0 if it is off, and may be computed using the tables of Example 4.10. For example, in the first row

$$\begin{aligned}(x + y + z')(x' + y + z) &= (1 + 1 + 0)(0 + 1 + 1)\\ &= (1)(1)\\ &= 1\end{aligned}$$

We see that current flows in the circuit except when x is closed and y and z are open or when x and y are open and z is closed.

Two circuits involving switches $x, y, z, \ldots$ are equivalent if and only if their closure properties are the same. Two boolean algebraic statements are **equal** if and only if they represent equivalent circuits. Of course two algebraic statements are equal if each one of them can be derived from the other one using the axioms and theorems of boolean algebra.

Exercise Set 4.5

Find the expression from boolean algebra which describes each of the circuits in Figures 4–9 through 4–14.

1.

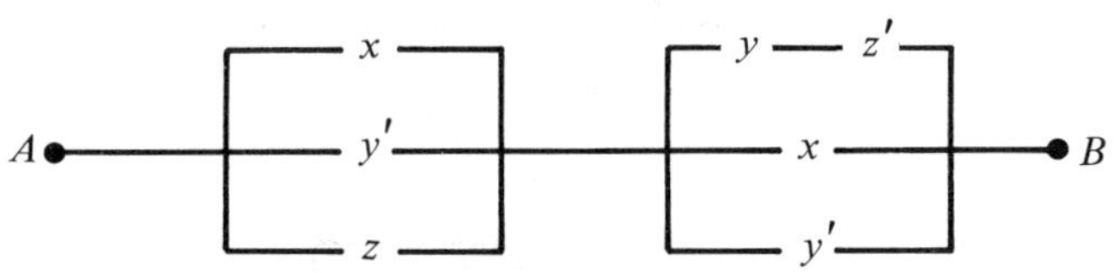

FIGURE 4–9

2.

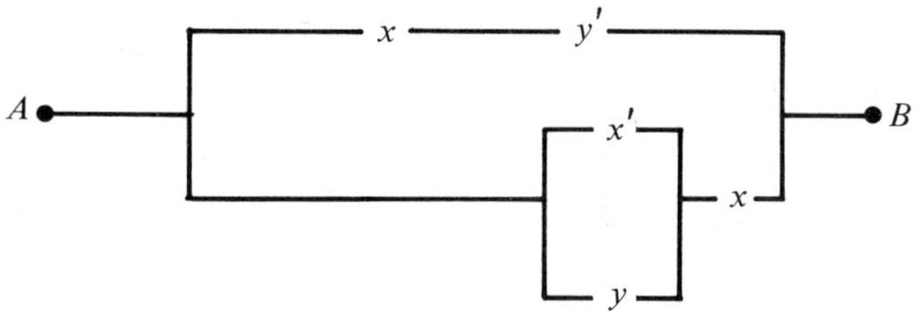

FIGURE 4–10

3.

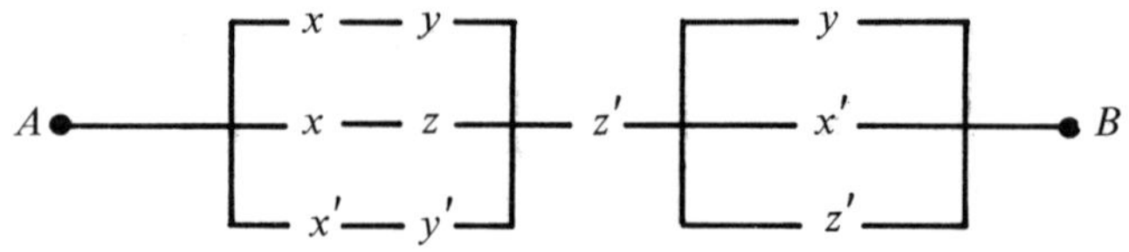

FIGURE 4–11

4.

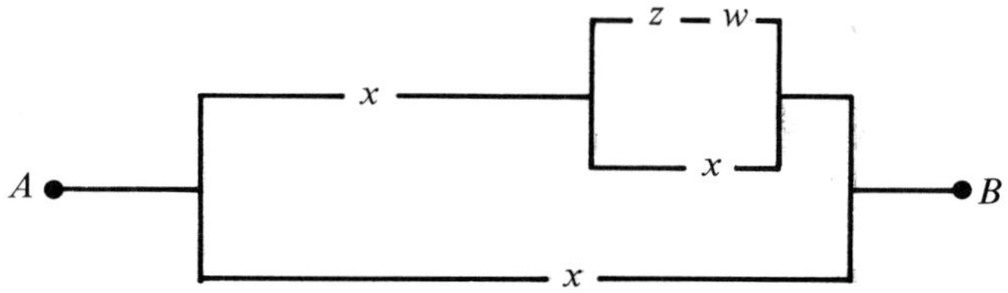

FIGURE 4–12

5.

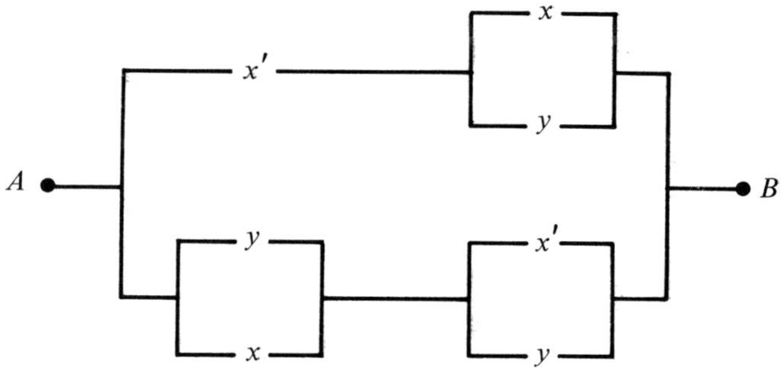

FIGURE 4–13

6.

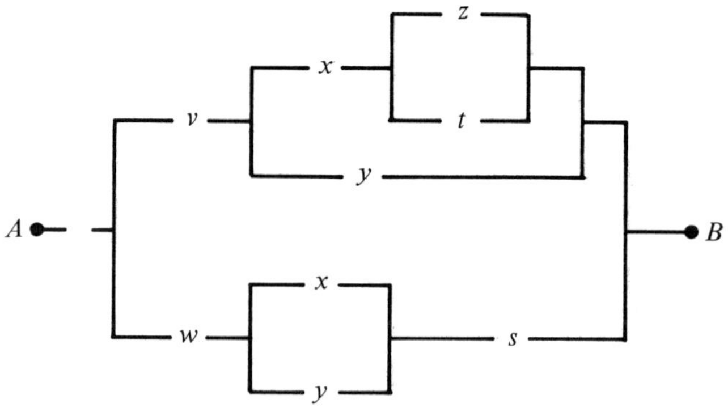

FIGURE 4–14

Draw circuits which represent these expressions:

7. $x(x + y)$
8. $x + y(z + w)$
9. $(x + y)(x + z)$
10. $x + (x + z)$
11. $xyz' + x'(y + z')$
12. $xyz + xy(wz + rs)$
13. $x[y(z + w) + z(u + v)]$

Construct a table of closure properties for the circuits of:

14. Problem 1
15. Problem 2
16. Problem 3
17. Problem 4
18. Problem 5
19. Problem 7
20. Problem 9
21. Problem 11

22. Show that the circuits of Problems 2 and 4 are equivalent.

23. Show that $x + xz + x'yz = x + x'yz$.

4.6 DESIGNING CIRCUITS

The problem of designing a circuit with certain properties is essentially that of writing a boolean algebraic expression which satisfies a table of closure properties for the circuit.

EXAMPLE 4.16 Write the algebraic expression which represents the circuit **f** whose closure properties are given in Table 4–9.

TABLE 4–9

	x	y	z	f
Row 1	1	1	1	0
Row 2	1	1	0	0
Row 3	1	0	1	1
Row 4	1	0	0	0
Row 5	0	1	1	1
Row 6	0	1	0	0
Row 7	0	0	1	0
Row 8	0	0	0	0

Solution We note that **f** has 1's in only row 3 and row 5. That is, current flows in the circuit only when switches x and z are closed and y is open or when x is open and y and z are closed. Recalling that current does not flow through x' when it flows through x, we write for row 3,

$$xy'z$$

and for row 5 we write

$$x'yz$$

Now we wish circuit f to be one which is closed when either (or both) of these circuits is closed and open otherwise. This means that we can write

$$f = xy'z + x'yz$$

EXAMPLE 4.17 Write the algebraic expression satisfying Table 4–10.

TABLE 4–10

x	y	z	f
1	1	1	1
1	1	0	1
1	0	1	0
1	0	0	0
0	1	1	0
0	1	0	0
0	0	1	1
0	0	0	0

Solution There are 1's only in rows 1, 2, and 7. Row 1 is represented by xyz, row 2 by xyz', and row 7 by $x'y'z$. The entire table is represented by the expression

$$\mathbf{f} = xyz + xyz' + x'y'z$$

The method employed in Examples 4.16 and 4.17 may be summarized:

To write the algebraic expression corresponding to a given table of closure properties:

1. Find the first row of the table which has a 1 in the right-hand column. Write a product of all the variables involved, priming those variables which have value 0 in that row.

2. Repeat step 1 for each succeeding row having a 1 as its final entry.
3. Add the products obtained in steps 1 and 2.

The circuit designed by the method suggested above may not be the simplest one that could be designed to realize the table. In the next section we will consider some of the problems involved in the simplification of circuits.

EXAMPLE 4.18 Write an expression **f** whose closure properties are shown in Table 4–11.

TABLE 4–11

x	y	z	f
1	1	1	1
1	1	0	1
1	0	1	1
1	0	0	0
0	1	1	1
0	1	0	1
0	0	1	1
0	0	0	0

Solution The method of the previous two examples could be used here, but because there are more 1's than 0's under **f** we shall use a variation. The expression

$$xy'z' + x'y'z'$$

represents the fourth and eighth rows of the table, and since each of those rows has value 0 we have

$$xy'z' + x'y'z' = 0$$

Now, complement both sides of this equation:

$$(xy'z' + x'y'z')' = 1 \qquad (\textit{since } 0' = 1 \textit{ by } T_6)$$

or

$$(x' + y + z)(x + y + z) = 1 \qquad (\textit{by } T_8, \textit{ DeMorgan's Laws})$$

One expression that has the given closure properties is

$$\mathbf{f} = (x' + y + z)(x + y + z)$$

This method is especially useful when more of the closure properties in a table are 0's than 1's.

EXAMPLE 4.19 Design a circuit containing two switches and a light bulb in such a way that either switch may be used to operate the light independently of the other.

Solution Represent the switches by x and y. Suppose that initially both switches are closed and the light is on. Open y, and the light goes off. Open x and close y, and the light stays off. Open y, and the light goes on. The closure properties of the circuit are shown in Table 4–12.

TABLE 4–12

x	y	ℓ
1	1	1
1	0	0
0	1	0
0	0	1

The required circuit is represented by

$$\ell = xy + x'y'$$

Its diagram is shown in Figure 4–15(a). Figure 4–15(b) shows more accurately the way the actual wiring would be done. The switches in this figure are called **double throw** switches because each switch is in reality two switches, one of which is closed when the other is open.

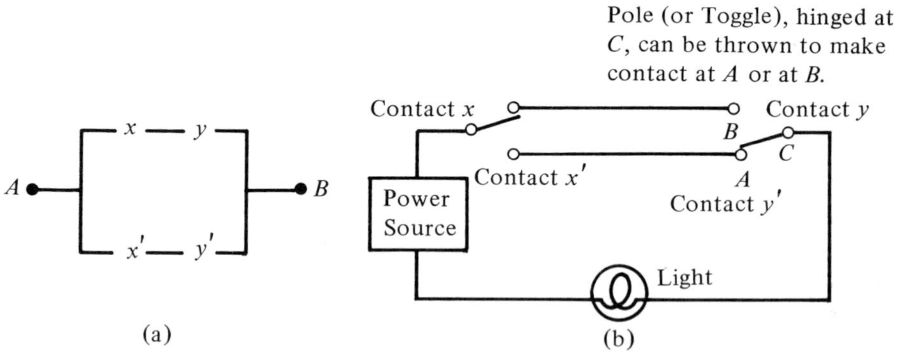

FIGURE 4–15

Exercise Set 4.6

1–4. Design circuits with the properties given by each of the columns headed 1, 2, 3, and 4 in this table.

x	y	z	1	2	3	4
1	1	1	1	0	1	0
1	1	0	0	1	0	0
1	0	1	0	1	1	1
1	0	0	1	0	1	1
0	1	1	0	0	0	1
0	1	0	1	0	0	1
0	0	1	0	0	1	1
0	0	0	0	1	1	1

5. Design a circuit to operate a light independently from any one of three switches.

6. A furnace is operated independently by two switches y and z. A third switch x is operated by a thermocouple so that it is turned off if the pilot light ever goes off. Design a circuit so that either switch y or z can be used to turn the furnace off if it is on or on if it is off, except that the furnace can never be on when the pilot is off.

7. A game of electronic "coin matching" is played by two players x and y. Each person either pushes or does not push a button-type switch which is wired into a circuit with a light. The light comes on if one player pushes his button but the other does not. If each player pushes his button or if neither one pushes his button, the light stays off. If the light comes on y wins; otherwise x wins. Design the circuit for this game.

4.7 SIMPLIFYING CIRCUITS

A basic consideration in the design of circuitry is that of designing the simplest circuit which has a particular set of closure properties. What constitutes a "simplest" circuit in a given situation depends upon a variety of factors (the relative cost of switches and wiring and the type of switch available, for instance). We will assume that the simplest circuit is always a series-parallel circuit which contains the least number of switches possible. We shall also assume that a series-parallel circuit is the simplest possible circuit, although this is not always the case.

EXAMPLE 4.20 Simplify the circuit shown in Figure 4–16.

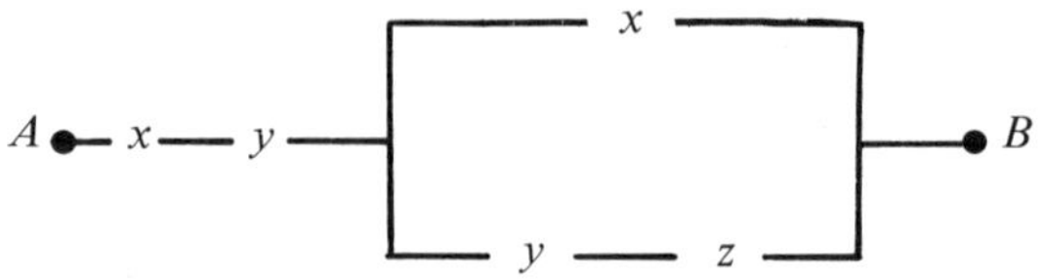

FIGURE 4–16

Solution The algebraic representation of the circuit is

$$xy(x + yz)$$

Using the axioms and theorems of boolean algebra, this becomes

$$\begin{aligned}
xy(x + yz) &= xy(x) + (xy)(yz) && (\textit{Distributive property } A_4)\\
&= x(yx) + x(yy)z && (\textit{Associative property } T_5)\\
&= x(xy) + x(yy)z && (\textit{Commutative property } A_1)\\
&= (xx)y + x(yy)z && (\textit{Associative property } T_5)\\
&= xy + xyz && (\textit{Idempotent property } T_2)\\
&= xy(1 + z) && (\textit{Distributive property } A_4)\\
&= xy(1) && (\textit{Property } T_3)\\
&= xy && (\textit{Multiplicative identity } A_3)
\end{aligned}$$

Thus the circuit which has x and y in series will have the same closure properties as the original circuit.

Sometimes the dual (Property T_1) of an expression is easier to simplify than the original expression.

EXAMPLE 4.21 Simplify the circuit which represents the expression

$$(x + y)(x' + xy)(x' + y')$$

Solution Write the dual of $(x + y)(x' + xy)(x' + y')$.

$$\begin{aligned}
xy + x'(x + y) + (x'y') &= xy + x'x + x'y + x'y' && (A_4)\\
&= xy + 0 + x'y + x'y' && (A_3)\\
&= xy + x'y + x'y' && (A_2)\\
&= xy + x'(y + y') && (A_4)\\
&= xy + x'(1) && (A_3)\\
&= xy + x' && (A_2)
\end{aligned}$$

The dual of this expression

$$(x + y)x'$$

is an expression which is equivalent to the original one. But

$$
\begin{aligned}
(x + y)x' &= x'(x + y) && (A_1)\\
&= x'x + x'y && (A_4)\\
&= 0 + x'y && (A_3)\\
&= x'y && (A_2)
\end{aligned}
$$

Thus a circuit which has switches x' and y in series will do the same task as the original circuit. Because x' is half of a double-throw switch, when the actual wiring is done only two single-throw switches are used, one designated x' and the other designated y.

Exercise Set 4.7

Simplify the circuits:

1.

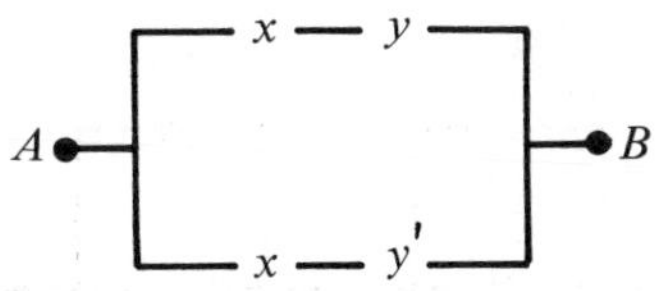

FIGURE 4–17

2.

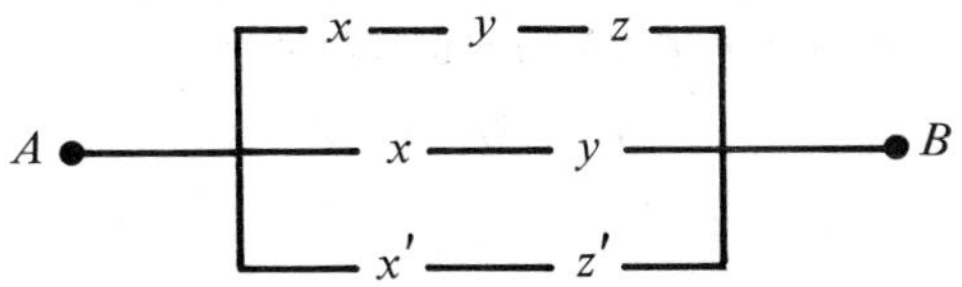

(Use the method of Example 4.21.)

FIGURE 4–18

3.

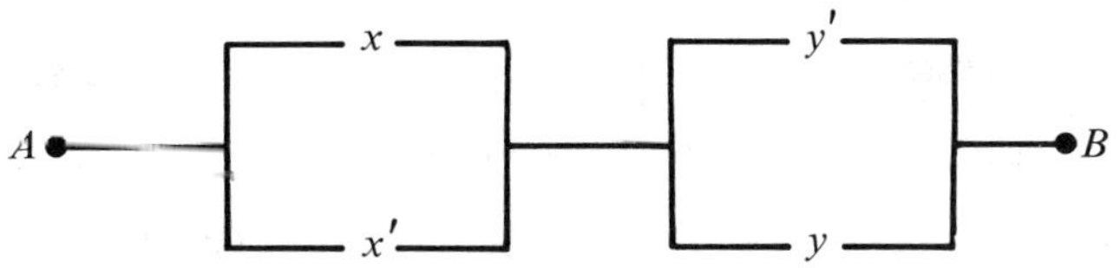

FIGURE 4–19

4.

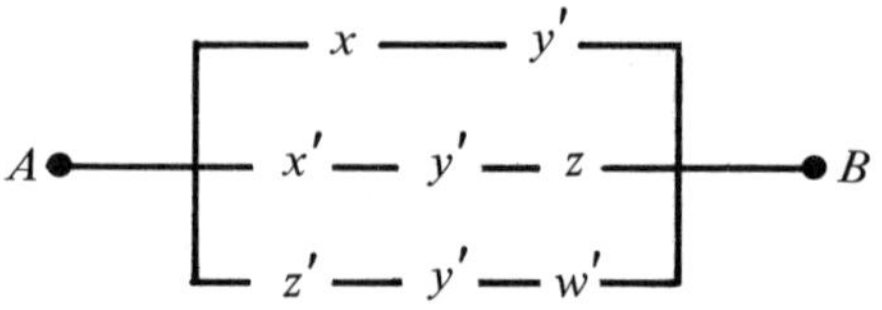

FIGURE 4–20

5.

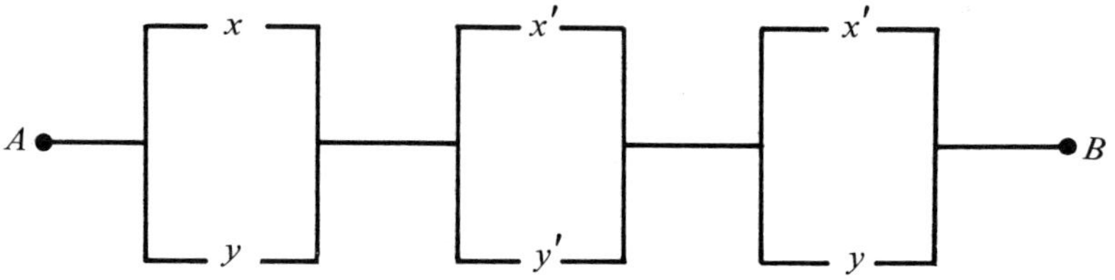

FIGURE 4–21

6.

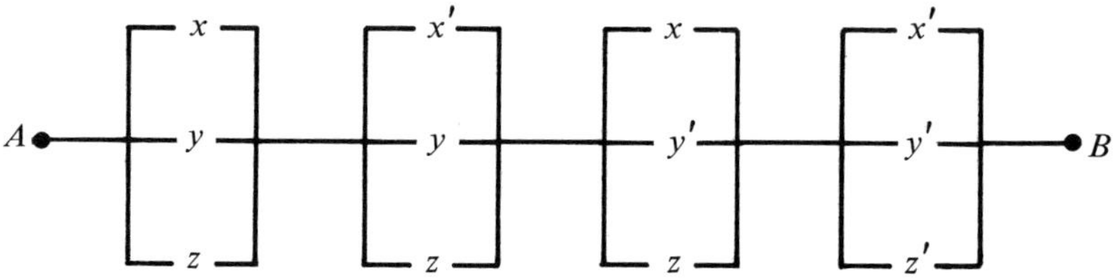

(Simplify the first two parts and the second two parts independently, and then combine and simplify again.)

FIGURE 4–22

SUMMARY

Terms you should know:

1. Mathematical system
2. Binary operation
3. Closure
4. Commutative
5. Group
6. Associative
7. Identity
8. Inverse
9. Distribute
10. Boolean algebra

REVIEW EXERCISES

1. Consider this table:

#	1	2	3	4	5
1	1	2	3	4	5
2	2	3	4	5	1
3	3	1	5	2	4
4	4	5	2	1	3
5	5	4	1	3	2

   (a) Compute 5 # 4 and 4 # 5. Do the results show that # is commutative? Explain.
   (b) Compute (3 # 2) # 5 and 3 # (2 # 5). Do the results tell you anything of importance? If so, what?
   (c) What is the identity element for #?
   (d) Which elements have inverses, and what are they?

2. Suppose the four tires on your car (not including the spare) can be rotated in only one of the four ways indicated in Figure 4–23.

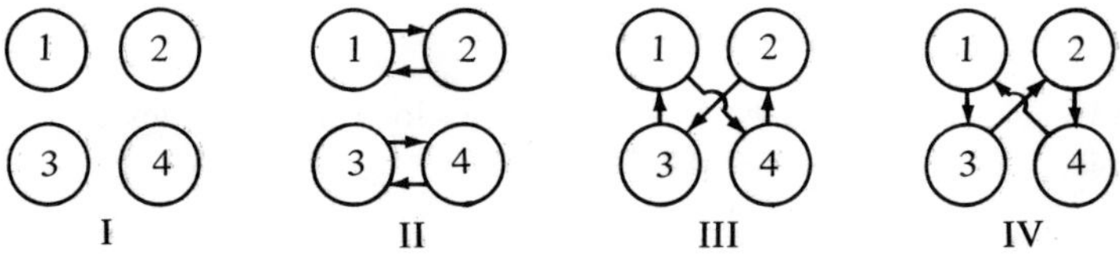

FIGURE 4–23

Make an operation table for the operation +, if + means "followed by," for example

$$\text{II} + \text{III} = \text{IV}$$

and

$$\text{IV} + \text{IV} = \text{II}$$

How many of the group properties does this system exhibit?

3. Draw a switching circuit to represent each of these expressions:

   (a) $x + y(z + wv) + rs$

   (b) $(x + y' + z)(x + yz') + zw + w(y' + z)$

4. Write the boolean algebraic expression which represents each of the circuits in Figure 4–24.

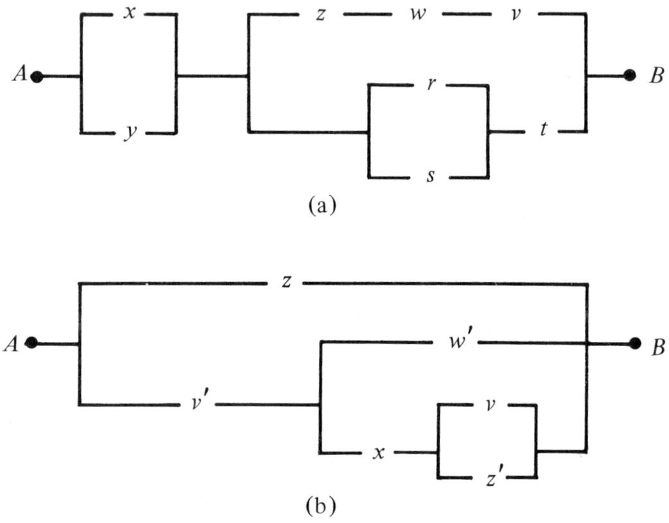

FIGURE 4–24

5. Draw a circuit diagram with the same closure properties as a and b in this table.

x	y	z	a	b
1	1	1	0	1
1	1	0	0	1
1	0	1	0	1
1	0	0	1	1
0	1	1	1	0
0	1	0	0	1
0	0	1	0	0
0	0	0	0	1

6. Simplify each of the circuits in Figure 4–25.

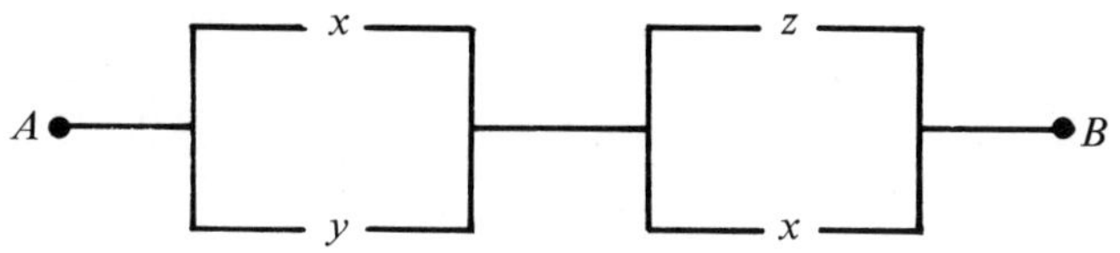

(a)

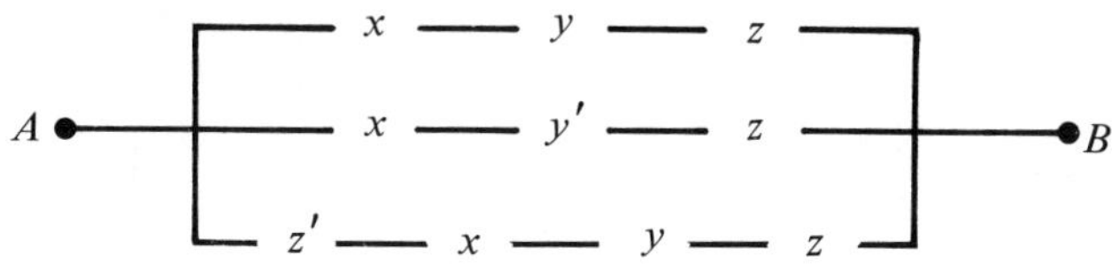

(b)

FIGURE 4–25

5 NUMBER THEORY

There is probably nothing in mathematics which is taken more for granted than the system of numbers used in arithmetic, the real numbers. Our first introduction to this system comes at a very early age, when we first learn to count. As we pass through the primary grades we learn addition, subtraction, multiplication, and division, and we struggle with the complexities of fractions and long division. By the time we have finished high school the real numbers have become, if not our friends, at least acquaintances whom we grudgingly admit can be useful.

In fact, our familiarity with the properties of the real number system has become so strong by this time that we find it difficult to imagine that there can be other systems. When we are told that addition is a commutative operation, our reaction is "so what?" and when we are shown an example of a system in which multiplication does not commute we are skeptical and say "Oh, sure, but that system is just one which is made up, it isn't real." (Perhaps you can guess how the real number system got its name?)

In this chapter we shall trace the evolution of the real number system. We shall see that at a certain point in the history of mathematics, the existence of negative numbers was probably viewed with as much skepticism as today's man on the street might be expected to have about the possibility of there being a noncommutative arithmetic.

We shall also, in this chapter, investigate some of the properties of the numbers of the real system. Some of these properties are probably unknown to you.

5.1 THE REAL NUMBER SYSTEM

In the beginning man learned to count. The numbers which he used, $\{1, 2, 3, \ldots\}$, are called the **natural** numbers. The natural numbers were sufficient so long as all that man needed to do was to count. But after a while he found it desirable to describe quantities which are part of a whole thing. So he invented numbers such as $\frac{1}{5}$, $\frac{1}{2}$, $\frac{2}{3}$, etc. These numbers he called **fractions**, and the natural numbers he called **whole** numbers. (*Natural number* is a modern designation; there was no need to call the natural numbers anything but *numbers* before the invention of fractions.)

Centuries after the invention of fractions (by sometime after 600 B.C.) Greek mathematicians were involved in finding the properties of the geometric figures called polygons. One of these is the square, a four-sided polygon with sides of the same length and four right angles. The Greeks were able to prove that for a square with sides one unit long, the length of a diagonal (see Figure 5–1) must be some number x such that $x^2 = 2$. But try as they might, they could find no known number whose square was 2. Finally they proved that the

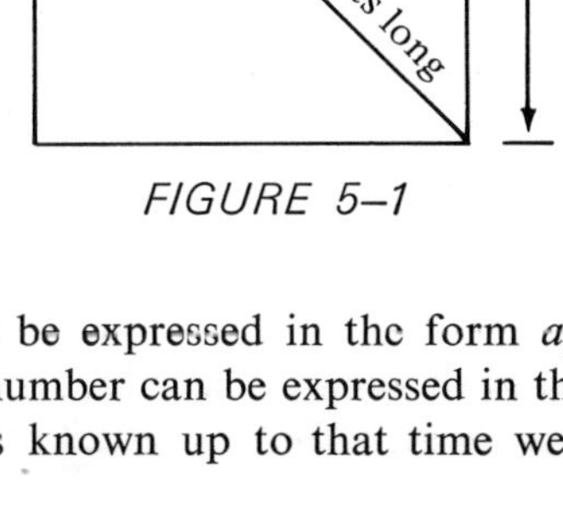

FIGURE 5–1

required number cannot be expressed in the form a/b with a and b natural numbers. Any natural number can be expressed in the form a/b, for example, 6 is $\frac{6}{1}$, and the fractions known up to that time were already in that form.

Because the diagonal of a square clearly exists and it must have a length, the Greeks were forced by their discovery to invent new numbers which they called **irrational** numbers. The older kind of numbers, which could be expressed in the form a/b with a and b natural numbers, they called **rational** numbers. Today we recognize many irrational numbers. Some of these are:

1. Square roots of numbers which are not perfect squares, that is, $\sqrt{2}$, $\sqrt{3}$, and $\sqrt{5}$.

2. Cube roots of numbers which are not perfect cubes, that is, $\sqrt[3]{2}$, $\sqrt[3]{4}$, and $\sqrt[3]{10}$.

3. π, the ratio of the circumference of a circle to its diameter.

EXAMPLE 5.1 Which of these numbers are rational and which are irrational?
(a) $\frac{3}{4}$ (b) 2 (c) $\sqrt[3]{8}$ (d) $\sqrt{6}$

Solution (a) $\frac{3}{4}$ is in the form a/b with $a = 3$ and $b = 4$. Therefore $\frac{3}{4}$ is rational.

(b) 2 may be written $\frac{2}{1}$ or $\frac{6}{3}$ or $\frac{10}{5}$ to name only three ways. Each of these is in the form a/b with a and b natural numbers. Thus 2 is rational.

(c) $\sqrt[3]{8} = 2$ since $2 \times 2 \times 2 = 8$. Therefore $\sqrt[3]{8}$ is rational.

(d) $\sqrt{6}$ is irrational since 6 is not a perfect square.

Although the Greeks were much involved in the study of number theory, they apparently felt no need to expand the number system beyond the irrational numbers. Figure 5–2 shows the numbers used by the Greeks and the relationship between them.

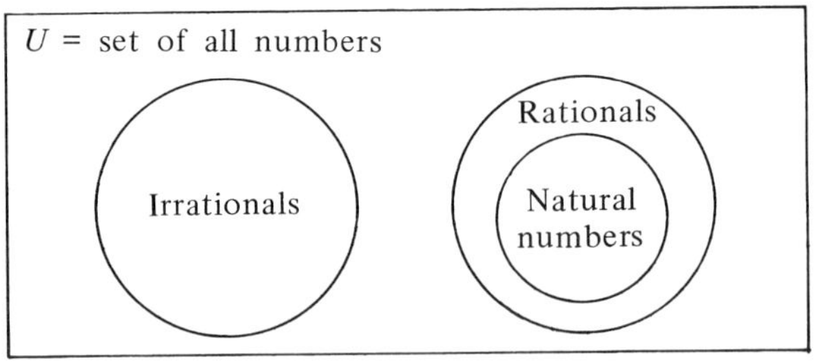

FIGURE 5–2 Numbers known to the Greeks.

Note that zero does not occur in any of the sets of Figure 5–2. The Greeks may have been aware of zero, for Greek traders had traveled in Babylonia and the Babylonians had a symbol for zero, but apparently the Greeks saw no reason for elevating zero to the status of number. Because of the lack of a zero, questions such as "$5 + (\text{what number}) = 0$?" could not even be asked. If someone had shown a Greek schoolboy how to solve the problem $2 - 5 = ?$ he would very likely have replied "Oh sure, but -3 isn't really a number."

Sometime around the second century A.D., some unknown Hindu mathematician had the perception to accept zero as a number. The immediate consequence of this was a new system of numeration, the Hindu-Arabic system which we use today.† The Hindu-Arabic system is a decimal system with place value.‡ Now decimal systems were not new; the Egyptians had used one and so had the Greeks. Place-value systems were more rare; only the Babylonians had had one of these.§ Without a symbol for zero, a place-value system is awkward to use. To get an idea of the difficulty, try to write the numerals for *one hundred twenty* and the one for *one hundred two* without using zeros. How can you distinguish between the two numerals?

With the inclusion of a symbol for zero it became possible to write any natural number by using a finite arrangement of only 10 symbols, {0, 1, 2, 3, 4, 5, 6, 7, 8, 9}.‖ It was soon discovered that any rational number could be expressed as a decimal fraction (fractions having as denominators powers of 10 such as 10, 100, 1000, etc.). The process used to change a rational number to a decimal is one the reader has undoubtedly seen before. It is reviewed in the following Example.

EXAMPLE 5.2 Express as decimals the numbers:

(a) $\frac{2}{5}$
(b) $\frac{3}{8}$
(c) $\frac{1}{6}$
(d) $\frac{5}{7}$

† A numeration system is a method of expressing numbers. There have been a number of these; one of them which you undoubtedly remember is the system of Roman numerals.

‡ A decimal system is one which has names for the first 10 natural numbers and names for each multiple of 10 up to 100, then names for each multiple of 100 up to 1000 and so on.

§ A place-value system is one in which the position of a number symbol in the numeral affects the value represented by that number symbol. Thus in our numeral 44, the 4 on the left represents the number forty and the 4 on the right represents the number four.

‖ Of course these symbols are not the ones used by the early Hindus but are refinements of the ones they actually used.

Solution (a) To change $\frac{2}{5}$ to a decimal, divide 2 by 5.

$$\begin{array}{r} 0.4 \\ 5\,\overline{)2.0} \\ 2\,0 \\ \hline \end{array}$$

Therefore $\frac{2}{5} = 0.4$.

(b) Divide 3 by 8.

$$\begin{array}{r} 0.375 \\ 8\,\overline{)3.000} \\ 2\,4 \\ \hline 60 \\ 56 \\ \hline 40 \\ 40 \\ \hline \end{array}$$

Thus $\frac{3}{8} = 0.375$.

(c) Divide 1 by 6

$$\begin{array}{r} 0.16 \\ 6\,\overline{)1.000} \\ 6 \\ \hline \rightarrow 40 \\ 36 \\ \hline (4) \end{array}$$

The remainder 4 (circled in the computation) is exactly the same as the remainder in the preceding step of the division (indicated by the arrow). Thus the next digit in the quotient will be a 6 and the next remainder will be a 4; the fourth digit in the quotient will be a 6 and the remainder in that step will be a 4; and so on. Thus

$$\frac{1}{6} = 0.166666\ldots.$$

We may represent the fraction $\frac{1}{6}$ by means of a finite number of decimal digits in this way:

$$\frac{1}{6} = 0.1\overline{666}$$

The bar over the 6's indicates that there are supposed to be an infinite number of digits after the last 6 and all of them are 6's.

(d) Divide 5 by 7.

```
     0.714285
 7 )5.0000000000
    49
    --
     10
      7
     --
      30
      28
      --
       20
       14
       --
        60
        56
        --
         40
         35
         --
          5
```

The last remainder 5 is the same as the number which was originally to be divided by 7. Therefore the next six digits in the quotient will be 714285 in that order, and the remainder after 12 digits will again be 5. We see that the quotient repeats in blocks of six digits. We write

$$\tfrac{5}{7} = 0.\overline{714285}$$

Once again the bar over the six digits indicates that the decimal expression of the rational number is an infinite sequence of digits and that the sequence is made up of blocks of the digits 714285.

The decimals in parts (a) and (b) of Example 5.2 are called **terminating** decimals; those of parts (c) and (d) are called **repeating** decimals. Any rational number can be expressed as either a terminating or a repeating decimal. An irrational number can be expressed as a decimal, but the decimal representation of an irrational number will never terminate nor will any sequence of digits in the decimal repeat. As examples, $\sqrt{2}$ is approximately 1.414213562373, and π is approximately 3.14159265358979323846.†

With the recognition of zero as a number, the means were at last at hand for considering the additive inverses of the known numbers. The number -1 was defined to be one such that $1 + (-1) = 0$; -2 was defined as a number such that $2 + (-2) = 0$; $-\sqrt{2}$ was defined as a number such that $\sqrt{2} + (-\sqrt{2}) = 0$;

† The value of π has been computed to several million decimal places.

etc. These new numbers were called **negative** numbers, and the older numbers, except for zero, became the **positive** numbers.

The set of numbers consisting of 0 and the natural numbers were named **whole** numbers. The whole numbers and the additive inverses of the natural numbers became known as **integers**, and the set of all numbers which can be represented in the form a/b where a and b are *integers* with b not equal to 0 became known as the rational numbers. A breakdown of the numbers into these various categories is given below.

NATURAL NUMBERS $\{1, 2, 3, \ldots\}$

WHOLE NUMBERS $\{0, 1, 2, 3, \ldots\}$

INTEGERS $\{\ldots, -3, -2, -1, 0, 1, 2, 3, \ldots\}$

RATIONAL NUMBERS {Numbers which can be expressed in the form a/b in which a and b are integers with $b \neq 0$} Examples of rational numbers are $1/2$, $3/5$, 8, $-3/4$, -5. Rational numbers can be written as either terminating or repeating decimals.

IRRATIONAL NUMBERS {Numbers which cannot be expressed in the form a/b with a and b integers and $b \neq 0$} Irrational numbers can be approximated by nonrepeating nonterminating decimals. Examples of irrational numbers are $\sqrt{5}$, $\sqrt[5]{3}$, $-\sqrt[3]{7}$, and $-\pi$.

A Venn diagram showing the relations between these sets of numbers is shown in Figure 5–3.

The numbers described above make up the set we call the **real** numbers. It is not uncommon for the real numbers to be defined as all those numbers which can be expressed as decimals. Numbers which cannot be expressed as

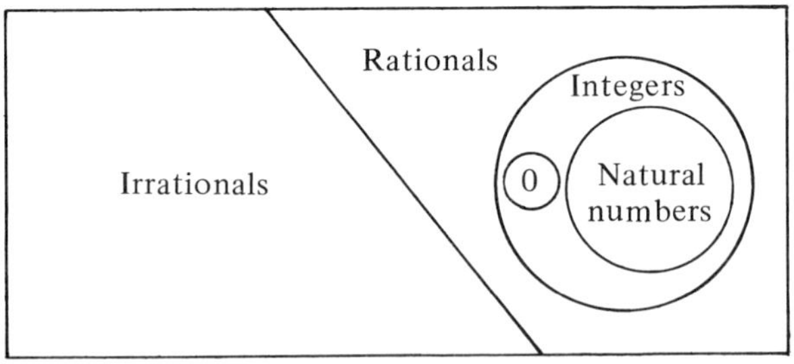

FIGURE 5–3 The real numbers.

decimals, and there are many that cannot, are not real numbers. Some of the latter are called **imaginary** numbers; there are others called **quaternions** and still others called **vectors**. Vectors will be discussed in Chapter 6.

The important lesson to be learned from the history of the development of the real number system is this: Number systems evolve as man finds a need for them. Whenever a set of numbers (or an entire system) fails to do a job, man simply expands the system by adding new numbers or he discards the system and starts all over with numbers that will do what he wants them to. The number of new kinds of numbers and the number of new systems is limited only by man's lack of ingenuity in creating them.

Exercise Set 5.1

1. Classify each of the following by naming the subset of the real numbers to which it belongs.

   (a) 23 (c) $\frac{15}{16}$ (e) $\sqrt{4}$ (g) -1.21

   (b) $\frac{2}{5}$ (d) 0 (f) $-\sqrt{5}$ (h) -10

2. Classify each of the following as either rational or irrational.

   (a) $\frac{5}{6}$ (c) $\sqrt{\frac{1}{9}}$ (e) $\pi - 1$

   (b) 3 (d) $\sqrt[4]{5}$

3. Find the real number replacement for n in each problem.

   (a) $n + 4 = 0$ (c) $\dfrac{n}{2} = 6$ (e) $n^2 = 3$

   (b) $\dfrac{3}{4} + n = 0$ (d) $\dfrac{10}{n} = 5$ (f) $n^3 = 7$

4. Write each of the following as a decimal.

   (a) $\frac{3}{5}$ (b) $\frac{5}{8}$ (c) $\frac{2}{9}$ (d) $\frac{4}{7}$

5. What is another name for the union of 0, the natural numbers, and the negatives of the natural numbers?

6. What is another name for the union of the rational numbers and the irrational numbers?

7. Consider the decimal

   $$0.01001000100001\cdots$$

   If the pattern is continued infinitely, does the decimal represent a rational or an irrational number?

5.2 ORDER RELATIONS

There are two **order relations, less than** and **greater than**, for the set of real numbers. In this section we consider these relations only on the set of natural numbers $N = \{1, 2, 3, \ldots\}$. Some of the properties of order, which we shall derive, can be extended, by suitable changes in the definitions, to apply to all the real numbers. The reason we confine our discussion to the natural numbers rather than considering the more general case is that the properties are easy to verify when only the natural numbers are involved.

5.2.1 Less Than

We shall first define the relation **less than**.

LESS THAN If A and B are nonempty finite sets with $N(A) = a$ and $N(B) = b$, then a is **less than** b if and only if A is equivalent to some proper subset of B. We write $a < b$.

EXAMPLE 5.3 Suppose $A = \{x, y, z\}$ and $B = \{r, s, t, u, v\}$. We can make a one-to-one correspondence between the elements of A and some of the elements of B. One such correspondence is

$$\begin{array}{rl} A = & \{x,\ y,\ z\} \\ & \ \ \updownarrow\ \ \updownarrow\ \ \updownarrow \\ B = & \{r,\ s,\ t,\ u,\ v\} \end{array}$$

Thus $A \sim \{r, s, t\}$, and since $\{r, s, t\} \subset B$, it follows that $N(A) < N(B)$.

EXAMPLE 5.4 Suppose $N(A) = 4$ and $N(B) = 6$. Then A and B may be represented by the Venn diagram in Figure 5–4 and the elements of A can be put into a one-to-one correspondence with a proper subset of B as indicated. Therefore we conclude that $4 < 6$.

FIGURE 5–4

The definition of **less than** says in formal language what our intuition tells us must be true: If when counting the elements of two sets, A and B, we have to count "further" to count B than we do to count A, then A has fewer (or less) elements than B.

5.2.2 Greater Than—Properties of Order

The second order relation is defined in the following way.

GREATER THAN If a and b are natural numbers with a less than b, then b is **greater than** a. We write $b > a$.

If A and B are two finite nonempty sets, then exactly one of these propositions must be true:

1. $A = B$
2. A is equivalent to some proper subset of B
3. B is equivalent to some proper subset of A

If $A = B$, then $N(A) = N(B)$. If A is equivalent to a proper subset of B, then $N(A) < N(B)$. If B is equivalent to a proper subset of A, then $N(B) < N(A)$, and thus $N(A) > N(B)$. This proves the following **theorem**.†

PROPERTY 1 (*Trichotomy*) If a and b are natural numbers, then exactly one of the following is true:

1. $a = b$
2. $a < b$
3. $a > b$

The following order properties are also theorems.

PROPERTY 2 If a and b are natural numbers, then $a < b$ if and only if there is a natural number c such that $a + c = b$.

Before we prove Property 2, let us look at some examples which illustrate it.

† A theorem is a proposition which is a logical consequence of other propositions. That is, a theorem is a proposition which can be proved by a logical argument.

EXAMPLE 5.5 We know that $3 < 5$; Property 2 says there is some natural number x such that $3 + x = 5$. Clearly x must be 2.

EXAMPLE 5.6 Suppose that we know that $x + 4 = y$. Property 2 tells us that $x < y$.

To prove Property 2, we recall that a proposition of the form "p if and only if q" can be rewritten as the compound proposition "if p then q, and if q then p." Thus Property 2 can be broken into two simpler propositions.

PART 1 If $a < b$, then there is a natural number c, such that $a + c = b$.

PART 2 If a, b, and c are natural numbers with $a + c = b$, then $a < b$.

Proof of Part 1 Let A and B be two nonempty finite sets with $N(A) = a$ and $N(B) = b$. Then because $a < b$, A is equivalent to C, a proper subset of B, and $N(C) = a$. We know that $B = C \cup (\tilde{C} \cap B)$ (see Figure 5–5). Thus $N(B) = N(C \cup [C \cap B])$. But C and $(\tilde{C} \cap B)$ are disjoint sets, and so

$$N(B) = N(C) + N(\tilde{C} \cap B)$$

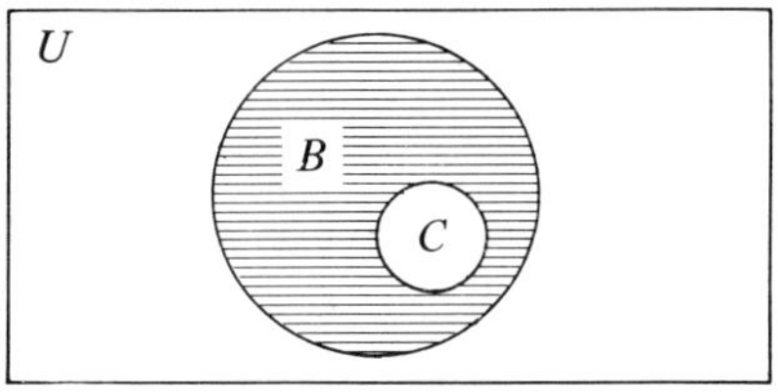

FIGURE 5–5

that is,

$$b = a + N(\tilde{C} \cap B)$$

But C is a proper subset of B; therefore $\tilde{C} \cap B$ is not empty, and $N(\tilde{C} \cap B)$ is some natural number. Call that number c; then we have

$$b = a + c$$

which is what we set out to prove.

Proof of Part 2 To prove that whenever a, b, and c are natural numbers with $a + c = b$, and then $a < b$, we recall that by Property 1, the trichotomy property, only one of these propositions can be true:

$$a = b \qquad a > b \qquad \text{or} \qquad a < b$$

We shall show that if either of the first two propositions is true, it is impossible for the proposition "a, b, and c are natural numbers with $a + c = b$" to be true. Since we are starting our proof with this last proposition as one of the premises, we will conclude that $a = b$ and $a > b$ must be false, and thus the only possibility left is that $a < b$ is true.

Suppose $a = b$ and $a + c = b$ are both true. Then c must be 0 because the only number which can be added to a and have the sum a is 0. The number 0 is not a natural number. But we started with "$a + c = b$ *and* c is a natural number" as a true proposition. Therefore "$a = b$" must be false.

Suppose $a > b$. Then $b < a$, and by Part 1 of this theorem we know there is some natural number x such that $b + x = a$. We also know that the sum of two natural numbers is a unique natural number; that is, if we add the natural number c to both $b + x$ and a, the two sums will be equal. We have

$$\begin{aligned} (b + x) + c &= a + c \\ b + (x + c) &= a + c \qquad \textit{(since addition is associative)} \\ b + (c + x) &= a + c \qquad \textit{(since addition is commutative)} \end{aligned}$$

Now we are starting with the assumption that $a + c = b$, and if this is the case, it must be true that $c + x = 0$ in the last equation above. This cannot be true since c and x are both natural numbers and the sum of two natural numbers is another natural number. We must conclude that it is impossible for a to be greater than b when $a + c = b$ and c is a natural number.

We have shown that $a = b$ and $a > b$ are both impossible under the given conditions. We must conclude that whenever a, b, and c are natural numbers with $a + c = b$, it must be true that $a < b$.

PROPERTY 3 (Transitivity) If a, b, and c are natural numbers with $a < b$ and $b < c$, then $a < c$.

The proof of Property 3 is left as an exercise.

PROPERTY 4 If a, b, and c are natural numbers with $a < b$, then $a + c < b + c$.

The proof makes use of a property of the real numbers called the Additive Property of Equality. This property is:

ADDITIVE PROPERTY OF EQUALITY If a, b, and c are real numbers with $a = b$, then $a + c = b + c$.

Proof of Property 4 If $a < b$, then by Property 2, there is some natural number x such that

$$a + x = b$$

Then by the Additive Property of Equality,

$$(a + x) + c = b + c$$

Then

$$\begin{aligned} a + (x + c) &= b + c \qquad (\textit{Why?}) \\ a + (c + x) &= b + c \qquad (\textit{Why?}) \\ (a + c) + x &= b + c \qquad (\textit{Why?}) \end{aligned}$$

and

$$a + c < b + c \qquad (\textit{by Property 2})$$

There are times when all that we know about two natural numbers a and b is that a is not greater than b. When this is so, we say that a is less than or equal to b. The "or" in this proposition is the *exclusive or* since it is not possible for $a < b$ and $a = b$ to be true at the same time. The symbol for "a is less than or equal to b" is $a \leq b$. Similarly, the proposition "a is greater than or equal to b" is written $a \geq b$.

We conclude with a property which is very useful in theoretical discussions. We shall use this property in the next section.

PROPERTY 5 (Well-Ordering Axiom)† Every nonempty set of natural numbers contains a least number.

EXAMPLE 5.7 Consider these sets:

$$\begin{aligned} A &= \{3, 5, 7, 9, \ldots\} \\ B &= \{8, 14\} \\ C &= \{145\} \end{aligned}$$

Set A is the set of natural numbers greater than 2. It contains an element 3 which is less than any other number in the set. Set B contains only two numbers, 8

† An axiom is a proposition the truth of which is self-evident.

and 14. Since $8 < 14$, B contains a least element. Set C contains only one element, the natural number 145. Since there are no other numbers in the set, 145 is smallest, and C contains a least element.

We have considered the two order relations *less than* and *greater than* in this section. We have also studied five properties of the order relation. The first four of these are true when the order relations are defined to apply to the set of all real numbers. The fifth property, the Well-Ordering Axiom, cannot be made to apply to every nonempty set of real numbers (see Problem 3 of the following exercise set).

Exercise Set 5.2

1. Use the definitions of order and the order properties to determine whether each of the following inequalities is a true proposition. In each case, x represents some natural number.

   (a) $3 < 8$
   (b) $4 > 3$
   (c) $1 \leq 1$
   (d) $x \geq x$
   (e) $x + 2 < 3 + x$
   (f) If $x > 4$, then $4 < x$.
   (g) If $2 < x$ and $x < y$, then $2 < y$.
   (h) If $x < y$, then $x - 2 < y$.
   (i) If $x < 4$, then $x \leq 3$.
   (j) If $x > 2$, then $x \geq 1$.
   (k) If $x \neq 2$ and $x \nless 2$, then $x > 2$.
   (l) If $x \nless 4$ and $x \ngtr 4$, then $x = 4$.

2. Use a Venn diagram to prove the transitive property: If a, b, and c are natural numbers and $a < b$ and $b < c$, then $a < c$.

3. Show that each of the following sets does not have a least element. Explain in each case why this does not contradict the Well-Ordering Axiom.

   (a) $\{\ldots, -3, -2, -1\}$
   (b) $\{1, \frac{1}{2}, \frac{1}{4}, \frac{1}{8}, \frac{1}{16}, \ldots\}$
   (c) $\{x$ such that x is a natural number and $x \times 1 \neq x\}$

5.3 THE DIVISION ALGORITHM

5.3.1 Divides

In all the discussions which have preceded this one, in this book, it has been assumed that the reader was familiar with that operation of arithmetic called division. At this time we shall present some elementary theorems on division on the natural numbers. The most important of these is the theorem called the division algorithm which justifies the process called long division. We first define a relation called **divides**:

DIVIDES If a and b are natural numbers and there is some natural number n such that $a = bn$, then we say that b **divides** a.

EXAMPLE 5.8

(a) 3 divides 6, since $3 \times 2 = 6$.

(b) 4 divides 12, since $4 \times 3 = 12$.

(c) 7 does not divide 10 because there is no natural number n such that $7 \times n = 10$.

Note that although 2×3 is a numeral representing the number 6, "2 divides 8" does not represent a number, but it does represent a relation between 2 and 8. We must take care not to write "2 divides 8 = 4." The operation **division** which we shall discuss later is related to **divides** but the two are not identical. For example, $3 \div 6 = \frac{1}{2}$ but 6 does not divide 3 since $\frac{1}{2}$ is not a natural number.

DIVISORS, FACTORS, MULTIPLES If b divides a, then b is called a **divisor** or **factor** of a and a is said to be a **multiple** of b.

EXAMPLE 5.9

(a) 2 divides 4 so 2 is a divisor of 4. 4 is a multiple of 2.

(b) 3 is a divisor of 18 and 18 is a multiple of 3.

(c) 2 and 5 are factors of 10. 10 is a multiple of 2 and 10 is also a multiple of 5.

EVEN AND ODD NUMBERS If 2 is a divisor of a natural number n, then n is said to be an **even** number. If a natural number is not **even**, it is said to be **odd**.

Examples of even numbers are 4, 8, 26, and 1235472. Examples of odd numbers are 1, 7, 17, and 394423.

The following theorem will be useful in the next section.

THEOREM If b divides a and b divides c, then b divides $a + c$.

Proof If b divides a, then $a = bn$ where n is a natural number. If b divides c, then $c = bm$ where m is a natural number. Then

$$a + c = bn + bm$$

Then

$$a + c = b(n + m) \qquad \text{(since multiplication distributes over addition)}$$

But if m and n are natural numbers, then $n + m$ is a natural number. Therefore by the definition of divides, we know that b divides $a + c$.

5.3.2 The Division Algorithm

THE DIVISION ALGORITHM If a and b are natural numbers with $a > b$, then there are unique non-negative integers q and r such that $a = bq + r$ with $0 \leq r < b$. The number q is called the **quotient**, and r is called the **remainder**.

The Division Algorithm is a theorem. A proof is given following Example 5.10.

EXAMPLE 5.10 Let $a = 10$ and $b = 5$; then $10 = 5 \times 2 + 0$, that is, $q = 2$ and $r = 0$.

Let $a = 17$ and $b = 3$; then $17 = 3 \times 5 + 2$, that is, $q = 5$ and $r = 2$.

Proof of the Division Algorithm Either b divides a or b does not divide a. If b divides a, then $a = bk$ where k is a natural number. In this case, the q and r of the Division Algorithm are $q = k$ and $r = 0$.

If b does not divide a, then there is no natural number k such that $a = bk$. Let n be any one of the natural numbers which make $a - bn > 0$ and consider the set $S = \{a - bn\}$. If $n = 1$, $a - bn = a - b$, and since $a > b$, $a - b > 0$; thus S is not empty. Therefore, by the Well-Ordering Axiom, S contains a least number. Designate this least number by r, and take the value of n which yields this least value as q. Then $a - bq = r$ and $a = bq + r$.

We have shown that for any pair of natural numbers a and b, with $a > b$, there exist non-negative integers q and r such that

$$a = bq + r$$

Now we must show that $0 \leq r < b$. We know $r = 0$ when b divides a. If b does not divide a, then r is the least natural number in S and hence $r > 0$ since

S contains only natural numbers. Thus we have $r \geq 0$. To prove that $r < b$, we recall that the trichotomy property says one and only one of the following is true:

$$r = b \qquad r < b \qquad \text{or} \qquad r > b$$

We shall prove that neither $r = b$ nor $r > b$ is true. First we suppose that $r = b$. Then since

$$a = bq + r$$
$$a = bq + b \qquad \textit{(since by assumption } r = b\textit{)}$$
$$a = b(q + 1) \qquad \textit{(since multiplication distributes over addition)}$$

Then because q is a natural number, $q + 1$ is also a natural number, and it must be true that b divides a. But this means that we could have written $a = bq_1 + 0$ where $q_1 = q + 1$ and $r = 0$. Therefore if $r = b$, it must be true that $b = 0$ also. However we know b is a natural number, and so b cannot be 0. We must conclude that r is not equal to b.

Next let us see what happens if $r > b$. Then $r = b + k$ where k is a natural number and since

$$a = bq + r$$
$$a = bq + (b + k)$$
$$a = (bq + b) + k \qquad \textit{(since addition is associative)}$$
$$a = b(q + 1) + k \qquad \textit{(since multiplication distributes over addition)}$$

and

$$a - b(q + 1) = k$$

Now k is an element of S since $a - b(q + 1)$ is of the form $a - bn$ with $n = q + 1$. However, $a - b(q + 1) < a - bq$ and since $a - b(q + 1) = k$ and $r = a - bq$, $k < r$. Thus if $r > b$, r is not the least element in set S, but we know that this is not the case since we picked the least element and named it r. Therefore r is not greater than b.

We have shown that r is not equal to and not greater than b. Therefore $r < b$, and this is what we wished to prove. We have now proved that when a and b are natural numbers with $a > b$, there exists a natural number q and a non-negative integer r such that

$$a = bq + r \qquad \text{where } 0 \leq r < b$$

The Division Algorithm says that q and r are unique; that is, for any given a and b, there is but one q and one r which satisfy the statement of the algorithm. We shall accept this fact without proof in this book.

Recall that *divides* is a relation rather than an operation. Although we have developed the Division Algorithm, we have not defined the operation of division between all pairs of natural numbers. In fact, we cannot define an operation of division that is closed relative to the natural numbers; it is necessary to extend the set to include the rational numbers, except 0, before that can be done.

Whenever we are asked to do a long division in arithmetic, the problem is one of finding a q and an r guaranteed to us by the Division Algorithm. The actual method used may vary from problem to problem and from person to person because there are many different methods which may be used. The one used here is a rather common one.

EXAMPLE 5.11 Find the quotient and remainder when $a = 471$ and $b = 23$.

Solution

$$\begin{array}{r} 20 \\ 23\,\overline{)471} \\ 460 \\ \hline 11 \end{array}$$

Thus $q = 20$ and $r = 11$.

In this section we have considered a third relation on the natural numbers, the relation *divides*. We have also considered a special theorem called the *Division Algorithm*. The relation *divides* can be extended to the set of integers if we make special provisions that 0 can never be a divisor. For example, we say that integer b divides integer a when $b \neq 0$ if there exists an *integer* n such that $a = bn$. Thus 5 divides -10 because $5 \times (-2) = -10$.

Exercise Set 5.3

1. Test each of the following pairs of numbers to determine whether b divides a.

   (a) $b = 6,\ a = 18$ (e) $b = 3,\ a = 21$

   (b) $b = 7,\ a = 35$ (f) $b = 15,\ a = 5$

   (c) $b = 4,\ a = 26$ (g) $b = 8,\ a = 10$

   (d) $b = 10,\ a = 87$ (h) $b = 10,\ a = 5$

2. Express 24 as a multiple of each of the natural numbers 1, 2, 8, and 12.
3. Express 96 as a multiple of each of the natural numbers 4, 6, 16, and 32.
4. Find all the factors of 30.

5. Find all the factors of 32.

6. Compute the natural numbers q and r of the division algorithm ($a = bq + r$, $0 \leq r < b$) corresponding to each of the following pairs of natural numbers.

   (a) $a = 18, b = 7$ (c) $a = 171, b = 133$

   (b) $a = 28, b = 4$ (d) $a = 97, b = 300$

7. What are the possible remainders if a number is divided by:

   (a) 5 (b) 8 (c) 23 (d) 156

5.4 PRIME NUMBERS

One of the most interesting of the subsets of the natural numbers is the set of prime numbers. These are numbers which have only themselves and 1 as divisors. The relation which exists between these prime numbers and the remaining natural numbers is of such importance that it has been given the name *The Fundamental Theorem of Arithmetic*. In the section which follows this one we shall discuss the Fundamental Theorem; in this section we discuss the prime numbers and a very few of the many interesting things which have been discovered about them.

PRIME AND COMPOSITE NUMBERS A natural number is called a **prime** number if it is not 1 and if it has no divisors except 1 and itself. A natural number is a **composite** number if it is neither 1 nor a prime.

The numbers 2, 3, and 5 are primes because they have no divisors except themselves and 1. The numbers 4, 9, and 35 are composite numbers because each of them has divisors other than itself and 1. The number 1, by definition, is neither a prime number nor a composite number. The definition of primes could have been worded in such a way that 1 would be a prime number; however, many proofs in higher mathematics are shorter when 1 is not called a prime.

The smallest prime number is 2. Any even number greater than 2 has 2 as a factor, and thus 2 is the only even prime number. If we write down the primes in order of increasing size, we have

$$2, 3, 5, 7, 11, 13, 17, 19, 23, 29, \ldots$$

The dots at the end of the sequence indicate that there are more primes in the sequence. Can you find the next three?

5.4.1 Identifying Prime Numbers

One method of finding all the primes less than a given number n is credited to Eratosthenes, a Greek mathematician who lived sometime around 200 B.C. The method is called the **Sieve of Eratosthenes** because it sifts the composite numbers from the natural numbers leaving only the primes. To use the sieve, one writes down, in order of size, all the odd numbers greater than 2 and less than or equal to n. One then crosses out every third number after 3. The next larger number after 3 in the listing is 5; thus the next step is to cross out every fifth number after 5. The next larger number after 5 is 7; therefore the next step is to cross out every seventh number after the 7. The next larger number after the 7 is 11; the next step is to cross out every eleventh number after the 11.

The next number which remains after the 11 is 13, and so every thirteenth number after the 13 is crossed out, and so on. In this process some numbers will be crossed off more than once. The numbers which remain, along with 2, constitute the set of prime numbers less than or equal to n. Example 5.12 shows the use of the sieve to find all the primes less than 100.

EXAMPLE 5.12 *Step 1* List all the odd numbers from 3 to 99 in order of increasing size.

3 5 7 9 11 13 15 17 19 21 23 25 27 29 31
33 35 37 39 41 43 45 47 49 51 53 55 57 59 61
63 65 67 69 71 73 75 77 79 81 83 85 87 89 91
93 95 97 99

Step 2 Cross out every third number after 3. (Done with | in the list above.)

Step 3 Cross out every fifth number after the 5. (Done with /).

Step 4 Cross out every seventh number after the 7. (Done with \.)

The next step is to cross out every eleventh number after the 11 but it turns out that all of these are already marked out. Also every thirteenth number after the 13 has been already crossed off and every seventeenth number after the 17, etc. The primes less than 100 are those numbers which are not crossed out in the list and the number 2.

It can be shown that to use the sieve to sift out the primes less than n, we may stop crossing out numbers after crossing out every kth number after the number k, where k is the largest prime less than or equal to $\sqrt{n}$. In the above example $\sqrt{100} = 10$ and 7 is the largest prime less than or equal to 10. Therefore we were justified in stopping the crossing-out process after step 4.

5.4.2 Theorems and Conjectures About Primes

One of the earliest theorems which was proved about prime numbers is this one:

THEOREM The set of prime numberr is infinite.

The first proof was given by Euclid (about 300 B.C.) and it was done in this manner: Either the set of primes is infinite or it is finite. If we can show that it is not finite, then it must be infinite.

Suppose there are only n primes where n is some finite number. Let these n primes be $p_1, p_2, p_3, \ldots, p_n$. Now let $A = p_1 p_2 p_3 \ldots p_n + 1$; that is, A is the natural number formed by multiplying together all the prime numbers and then adding 1. Now A is either prime or it is composite. If it is prime, we have a prime not in our original set. (Do you see why?) If A is not prime, then A is divisible by a prime p. Can p be equal to p_1? No, for if we divide A by p_1, we get a remainder of 1. In the same way, p cannot be p_2 or p_3 or any other of the primes in the set $p_1, p_2, p_3, \ldots, p_n$. Therefore whether A is prime or composite, there is at least one prime which is not in the original list. But that list was supposed to contain all the primes. Therefore the set of all primes must be infinite.

We note that in the sequence of primes

$$2, 3, 5, 7, 11, 13, 17, 19, 23, 29, 31, 37, \ldots$$

there does not seem to be any pattern in the distribution of the primes. Is this true or does it only seem so? This question has been answered; there is no pattern. In fact, it is possible to write arbitrarily long sequences of consecutive composite numbers. To do so, it is first necessary to introduce a new notation, the **factorial** n, written $n!$. In general $n!$ is the product of the first n natural numbers. Thus $2! = 2 \times 1$, $3! = 3 \times 2 \times 1$, $4! = 4 \times 3 \times 2 \times 1$ and $5! = 5 \times 4 \times 3 \times 2 \times 1$. Now consider the sequence

$$(21! + 2), (21! + 3), (21! + 4), \ldots, (21! + 21)$$

There are 20 numbers in this sequence. The first one, $(21! + 2)$, is divisible by 2 because both $21!$ and 2 are divisible by 2; the second number, $(21! + 3)$, is divisible by 3; the third number, $(21! + 4)$, is divisible by 4; and so on. We have demonstrated a sequence of 20 consecutive composite numbers. You can easily see how this technique can be modified to give a sequence of consecutive composite numbers of any desired length.

Primes can be arbitrarily far apart. Can they be close together? There are two primes, 2 and 3, which are consecutive. Are there any others? No, for one number in each pair of consecutive numbers must be even, and 2 is the only even prime; therefore 2 and 3 are the only consecutive primes. There are many primes which differ by 2. Some of these are 3 and 5, 5 and 7, 11 and 13, 29 and 31. Are there an infinite number of such pairs? No one knows, although many people have tried to find the answer.

Another problem which has defied solution is that of finding a formula which will produce nothing but primes. The expression $n^2 - 79n + 1601$ represents a prime number if n is replaced with any natural number up to and including 79; but if $n = 80$, $n^2 - 79n + 1601$ is a composite number. Formulas have been written which will yield as many primes as desired, but the formula which will give nothing but primes is yet to be found.

One of the most famous unsolved problems about prime numbers is the one called **Goldbach's Conjecture**. Charles Goldbach, an eighteenth-century mathematician, surmised that every even number larger than 4 is the sum of two primes. For example, $8 = 5 + 3$ and $10 = 3 + 7$. This conjecture has been verified for all even integers less than or equal to 100,000, but it has not been proved for all even integers.

The problems mentioned above are only a few of those which have intrigued mathematicians and nonmathematicians alike for centuries. Although some of the problems may seem trivial, it is surprising how many have contributed to advances in other branches of mathematics.

Exercise Set 5.4

1. Find all the prime numbers less than 200.
2. Does the formula $n^2 - n + 41$ yield a prime for $n = 1, 2, 3, 4, 5, 6, 7$?
3. Verify Goldbach's conjecture for every even number between 14 and 24.
4. Find a set of 5 consecutive composite numbers.
5. Find eight primes of the form $4n - 1$ where n is a natural number. (EXAMPLE: $3 = 4n - 1$ with $n = 1$.)

5.5 FUNDAMENTAL THEOREM OF ARITHMETIC

5.5.1 Fundamental Theorem—Prime Factorization

We now know enough about the prime numbers that we may discuss one of the most essential of the theorems of number theory.

FUNDAMENTAL THEOREM OF ARITHMETIC Every composite natural number can be written as a product of primes in a unique way except for the order of the factors.

EXAMPLE 5.13

(a) $12 = 2 \times 2 \times 3$. These factors may be rearranged but they are the only set of prime numbers whose product is 12.

(b) $715 = 5 \times 11 \times 13$

(c) $798 = 2 \times 3 \times 7 \times 19$

Proof of the Fundamental Theorem of Arithmetic If n is a composite number, it contains a prime factor p_1 where $1 < p_1 < n$, and we may write

$$n = p_1 n_1$$

where n_1 is some natural number. If n_1 is prime, then the prime factors of n are p_1 and n_1; but if n_1 is composite, then it can be written

$$n_1 = p_2 n_2$$

where p_2 is a prime and n_2 is some natural number. Also, if n_1 is composite and p_2 is a prime, $n_2 \neq 1$. Therefore $n_2 < n_1$ since the product of two natural numbers neither of which is 1 is a natural number larger than either of its factors.

Then

$$n = p_1 p_2 n_2$$

and if n_2 is prime, p_1, p_2, and n_2 are the prime factors of n. If n_2 is not prime then it contains a prime factor p_3, and we may write

$$n = p_1 p_2 p_3 n_3$$

where n_3 is a natural number with $n_3 < n_2$. We continue this process until some one of the *n's* is a prime. We know that this will happen sooner or later since

$$n > n_1 > n_2 > n_3 > \cdots > n_k$$

is a sequence of natural numbers and for any given n there can be at most n numbers in this sequence. Therefore one of the n's will be a 2 if none of them is a prime before that one.

Before we prove that the factorization of any composite number is unique, let us find prime factors of a number.

EXAMPLE 5.14 Find prime factors of 952.

Solution Consider the primes 2, 3, 5, 7, 11, 13, 17, 19, 23, 29, ... Starting with the smallest, test to see if it divides 952. It does, and so we write

$$952 = 2 \times 476$$

Next divide 476 by the smallest prime which divides it. That one is 2 and

$$476 = 2 \times 238$$

and so we replace the 476 by 2 × 238 and then we have

$$952 = 2 \times 2 \times 238$$

But 2 divides 238, and so we can write

$$952 = 2 \times 2 \times 2 \times 119$$

Since 119 is odd, we know that 2 will not divide it, and so we try the next larger prime as a divisor. That one is 3, and 3 does not divide 119. Neither does 5, the next larger prime after 3. The next prime larger than 5 is 7, and 7 does divide 119 because 119 = 7 × 17. We may write

$$952 = 2 \times 2 \times 2 \times 7 \times 17$$

and 17 is a prime; so we are finished.

5.5.2 Proof of Uniqueness of Prime Factorization

Now we return to the proof of the Fundamental Theorem. We have to show that there is only one set of prime factors of a composite number. Let us assume that there might be two prime factorizations of the same composite number, and let us designate these as $p_1, p_2, p_3, \ldots, p_r$ and $q_1, q_2, q_3, \ldots, q_s$ where the p's and q's are primes and

$$p_1 \leq p_2 \leq p_3 \leq \cdots \leq p_r \qquad \text{and} \qquad q_1 \leq q_2 \leq q_3 \leq \cdots \leq q_s$$

Since $p_1p_2p_3 \cdots p_r = q_1q_2q_3 \cdots q_s$, p_1 divides $q_1q_2q_3 \cdots q_s$. But none of the q's can have p_1 as a divisor; therefore p_1 must be equal to one of the q's. By the same line of reasoning, q_1 must be equal to one of the p's. But if the smallest p is equal to one of the q's and the smallest q is equal to one of the p's, it must be true that the smallest p is equal to the smallest q, that is, $p_1 = q_1$. Therefore it must be true that

$$p_2p_3 \cdots p_r = q_2q_3 \cdots q_s$$

Proceeding as before, we conclude that

$$p_2 = q_2, \quad p_3 = q_3, \quad \ldots$$

Now r cannot be less than s, for if it is, eventually we will reach a step

$$p_r = q_r$$

and then we will have

$$1 = q_{r+1}q_{r+2}\cdots q_s$$

and this is not possible since all the q's are primes. Similarly, r cannot be greater than s, for if it is, we obtain

$$p_{s+1}p_{s+2}\cdots p_r = 1$$

and this is not possible because all the p's are primes. Thus by the trichotomy, $r = s$.

We have shown that there are the same number of factors in the two factorizations and that

$$p_1 = q_1, \quad p_2 = q_2, \quad p_3 = q_3, \quad \ldots, \quad p_r = q_r$$

Therefore there is only one prime factorization of any composite number.

We conclude this chapter with two practical applications of the Fundamental Theorem.

5.5.3 GCF *and* LCM

GREATEST COMMON FACTOR (GCF) The **greatest common factor** of two natural numbers is the largest number which divides both of them.

EXAMPLE 5.15 The greatest common factor of 12 and 18 is 6 since 6 is the largest number which divides both 12 and 18.

We may use the Fundamental Theorem to find the greatest common factor of two natural numbers. First we express each number by its prime factorization. In doing this we use exponents. For example, $36 = 2 \times 2 \times 3 \times 3$, and we write this as

$$36 = 2^2 \times 3^2$$

After we have expressed the two numbers by their prime factorizations, we write down each of the numbers which is a prime factor of *both* of the original numbers, and we give that prime factor the smallest exponent it had in either of the two prime factorizations. The product of these numbers is the greatest common factor of the original numbers.

EXAMPLE 5.16 Find the greatest common factor of 36 and 60.

Solution

$$36 = 2^2 \times 3^2 \qquad \text{and} \qquad 60 = 2^2 \times 3 \times 5$$

Because 2 and 3 are factors of both 36 and 60, they will be factors of the greatest common factor of those numbers. The factor 2 occurs twice in each of the numbers (its exponent is 2 in each factorization); thus it will have an exponent of 2 in the greatest common factor. The factor 3 occurs twice in the factorization of 36 but only once in the factorization of 60. Therefore 3 will be a factor only once in the greatest common factor. We write

$$\text{Greatest common factor} = 2^2 \times 3 = 12$$

EXAMPLE 5.17 Find the greatest common factor of 90 and 750.

Solution

$$90 = 2 \times 3^2 \times 5 \qquad \text{and} \qquad 750 = 2 \times 3 \times 5^3$$

therefore the greatest common factor of these numbers is

$$2 \times 3 \times 5 = 30$$

In adding rational numbers such as $\frac{5}{12}$ and $\frac{7}{36}$, we find it desirable to use a least common denominator, that is, the smallest number which is divisible by both 12 and 36. The least common denominator is an example of a kind of number called the least common multiple.

LEAST COMMON MULTIPLE (LCM) The least common multiple of two natural numbers a and b is the smallest number divisible by both a and b.

To find the LCM of two numbers, first find the prime factorization of each of the numbers and write the factorizations using exponents. The LCM will have as its prime factors every factor which appears in either of the two original numbers, and the exponent of each factor will be the largest exponent that that factor had in either of the original numbers.

EXAMPLE 5.18 Find the LCM of 12 and 18.

Solution

$$12 = 2^2 \times 3 \qquad \text{and} \qquad 18 = 2 \times 3^2$$

Therefore the LCM of 12 and 18 will have as its prime factors, 2's and 3's. The largest exponent of 2 in either of the factorizations is 2, and so 2 will occur twice as a factor in the least common multiple. Similarly, 3 will occur twice as a factor in the LCM because the largest exponent it has in the factorizations of 12 and 18 is 2. We write

$$\text{LCM} = 2^2 \times 3^2 = 36$$

EXAMPLE 5.19 Find the smallest natural number which is divisible by both 24 and 30.

Solution

$$24 = 2^3 \times 3 \qquad \text{and} \qquad 30 = 2 \times 3 \times 5$$

The required number, the LCM of 24 and 30, contains every factor which occurs in either 24 or 30; therefore it contains 2's and 3's and 5's. In the factorizations of 24 and 30, the largest exponent of 2 is 3; thus 2 will be a factor 3 times in the LCM. The other factors, 3 and 5, occur at most once in either of the original numbers, and they will therefore occur but once each in the LCM. Therefore the least common multiple of 24 and 30 is

$$2^3 \times 3 \times 5 = 120$$

Exercise Set 5.5

1. Express each of the following as a product of primes.

   (a) 330 (d) 315 (g) 5940
   (b) 273 (e) 900 (h) 2790
   (c) 140 (f) 504

2. For each of the following, find the GCF.

   (a) 27 and 15 (d) 286 and 392
   (b) 16 and 40 (e) 255 and 406
   (c) 75 and 45 (f) 357 and 437

3. Find the LCM for each of the pairs of numbers in Problem 2.

5.6 MODULAR ARITHMETIC

5.6.1 Residues

In Chapter 4, arithmetic on a four-hour clock was defined. At that time it was mentioned that this arithmetic is an example of a modular system. We shall now consider modular systems in a more general way.

MODULUS AND RESIDUE Let m be any natural number except 1. Then any non-negative integer a may be written in the form

$$a = mq + r$$

in which q is a non-negative integer and r is an integer with $0 \leq r < m$. We call m a **modulus**, and we call r the **residue of a with respect to m.**

EXAMPLE 5.20 Suppose m is 5. What are the residues of 3, 5, 7, and 12 with respect to m?

Solution $3 = 5 \times 0 + 3$, and therefore the residue of 3 with respect to 5 is 3.
$5 = 5 \times 1 + 0$, and therefore the residue of 5 with respect to 5 is 0.
$7 = 5 \times 1 + 2$, and therefore the residue of 7 with respect to 5 is 2.
$12 = 5 \times 2 + 2$, and therefore the residue of 12 with respect to 5 is 2.

We may classify any non-negative integer by its residue with respect to some modulus. For example, if 3 is taken as a modulus, any non-negative integer can be placed into one of three distinct categories according to the scheme in the following table:

Category	Number
Residue 0	0, 3, 6, 9, 12, . . .
Residue 1	1, 4, 7, 10, 13, . . .
Residue 2	2, 5, 8, 11, 14, . . .

To indicate this categorization, we invent a new relation on the non-negative integers.

CONGRUENCE Two non-negative integers are **congruent modulo** (mod m) if they have the same residue with respect to m. The symbol for congruence is $\equiv$.

On the basis of the above definition we say

0, 3, 6, 9, 12, . . . are congruent to 0 (mod 3)
1, 4, 7, 10, 13, . . . are congruent to 1 (mod 3)
2, 5, 8, 11, 14, . . . are congruent to 2 (mod 3)

and we write $3 \equiv 0 \pmod 3$, $4 \equiv 7 \pmod 3$, $11 \equiv 8 \pmod 3$, and $35 \equiv 17 \pmod 3$.

5.6.2 Modular Arithmetic

We may construct addition and multiplication tables for arithmetic mod m. For example see Table 5–1 for mod 5. Note that the only numbers

TABLE 5–1

+	0	1	2	3	4
0	0	1	2	3	4
1	1	2	3	4	0
2	2	3	4	0	1
3	3	4	0	1	2
4	4	0	1	2	3

·	0	1	2	3	4
0	0	0	0	0	0
1	0	1	2	3	4
2	0	2	4	1	3
3	0	3	1	4	2
4	0	4	3	2	1

which occur in the tables are the residues mod 5. We shall refer to these residues as the numbers of the modular system, and it should be understood that the "number 2," for instance, is any number which is congruent to 2.

It is a relatively easy matter to show, although we shall not do it, that whenever m is a prime number the set of numbers mod m with operations addition and multiplication satisfies all of the field properties. That is:

1. There is closure with respect to both operations.
2. If a and b are any two numbers of the system, then

$$a + b = b + a$$

and

$$a \cdot b = b \cdot a$$

3. *Associativity*: If a, b, and c are any numbers of the system, then

$$(a + b) + c = a + (b + c)$$

and

$$(a \cdot b) \cdot c = a \cdot (b \cdot c)$$

4. *Identity for* $+$: For every number a,

$$a + z = z + a = a$$

if and only if $z = 0$.

5. *Identity for* $\cdot$: If a is any number, then

$$a \cdot u = u \cdot a = a$$

if and only if $u = 1$.

6. *Inverses for* $+$: For every number a in the system, there is a unique number b such that

$$a + b = b + a = 0$$

The number b is called the additive inverse of a.

7. *Multiplicative Inverses*: For every number $a \neq 0$, there is a unique number x such that

$$a \cdot x = x \cdot a = 1$$

The number x is called the multiplicative inverse of a.

8. *Multiplication distributes over addition*: If a, b, and c are any numbers of the system, then

$$a \cdot (b + c) = (a \cdot b) + (a \cdot c)$$

If m is a composite number, field property 7 does not apply. (See Problem 2 in the following exercise set.)

In the discussion which follows, we shall be considering only modular systems in which m is a prime number, in order that the systems will exhibit all the field properties.

Let us first define symbols to use for the additive and multiplicative inverses of numbers in a modular system.

ADDITIVE INVERSES If a is any number in a modular system, we shall denote its additive inverse as $-a$. Then $a + (-a) = 0$ and $-a + a = 0$.

In arithmetic mod 5, $2 + 3 = 0$; therefore 2 is the additive inverse of 3; that is, $-3 = 2$. Similarly $-2 = 3$. What is -1? -4?

MULTIPLICATIVE INVERSES If $a \neq 0$ is any number in a modular system, then we shall denote the multiplicative inverse of a as $1/a$. Then

$$a \cdot \frac{1}{a} = 1 \qquad \text{and} \qquad \frac{1}{a} \cdot a = 1$$

In arithmetic mod 5, $2 \cdot 3 = 1$ and $3 \cdot 2 = 1$. Therefore $\frac{1}{2} = 3$ and $\frac{1}{3} = 2$. What is $\frac{1}{4}$ in this system?

We may now define two more operations on a modular system.

SUBTRACTION If a and b are mod m numbers, then b subtracted from a, written $a - b$, means $a + (-b)$.

DIVISION If a and b are numbers in a mod m system, and $b \neq 0$, then a divided by b, written $a \div b$, means $a \cdot 1/b$.

EXAMPLE 5.21 Perform these computations in arithmetic mod 5.

(a) $3 - 4$ (c) $3 \div 2$

(b) $2 - 3$ (d) $4 \div 3$

Solution (a) $3 - 4$ means $3 + (-4)$ where -4 is a number which can be added to 4 to obtain 0. From the addition table for mod 5 (see Tables 5–1), we find that $4 + 1 = 0$; therefore $-4 = 1$. The problem then becomes

$$3 + -4 = 3 + 1 = 4$$

(b)
$$\begin{aligned} 2 - 3 &= 2 + (-3) \\ &= 2 + 2 \\ &= 4 \end{aligned}$$

(c) $3 \div 2$ means $3 \cdot \frac{1}{2}$ where $\frac{1}{2}$ is a number such that when it is multiplied by 2, the result is 1. From the multiplication table (mod 5), it is clear that $\frac{1}{2} = 3$. Thus

$$3 \div 2 \qquad \text{means} \qquad 3 \cdot 3$$

and $3 \cdot 3 = 4$. Therefore

$$3 \div 2 = 4$$

(d)
$$\begin{aligned} 4 \div 3 &= 4 \cdot \tfrac{1}{3} \\ &= 4 \cdot 2 \\ &= 3 \end{aligned}$$

In using the Tables 5–1 for mod 5 arithmetic, we used without stating them the relations expressed in this theorem.

THEOREM If $a \equiv x \pmod{m}$ and $b \equiv y \pmod{m}$, then

$$a + b \equiv x + y \quad \text{and} \quad a \cdot b \equiv x \cdot y \pmod{m}$$

5.6.3 An Application of Modular Arithmetic

We shall use the facts expressed by the theorem above to show one practical application of modular arithmetic to number theory.

Any natural number can be written as a decimal in the form

$$a_0 10^n + a_1 10^{n-1} + \cdots + a_n$$

in which the a's are numbers from the set $\{0, 1, 2, \ldots, 9\}$ and n is a natural number. For example,

$$12 = 1 \times 10^1 + 2$$
$$247 = 2 \times 10^2 + 4 \times 10^1 + 7$$

Now we know that $10 \equiv 1 \pmod{9}$; therefore

$$10^2 \equiv 1 \times 1 \equiv 1 \pmod{9}$$
$$10^3 \equiv 1 \times 1 \times 1 \equiv 1 \pmod{9}$$
$$\cdots\cdots\cdots\cdots\cdots\cdots$$
$$10^n \equiv 1 \pmod{9}$$

Let X be any natural number in its decimal form. Then

$$X \equiv a_0 10^n + a_1 10^{n-1} + \cdots + a_n \pmod{9}$$

But by the theorem quoted above,

$$a_0 10^n \equiv a_0, \quad a_1 10^{n-1} \equiv a_1, \quad \ldots, \quad a_n \equiv a_n \pmod{9}$$

Therefore

$$X \equiv a_0 + a_1 + \cdots + a_n \pmod{9}$$

This means that the remainder of X, when X is divided by 9, is congruent to the remainder when the sum of the digits of X is divided by 9.

EXAMPLE 5.22 Find the remainder when each of the following is divided by 9.

(a) 134 (b) 2005 (c) 287493

Solution (a) $1 + 3 + 4 = 8$, and therefore the remainder when 134 is divided by 9 is 8.

(b) $2 + 0 + 0 + 5 = 7$, and therefore when 2005 is divided by 9, the remainder is 7.

(c) $2 + 8 + 7 + 4 + 9 + 3 = 33$, and therefore $287493 \equiv 33 \pmod 9$. But $3 + 3 = 6$, and so $33 \equiv 6 \pmod 9$. Therefore $287493 \equiv 6 \pmod 9$, and the remainder when 287493 is divided by 9 is 6.

If the remainder is 0 when a number is divided by 9, then 9 is a divisor of the number. From this we have the following important divisibility rule.

TEST FOR DIVISIBILITY BY 9 If the sum of the digits of a decimal number is congruent to 0 (mod 9), the number is divisible by 9.

There are other tests for divisibility by other numbers (11 and 13, for instance). These tests may also be derived by use of modular arithmetics.

Exercise Set 5.6

1. Perform the following computations (mod 5).

   (a) $4 + 2 + 3$
   (b) $3 \cdot 2 \cdot 4$
   (c) $3 \cdot (1 + 4)$
   (d) $2 + (-1)$
   (e) $3 + (-4)$
   (f) $4 - 2$
   (g) $2 - 4$
   (h) $2 \cdot \frac{1}{3}$
   (i) $2 + \frac{1}{2}$
   (j) $\frac{1}{3} + 4$
   (k) $2 \cdot -2$
   (l) $\frac{1}{3} - 3$
   (m) $\frac{1}{2} - \frac{1}{4}$
   (n) $3 + \left(\frac{-1}{3}\right)$

2. Make a multiplication table for arithmetic modulo 6. What is the identity for multiplication? Which elements have inverses?

3. Test each of the following for divisibility by 9.

   (a) 3456
   (b) 121212
   (c) 35731
   (d) 249076125

SUMMARY

Terms you should know:

1. Natural number
2. Integer
3. Rational
4. Irrational
5. Greater than
6. Less than
7. Divides
8. Divisor
9. Factor
10. Even and odd numbers
11. Prime number
12. Composite number
13. Sieve of Eratosthenes
14. Division Algorithm
15. Fundamental Theorem of Arithmetic
16. Greatest common factor
17. Least common multiple
18. Modulus
19. Residue
20. Congruent

REVIEW EXERCISES

1. Identify each of the following by naming each subset of the real numbers to which it belongs.

   (a) 3
   (b) -4
   (c) $\frac{2}{3}$
   (d) $\sqrt{8}$
   (e) 1.21
   (f) $1\frac{1}{3}$

2. Which of the following are true? In each case a and b are natural numbers.

   (a) $4 < 7$
   (b) $2 > 8$
   (c) $3 \leq 4$
   (d) $5 \geq 5$
   (e) If $a = b$, then $a + 3 > b$.
   (f) If $a < b$ and $b > 3$, then $3 < a$.

3. What is the truth value of each of the following?

   (a) 3 divides 24.
   (b) 2 is a divisor of 10.
   (c) 4 is a factor of 8.
   (d) 12 is a multiple of 36.
   (e) 1 is a prime number.
   (f) 10 is a composite number.
   (g) The LCM of 20 and 50 is 100.
   (h) The GCF of 20 and 50 is 10.
   (i) $8 \equiv 14 \pmod{6}$
   (j) $\frac{1}{2} + \frac{1}{4} \equiv 3 \cdot \frac{1}{4} \pmod{5}$
   (k) $\frac{1}{2} \cdot 4 \equiv 2 \pmod{6}$

4. Find the prime factors of 1020.

5. Compute

(a) $5 + 4 \pmod 7$

(b) $2 + 15 \pmod 5$

(c) $3 \cdot 7 \pmod{11}$

(d) $2 \cdot 15 \pmod 7$

(e) $3 - 4 \pmod 5$

(f) $5 \cdot \frac{1}{6} \pmod 7$

6

MATRIX ALGEBRA

Occasionally a mathematical system will be created to explain a particular science, but far more often a scientist finds the mathematics he needs in a system which was developed years before his problem was conceived. Matrix algebra is such a system. Developed by Caley in 1858, this system went virtually unnoticed until the development of high-speed computers and the resulting growth of technology caused a renewed interest in matrices. Now it is difficult to find a branch of science that does not use some matrix algebra.

Some of the applications are really "way out" as we shall show. First, however, we had better discuss the system itself.

6.1 VECTORS

A **vector** is an ordered set of numbers. If the numbers are written in a row, the vector is a **row vector**. Examples of row vectors are:

$$[3, 2, 1] \qquad [1, 0] \qquad [4, 6, 10, 7, 1]$$

The entries of a vector are called its **components**. The first vector above has three components, the second has two, and the third has five. The number of components of a vector is called its **dimension**.

In this book a vector will be denoted by a boldface letter, **a** for instance, and the components of **a** are $(a_1, a_2, a_3, \ldots, a_n)$. The lowered numbers $1, 2, 3, \ldots, n$ are called **subscripts**.

A vector may have its components listed in a column; if so it is a **column vector**. The following are column vectors:

$$\begin{bmatrix} 1 \\ 3 \end{bmatrix} \qquad \begin{bmatrix} 0 \\ 0 \\ 1 \\ 0 \end{bmatrix} \qquad \begin{bmatrix} 15 \\ 25 \\ 9 \end{bmatrix} \qquad \begin{bmatrix} a_1 \\ a_2 \\ a_3 \\ \vdots \\ a_n \end{bmatrix}$$

Two vectors **a** and **b** are **equal** if and only if they are both row vectors or both column vectors, have the same dimension, and $a_1 = b_1$, $a_2 = b_2$, $a_3 = b_3, \ldots, a_n = b_n$.

6.1.1 Vector Operations

Three operations are defined on vectors.

SCALAR MULTIPLICATION If k is any real number and **a** is a vector, the scalar product k**a** is a vector formed by multiplying each component of **a** by k. If **a** is a row vector, k**a** is a row vector. If **a** is a column vector, k**a** is a column vector.

EXAMPLE 6.1

$$2[3, 0, 2] = [6, 0, 4] \qquad \text{and} \qquad 3\begin{bmatrix} 1 \\ 4 \end{bmatrix} = \begin{bmatrix} 3 \\ 12 \end{bmatrix}$$

VECTOR ADDITION If $\mathbf{a} = (a_1, a_2, \ldots, a_n)$ and $\mathbf{b} = (b_1, b_2, \ldots, b_n)$ have the same dimension and both are row vectors or both are column vectors, their sum $\mathbf{c} = (c_1, c_2, \ldots, c_n)$ is a vector with $c_1 = a_1 + b_1$, $c_2 = a_2 + b_2$, $c_3 = a_3 + b_3, \ldots, c_n = a_n + b_n$. If **a** and **b** are row vectors, **c** is also a row vector; if **a** and **b** are column vectors, **c** is a column vector.

EXAMPLE 6.2

$$[1, 2, 0] + [3, 4, 7] = [4, 6, 7]$$

and

$$\begin{bmatrix}4\\3\end{bmatrix}+\begin{bmatrix}2\\8\end{bmatrix}=\begin{bmatrix}6\\11\end{bmatrix}$$

VECTOR MULTIPLICATION If **a** is a row vector and **b** is a column vector, both of the same dimension, their product is

$$\mathbf{a}\cdot\mathbf{b} = a_1b_1 + a_2b_2 + a_3b_3 + \cdots + a_nb_n$$

EXAMPLE 6.3

$$\begin{aligned}[2, 4]\begin{bmatrix}5\\3\end{bmatrix} &= 2\cdot 5 + 4\cdot 3\\ &= 10 + 12\\ &= 22\end{aligned}$$

$$[3, 1, 2, 0]\begin{bmatrix}1\\5\\4\\7\end{bmatrix} = 16$$

There are two things to notice concerning vector multiplication:

1. Vector multiplication is defined only for the case of a row vector written on the left of a column vector.
2. The product of two vectors is not a vector, but it is a number.

EXAMPLE 6.4 A firm employs three types of workers, skilled, semiskilled, and unskilled. Skilled workers receive $30 a day, semiskilled workers receive $20 and unskilled workers receive $14. On a particular day the firm employed 4 skilled workers, 8 semiskilled workers, and 3 unskilled workers. What was the firm's payroll for that day?

Solution To solve this problem using the method frequently taught in the primary grades, we would first find the wages paid to each classification of workers.

4 skilled workers earn	4 × $30 =	$120
8 semiskilled workers earn	8 × 20 =	160
3 unskilled workers earn	3 × 14 =	42
	Total	$322

The total of the wages paid is the day's payroll.

To make the same computation using vector multiplication, express the numbers of employees in a row vector and the wages in a column vector. The vector product is the amount of the payroll for that day. Thus

$$[4, 8, 3]\begin{bmatrix} 30 \\ 20 \\ 14 \end{bmatrix} = 322$$

Vector multiplication can be used to select a desired item from a given list. For instance, if

$$\mathbf{a} = \begin{bmatrix} 3 \\ 5 \\ 8 \end{bmatrix}$$

is multiplied by [0, 1, 0], the result is 5, the second component of **a**. If **a** is multiplied by (0, 0, 1), the result, 8, is the third component of **a**. In general, if **a** is a column vector with n components and i is a natural number with $1 \leq i \leq n$, then to select the ith component of **a**, one multiplies **a** by an n-component row vector which has a 1 in the ith position and 0's in every other position.

EXAMPLE 6.5 A grocery chain consists of 300 stores. Each store sends a report of its day's receipts to the main office of the chain. There the receipts from each store are recorded in a 300-component column vector in the company computer. The general manager of the chain may "lift" the recorded sales of any store from the record by causing the computer to multiply the sales vector by a 300-component row vector which has a 1 in that store's position and 0's in all the other positions.

In this section we have considered a number system called a vector system. The elements of this system are of two types, vectors and scalars. The vectors are ordered sets of numbers, and the scalars are real numbers. There are three operations defined for the system: addition, scalar multiplication, and vector multiplication. Examples 6.4 and 6.5 illustrate some applications of vectors; others will be discussed in later sections of this book.

Exercise Set 6.1

1. Write a row vector **a** having $a_1 = 2$, $a_2 = 4$, and $a_3 = a_4 = a_5 = 0$.
2. Write a column vector **b** of dimension 4 if $b_3 = 5$ and all other components are zero.

3. Which, if any, of the following vectors are equal?

(a) $[1, 2, 3]$
(b) $[3, 0, 0]$
(c) $[0, 3, 0]$
(d) $[4, 2, 0]$
(e) $[4, 2, 0, 0]$
(f) $\begin{bmatrix} 1 \\ 2 \\ 3 \end{bmatrix}$
(g) $\begin{bmatrix} 0 \\ 0 \\ 0 \end{bmatrix}$
(h) $\begin{bmatrix} 0 \\ 0 \\ 0 \\ 0 \end{bmatrix}$

4. Given the following vectors, do the indicated computations when possible. If a computation is impossible, explain why.

$\mathbf{a} = \begin{bmatrix} 1 \\ 2 \end{bmatrix}$ $\quad \mathbf{d} = [0, 0]$ $\quad \mathbf{g} = \begin{bmatrix} \frac{1}{3} \\ \frac{1}{4} \end{bmatrix}$

$\mathbf{b} = [3, 4]$ $\quad \mathbf{e} = [2, 1]$

$\mathbf{c} = [2, 0, 0]$ $\quad \mathbf{f} = [1, 0]$

(a) $\mathbf{a} + \mathbf{b}$
(b) $\mathbf{b} + \mathbf{c}$
(c) $\mathbf{b} + (\mathbf{f} + \mathbf{e})$
(d) $3\mathbf{a}$
(e) $\mathbf{a} + 4\mathbf{g}$
(f) $\mathbf{b} \cdot \mathbf{g}$
(g) $\mathbf{g} \cdot \mathbf{b}$
(h) $\mathbf{a} \cdot \mathbf{d}$
(i) $(\mathbf{f} \cdot \mathbf{g}) + \mathbf{a}$
(j) $\mathbf{f} \cdot (\mathbf{g} + \mathbf{a})$

5. A man made the following purchases at a grocery store:

3 quarts of milk @ 25¢ a quart
7 loaves of bread @ 35¢ a loaf
2 pounds of butter @ 83¢ a pound
10 pounds of potatoes @ 4¢ a pound
4 dozen eggs @ 52¢ a dozen

Write the purchases as a row vector and the prices as a column vector. Find the product of these vectors. What does the product represent?

6. Perform the following computations:

(a) $[2, 1] + [3, 4]$ and $[3, 4] + [2, 1]$

(b) $\begin{bmatrix} 1 \\ 2 \\ 0 \end{bmatrix} + \begin{bmatrix} 3 \\ 4 \\ 1 \end{bmatrix}$ and $\begin{bmatrix} 3 \\ 4 \\ 1 \end{bmatrix} + \begin{bmatrix} 1 \\ 2 \\ 0 \end{bmatrix}$

(c) $([2, 1] + [3, 4]) + [1, 5]$ and $[2, 1] + ([3, 4] + [1, 5])$

(d) $\begin{bmatrix} 1 \\ 2 \\ 1 \end{bmatrix} + \left(\begin{bmatrix} 3 \\ 8 \\ 2 \end{bmatrix} \quad \begin{bmatrix} 0 \\ 2 \\ 3 \end{bmatrix} \right)$ and $\left(\begin{bmatrix} 1 \\ 2 \\ 1 \end{bmatrix} + \begin{bmatrix} 3 \\ 8 \\ 2 \end{bmatrix} \right) + \begin{bmatrix} 0 \\ 2 \\ 3 \end{bmatrix}$

(e) $\left([2, 1, 3]\begin{bmatrix}3\\4\\1\end{bmatrix}\right) + \left([4, 2, 2]\begin{bmatrix}3\\4\\1\end{bmatrix}\right)$ and $([2, 1, 3] + [4, 2, 2])\begin{bmatrix}3\\4\\1\end{bmatrix}$

7. (a) From your answers to Problems 6(a) and 6(b), do you think vector addition is commutative?
   (b) From your answers to Problems 6(c) and 6(d), do you think vector addition is associative?
   (c) From your answer to Problem 6(e), do you think vector multiplication distributes over vector addition?
   (d) Is vector multiplication commutative? Explain.

6.2 MATRICES

In this section we consider a more general system than that of the vectors of the last section. The elements in this system will consist of rectangular arrays of numbers. These arrays we shall call **matrices**. Matrices are used in many practical situations; one novel application is illustrated in the following example.

EXAMPLE 6.6 Four points (these may be persons, locations, nations, etc.) are linked by communication lines as shown in Figure 6–1. Communications between two points takes place only in the direction of the arrows. Point 1 can send a message to point 2, but point 2 cannot send a message to point 1, and so on.

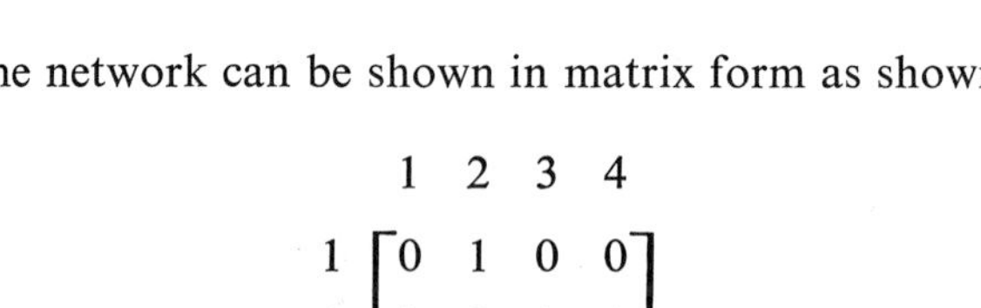

FIGURE 6–1 A communications network.

The same network can be shown in matrix form as shown below.

$$\begin{array}{c} \\ 1\\2\\3\\4\end{array}\begin{array}{c} \begin{array}{cccc}1&2&3&4\end{array}\\ \begin{bmatrix}0&1&0&0\\0&0&1&1\\1&1&0&1\\0&1&1&0\end{bmatrix}\end{array}$$

The matrix is the array of numbers inside the brackets. The horizontal row of numbers (1, 2, 3, 4) above the matrix and the vertical column of the same numbers to the left of the matrix are not parts of the matrix but are included only as references to the points in the communication network. Suppose that we wish to find whether point 1 can send a message to point 3. We look at the number in the first row and third column of the matrix. It is 0. This tells us that point 1 cannot send a message directly to point 3. The number in the third row and first column of the matrix is a 1. We interpret this to mean that point 3 can send a message directly to point 1. You should study both the matrix and Figure 6–1 until the meaning of each is quite clear to you. Remember that a 1 in the matrix indicates that communication can take place but a 0 means that it cannot.

Later we shall see how the matrix of the example above can be used to determine the number of ways that one point can communicate with another by relaying the message through other points. First, however, we must formally define a matrix and define some matrix operations.

MATRIX A **matrix** is a rectangular array of numbers. The numbers are called **components** of the matrix. Here a matrix will be denoted by an uppercase letter in boldface type. Components of the matrix will be identified by the same letter, in lowercase and with two subscripts. The first subscript indicates the row of the component, and the second subscript indicates the column of the component. As an example, suppose that matrix **A** is this one:

$$\mathbf{A} = \begin{bmatrix} 3 & 4 & 5 & 8 \\ 1 & 50 & 12 & 17 \\ 7 & 9 & 0 & 2 \end{bmatrix}$$

Then 3, the numeral in the first row and first column is a_{11}; 12, the numeral in the second row and third column is a_{23}; the numeral in the third row and fourth column is a_{34}; etc.

The **dimensions** of a matrix are determined by the numbers of its rows and columns. The matrix above is a three by four (3×4) matrix. A row vector is a $1 \times n$ matrix, and a column vector is an $n \times 1$ matrix.

Two $m \times n$ matrices **A** and **B** are **equal** if and only if their corresponding components are equal. That is,

$$\begin{array}{ccccc} a_{11} = b_{11} & a_{12} = b_{12} & a_{13} = b_{13} & \cdots & a_{1n} = b_{1n} \\ a_{21} = b_{21} & a_{22} = b_{22} & a_{23} = b_{23} & \cdots & a_{2n} = b_{2n} \\ \cdots\cdots & \cdots\cdots & \cdots\cdots & \cdots & \cdots\cdots \\ a_{m1} = b_{m1} & a_{m2} = b_{m2} & a_{m3} = b_{m3} & \cdots & a_{mn} = b_{mn} \end{array}$$

EXAMPLE 6.7 If

$$\mathbf{A} = \begin{bmatrix} 3 & 4 & 1 \\ 2 & 0 & 5 \end{bmatrix} \quad \text{and} \quad \mathbf{B} = \begin{bmatrix} x & y & z \\ r & s & t \end{bmatrix}$$

then $\mathbf{A} = \mathbf{B}$ if and only if $x = 3$, $y = 4$, $z = 1$, $r = 2$, $s = 0$, and $t = 5$.

6.2.1 Matrix Addition and Scalar Multiplication

As you might expect, since the set of vectors is a subset of the set of matrices, the operations on vectors may be extended to matrices.

SCALAR MULTIPLICATION The **scalar product** of a real number k and a matrix $\mathbf{A}$ is a matrix with the same dimensions as $\mathbf{A}$ but with every component k times as large as the corresponding component of $\mathbf{A}$.

EXAMPLE 6.8

$$2\begin{bmatrix} 1 & 2 \\ 0 & 3 \\ 5 & 6 \end{bmatrix} = \begin{bmatrix} 2 & 4 \\ 0 & 6 \\ 10 & 12 \end{bmatrix}$$

MATRIX ADDITION If $\mathbf{A}$ and $\mathbf{B}$ are two $m \times n$ matrices, their sum $\mathbf{A} + \mathbf{B}$ is an $m \times n$ matrix each of whose components is the sum of the corresponding components of $\mathbf{A}$ and $\mathbf{B}$. $\mathbf{A} + \mathbf{B}$ is not defined if $\mathbf{A}$ and $\mathbf{B}$ have different dimensions.

EXAMPLE 6.9

$$\begin{bmatrix} 2 & 3 & 4 \\ 1 & 5 & 0 \\ 1 & 1 & 2 \end{bmatrix} + \begin{bmatrix} 1 & 0 & 1 \\ 3 & 0 & 2 \\ 0 & 1 & 1 \end{bmatrix} = \begin{bmatrix} 3 & 3 & 5 \\ 4 & 5 & 2 \\ 1 & 2 & 3 \end{bmatrix}$$

A **zero matrix** is a matrix with all components zero. The symbol for a zero matrix is $\mathbf{0}$. The same symbol is used for any zero matrix regardless of its dimensions since the dimensions may be easily ascertained by context.

EXAMPLE 6.10 Compute $\mathbf{A} + \mathbf{0}$ if $\mathbf{A} = \begin{bmatrix} 3 & 2 & 1 \\ 1 & 4 & 5 \end{bmatrix}$.

Solution Since **A** is 2×3, **0** must also be 2×3, or addition will not be possible. Therefore,

$$\mathbf{0} = \begin{bmatrix} 0 & 0 & 0 \\ 0 & 0 & 0 \end{bmatrix} \quad \text{and} \quad \mathbf{A} + \mathbf{0} = \begin{bmatrix} 3 & 2 & 1 \\ 1 & 4 & 5 \end{bmatrix}$$

Let us pause, before we consider further properties of the matrix system, to summarize what we have accomplished thus far. We have defined a set of elements called matrices. These elements contain the set of vectors of Section 6.1 as a proper subset. On the matrices we have defined one relation, equality, and the two binary operations, scalar multiplication and matrix addition. We have found that for any given matrix there is a special matrix which will act as an identity element relative to addition. You may have guessed that matrix addition is commutative and associative and that each matrix has an inverse relative to addition. These facts are left for you to verify as exercises.

Exercise Set 6.2

1. What are the dimensions of each of the following matrices?

$$\mathbf{A} = \begin{bmatrix} 2 & 3 & 5 & 7 \\ 3 & 0 & 1 & 0 \\ 4 & 5 & 3 & 7 \end{bmatrix} \qquad \mathbf{B} = \begin{bmatrix} 5 & 6 \\ 7 & 8 \\ 6 & 5 \\ 4 & 2 \end{bmatrix} \qquad \mathbf{C} = \begin{bmatrix} 1 & 21 & 31 & 41 & 51 \\ 0 & 61 & 71 & 81 & 11 \\ 0 & 10 & 30 & 40 & 70 \\ 0 & 80 & 20 & 51 & 1 \\ 0 & 15 & 16 & 17 & 88 \end{bmatrix}$$

$$\mathbf{D} = \begin{bmatrix} 4 \\ 3 \\ 5 \\ 7 \\ 5 \end{bmatrix} \qquad \mathbf{E} = [3 \quad 6 \quad 7 \quad 1] \qquad \mathbf{F} = \begin{bmatrix} f_{11} & f_{12} & \cdots & f_{1s} \\ f_{21} & f_{22} & \cdots & f_{2s} \\ \cdot & \cdot & \cdots & \cdot \\ \cdot & \cdot & \cdots & \cdot \\ \cdot & \cdot & \cdots & \cdot \\ f_{w1} & f_{w2} & \cdots & f_{ws} \end{bmatrix}$$

2. For each of the matrices of Problem 1, what numbers, if any, correspond to:

(a) a_{12} (d) c_{21} (f) d_{13}

(b) b_{32} (e) d_{51} (g) e_{13}

(c) b_{43}

3. Perform these operations when possible. If a computation is not possible, tell why.

(a) $5\begin{bmatrix} 3 & 2 \\ 1 & 0 \\ 4 & 1 \end{bmatrix}$

(b) $\begin{bmatrix} 2 & 3 \\ 3 & 1 \end{bmatrix} + \begin{bmatrix} 1 & 1 \\ 2 & 2 \end{bmatrix}$

(c) $\begin{bmatrix} 3 & 1 & 4 \\ 1 & 2 & 0 \end{bmatrix} + 2\begin{bmatrix} 1 & 1 & 0 \\ 0 & 2 & 1 \end{bmatrix}$

(d) $\begin{bmatrix} 3 & 4 \\ 1 & 2 \\ 0 & 1 \end{bmatrix} + \begin{bmatrix} 1 & 2 & 3 \\ 2 & 4 & 5 \end{bmatrix}$

(e) $\begin{bmatrix} 3 & 6 & 8 \\ 6 & 2 & 1 \end{bmatrix} + \begin{bmatrix} 0 & 0 & 0 \\ 0 & 0 & 0 \end{bmatrix}$

4. Suppose four points are linked by the communications network of Figure 6–2. Show the network in matrix form.

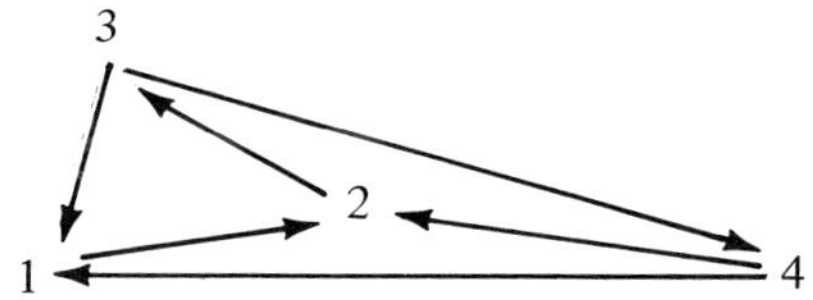

FIGURE 6–2

5. Which of the following do you think are true?
   (a) The set of matrices is closed under matrix addition.
   (b) Matrix addition is commutative.
   (c) Matrix addition is associative.

*6. Recall, from Chapter 4, that if $a + b = 0$, where 0 is the identity for $+$, a and b are inverses under $+$. Recall, also, that if a is a real number its additive inverse is $-a$. With these facts in mind, write a matrix **B** for each of the following problems if it is known that $\mathbf{A} + \mathbf{B} = \mathbf{0}$.

(a) $\mathbf{A} = \begin{bmatrix} 2 & 3 \\ 1 & 4 \end{bmatrix}$ (b) $\mathbf{A} = [3 \quad 2 \quad 2]$ (c) $\mathbf{A} = \begin{bmatrix} 1 \\ 8 \\ 6 \end{bmatrix}$

*7. Define $-\mathbf{A}$ to mean the additive inverse of the matrix **A**. Find:

(a) $-\begin{bmatrix} 2 & 4 & 6 \\ 1 & 2 & 0 \end{bmatrix}$ (b) $-\begin{bmatrix} 3 & 4 \\ 2 & 4 \\ 1 & 3 \end{bmatrix}$ (c) $-\begin{bmatrix} 0 & 2 & 3 \\ 1 & -3 & 2 \\ 4 & 8 & -7 \end{bmatrix}$

*8. Define **A** − **B** to mean **A** + (−**B**). Find:

(a) $\begin{bmatrix} 3 & 1 \\ 2 & 2 \end{bmatrix} - \begin{bmatrix} 2 & 1 \\ 1 & 2 \end{bmatrix}$

(b) $\begin{bmatrix} 3 \\ 4 \end{bmatrix} - \begin{bmatrix} 1 \\ 2 \end{bmatrix}$

(c) $[1 \quad 3 \quad 5] - [2 \quad 2 \quad 2]$

6.3 MATRIX MULTIPLICATION

6.3.1 Multiplication of a Matrix by a Matrix

The multiplication of a matrix by a matrix is defined only for very special cases. In order that the product **AB** exist, **B** must have the same number of rows as **A** has columns. The product is a matrix with as many rows as **A** and as many columns as **B**. The formal definition is:

MATRIX MULTIPLICATION If **A** is an $m \times n$ matrix and **B** is an $n \times r$ matrix, then **AB** is an $m \times r$ matrix and the component in the ith row and the jth column of the product matrix is the product of the ith row of **A** as a row vector with the jth row of **B** as a column vector.

The process of multiplication is illustrated in the next example.

EXAMPLE 6.11 Determine **C** if **AB** = **C** and

$$\mathbf{A} = \begin{bmatrix} 3 & 2 & 1 \\ 4 & 1 & 3 \end{bmatrix} \quad \text{and} \quad \mathbf{B} = \begin{bmatrix} 2 & 4 & 1 \\ 5 & 6 & 8 \\ 1 & 3 & 2 \end{bmatrix}$$

Solution Using vector multiplication, find the product of row 1 of **A** and column 1 of **B**. The product 17 is c_{11}.

$$\begin{bmatrix} 3 & 2 & 1 \\ 4 & 1 & 3 \end{bmatrix}\begin{bmatrix} 2 & 4 & 1 \\ 5 & 6 & 8 \\ 1 & 3 & 2 \end{bmatrix} = \begin{bmatrix} 17 & c_{12} & c_{13} \\ c_{21} & c_{22} & c_{23} \end{bmatrix}$$

Next multiply row 1 of **A** and column 2 of **B**; the product is c_{12}.

$$\begin{bmatrix} 3 & 2 & 1 \\ 4 & 1 & 3 \end{bmatrix}\begin{bmatrix} 2 & 4 & 1 \\ 5 & 6 & 8 \\ 1 & 3 & 2 \end{bmatrix} = \begin{bmatrix} 17 & 27 & c_{13} \\ c_{21} & c_{22} & c_{23} \end{bmatrix}$$

The component c_{13} is the product of row 1 of **A** and column 3 of **B**; c_{21} is the product of row 2 of **A** and column 1 of **B**; and so on until every row of **A** has been multiplied by every column of **B**.

$$\begin{bmatrix} 3 & 2 & 1 \\ 4 & 1 & 3 \end{bmatrix} \begin{bmatrix} 2 & 4 & 1 \\ 5 & 6 & 8 \\ 1 & 3 & 2 \end{bmatrix} = \begin{bmatrix} 17 & 27 & 21 \\ 16 & 31 & 18 \end{bmatrix}$$

Example 6.11, as well as 6.12 which follows, should be carefully studied to ensure that the technique of matrix multiplication is fully understood.

EXAMPLE 6.12 Given $\mathbf{A} = \begin{bmatrix} 1 & 3 \\ 2 & 4 \end{bmatrix}$ and $\mathbf{B} = \begin{bmatrix} 3 & 1 \\ 2 & 6 \end{bmatrix}$, compute **AB** and **BA**.

Solution

$$\mathbf{AB} = \begin{bmatrix} 1 & 3 \\ 2 & 4 \end{bmatrix} \begin{bmatrix} 3 & 1 \\ 2 & 6 \end{bmatrix} = \begin{bmatrix} 9 & 19 \\ 14 & 26 \end{bmatrix}$$

$$\mathbf{BA} = \begin{bmatrix} 3 & 1 \\ 2 & 6 \end{bmatrix} \begin{bmatrix} 1 & 3 \\ 2 & 4 \end{bmatrix} = \begin{bmatrix} 5 & 13 \\ 14 & 30 \end{bmatrix}$$

It should be clear from the above example that matrix multiplication is not commutative.

If matrix **A** has the same number of rows as it has columns, we call it a **square matrix**. If **A** is a square matrix, then the product of **A** by itself is possible. The product of **A** by itself is called the square of **A** and denoted $\mathbf{A}^2$. The product $\mathbf{A}^2 \cdot \mathbf{A}$ is denoted $\mathbf{A}^3$; $\mathbf{A}^4 = \mathbf{A}^3 \cdot \mathbf{A}$; etc.

6.3.1 Communications Networks

One interesting application of matrix multiplication can be made to the communication network of Example 6.6. The matrix representing that network is

$$\mathbf{A} = \begin{array}{c} \\ 1 \\ 2 \\ 3 \\ 4 \end{array} \begin{array}{c} \begin{array}{cccc} 1 & 2 & 3 & 4 \end{array} \\ \begin{bmatrix} 0 & 1 & 0 & 0 \\ 0 & 0 & 1 & 1 \\ 1 & 1 & 0 & 1 \\ 0 & 1 & 1 & 0 \end{bmatrix} \end{array}$$

Thus

$$\mathbf{A}^2 = \begin{array}{c} \\ 1 \\ 2 \\ 3 \\ 4 \end{array} \begin{array}{c} \begin{array}{cccc} 1 & 2 & 3 & 4 \end{array} \\ \begin{bmatrix} 0 & 0 & 1 & 1 \\ 1 & 2 & 1 & 1 \\ 0 & 2 & 2 & 1 \\ 1 & 1 & 1 & 2 \end{bmatrix} \end{array}$$

We may use $\mathbf{A}^2$ to determine the number of ways one point can send a message to a second point using only one relay point. To see how this is done, look at Figure 6–1 and trace a message from point 3 to point 2 if the message must be sent from point 3 to any other point except 2 and then relayed from that point to point 2. We see that there are two possible ways in which this may be done. Point 3 can send the message to point 1 and point 1 can send it on to point 2, or point 3 can send the message to point 4 and point 4 can relay it to point 2. Now look at the component in the third row and second column of $\mathbf{A}^2$. It is 2, the same as the number of ways that point 3 can send a message to point 2 using only one intermediate relay point. Similarly, the component in row 2, column 1 of $\mathbf{A}^2$ is a 1, and this tells us that there is but one way for point 2 to relay a message to point 1 through exactly one relay point. A check of Figure 6–1 shows that this way is from 2 to 3 and then to 1.

It can be shown that whenever a communications network is displayed as a matrix $\mathbf{A}$, then the matrix $\mathbf{A}^2$ shows the number of ways a message can be relayed from one point to another using exactly one relay point. The matrix $\mathbf{A}^3$ shows the number of ways a message can be relayed using exactly two intermediate points. The matrix $\mathbf{A} + \mathbf{A}^2$ shows the total number of ways a message can be sent from one point to another using no more than one relay point, and $\mathbf{A} + \mathbf{A}^2 + \mathbf{A}^3$ shows the total number of ways a message can be sent if no more than two relay points are used.

EXAMPLE 6.13 Let $\mathbf{A}$ be the communication matrix of Example 6.6. From

$$\mathbf{A} + \mathbf{A}^2 = \begin{array}{c} \\ 1 \\ 2 \\ 3 \\ 4 \end{array} \begin{array}{c} \begin{array}{cccc} 1 & 2 & 3 & 4 \end{array} \\ \begin{bmatrix} 0 & 1 & 1 & 1 \\ 1 & 2 & 2 & 2 \\ 1 & 3 & 2 & 2 \\ 1 & 2 & 2 & 2 \end{bmatrix} \end{array}$$

we find there are three ways in which point 3 can communicate with point 2 using no more than one relay point. There are two ways in which point 4 can send a message to point 2 using no more than one relay point. It is interesting to note that while no point can communicate directly with itself, point 1 is the only one which cannot send a message to itself via one intermediate point.

6.3.3 Dominance Relations

A sociometric problem to which matrices may be applied is that of dominance theory. A **dominance relation** is a relation between members of a group in which certain members exert an influence on other members. We shall consider only those dominance relations which satisfy these two requirements:

1. No member of a group dominates itself.
2. In each pair of members of the group, one member is dominant. That is, if x and y are members, either x dominates y or y dominates x but not both.

An example of this type of relation is found in one round of a round-robin contest between athletic teams if no ties are allowed.

EXAMPLE 6.14 In a local Little League there are four teams: the Tigers, the Cubs, the Sox, and the Cards. In the first round of games the Tigers beat the Cubs and the Sox but were beaten by the Cards. The Cubs beat the Sox and the Cards and the Sox beat the Cards. These dominances are shown in matrix **D**.

$$\mathbf{D} = \begin{array}{r} \\ \text{Tigers} \\ \text{Cubs} \\ \text{Sox} \\ \text{Cards} \end{array} \begin{array}{c} \begin{array}{cccc} \text{Tigers} & \text{Cubs} & \text{Sox} & \text{Cards} \end{array} \\ \begin{bmatrix} 0 & 1 & 1 & 0 \\ 0 & 0 & 1 & 1 \\ 0 & 0 & 0 & 1 \\ 1 & 0 & 0 & 0 \end{bmatrix} \end{array}$$

Notice that each row and column of **D** is associated with the name of one of the teams. A 1 component indicates that the team which names the row of the component beat the team which names the column of the component. A 0 component indicates that the team of that row did not beat the team of that column. The dominances of matrix **D** are **one-stage** dominances, that is, direct dominance of one member of a group by another member of the group. A **two-stage** dominance is one in which x dominates y and y dominates z; we say that x exerts a two-stage dominance over z. Two-stage dominances may be shown by squaring the one-stage matrix.

$$\mathbf{D}^2 = \begin{array}{r} \\ \text{Tigers} \\ \text{Cubs} \\ \text{Sox} \\ \text{Cards} \end{array} \begin{array}{c} \begin{array}{cccc} \text{Tigers} & \text{Cubs} & \text{Sox} & \text{Cards} \end{array} \\ \begin{bmatrix} 0 & 0 & 1 & 2 \\ 1 & 0 & 0 & 1 \\ 1 & 0 & 0 & 0 \\ 0 & 1 & 1 & 0 \end{bmatrix} \end{array}$$

The component 2 in the top right corner of $\mathbf{D}^2$ indicates that the Tigers exert a pair of two-stage dominances over the Cards. This does not mean that the Tigers beat the Cards; they did not. It does show that the Tigers beat two other teams each of which, in their turn, beat the Cards.

The **power** of a member of a group is the total number of one- and two-stage dominances which that member exerts. The matrix which is used to find the power of an individual is

$$\mathbf{P} = \mathbf{D} + \mathbf{D}^2$$

For the teams above,

$$\mathbf{P} = \begin{array}{l} \\ \text{Tigers} \\ \text{Cubs} \\ \text{Sox} \\ \text{Cards} \end{array} \begin{array}{c} \begin{array}{cccc} \text{Tigers} & \text{Cubs} & \text{Sox} & \text{Cards} \end{array} \\ \begin{bmatrix} 0 & 1 & 2 & 2 \\ 1 & 0 & 1 & 2 \\ 1 & 0 & 0 & 1 \\ 1 & 1 & 1 & 0 \end{bmatrix} \end{array}$$

The power of each team is the sum of the entries of its row in **P**. Thus the teams arranged in order of their powers are

Team	Power
Tigers	5
Cubs	4
Cards	3
Sox	2

In this section we have considered multiplication of a matrix by a matrix. We have discovered that matrix multiplication is not a commutative operation. Two applications of matrix multiplication were shown. One of these applications concerned communications networks and the other was to dominance relations. Another application will be discussed in Section 6.6.

Exercise Set 6.3

1. Perform these multiplications when possible.

(a) $\begin{bmatrix} 2 & 1 \\ 3 & 4 \end{bmatrix}\begin{bmatrix} 0 & 1 \\ 2 & 1 \end{bmatrix}$ (b) $\begin{bmatrix} 2 & 1 \\ 6 & 0 \end{bmatrix}\begin{bmatrix} 1 \\ 4 \end{bmatrix}$

(c) $[3 \quad 1 \quad 2]\begin{bmatrix} 4 & 2 \\ 2 & 1 \\ 1 & 1 \end{bmatrix}$ (f) $\begin{bmatrix} 3 & 2 \\ 1 & 5 \end{bmatrix}\begin{bmatrix} 0 & 1 \\ 1 & 0 \end{bmatrix}$

(d) $\begin{bmatrix} 9 & 6 & 2 \\ 4 & 3 & 1 \end{bmatrix}\begin{bmatrix} 2 & 4 \\ 0 & 2 \end{bmatrix}$ (g) $\left([1 \quad 2]\begin{bmatrix} 3 & 2 \\ 1 & 1 \end{bmatrix}\right)\begin{bmatrix} 2 \\ 3 \end{bmatrix}$

(e) $\begin{bmatrix} 3 & 2 \\ 1 & 5 \end{bmatrix}\begin{bmatrix} 1 & 0 \\ 0 & 1 \end{bmatrix}$ (h) $[1 \quad 2]\left(\begin{bmatrix} 3 & 2 \\ 1 & 1 \end{bmatrix}\begin{bmatrix} 2 \\ 3 \end{bmatrix}\right)$

2. Compute

(a) $\begin{bmatrix} 1 & 3 \\ 1 & 1 \end{bmatrix}\left(\begin{bmatrix} 3 \\ 4 \end{bmatrix} + \begin{bmatrix} 2 \\ 3 \end{bmatrix}\right)$ (b) $\left(\begin{bmatrix} 1 & 3 \\ 1 & 1 \end{bmatrix}\begin{bmatrix} 3 \\ 4 \end{bmatrix}\right) + \left(\begin{bmatrix} 1 & 3 \\ 1 & 1 \end{bmatrix}\begin{bmatrix} 2 \\ 3 \end{bmatrix}\right)$

3. Compute $\mathbf{A}^2$, $\mathbf{A}^3$, and $\mathbf{A}^4$ if

$$\mathbf{A} = \begin{bmatrix} 3 & 1 \\ 2 & 1 \end{bmatrix}$$

4. A trucking company has three types of trucks, types 1, 2, and 3. Each truck is equipped to haul three different types of heavy machines per load. The number of each type of machine handled by each type of truck is given by the following matrix.

$$\begin{array}{r c} & \text{Type Truck} \\ & \begin{array}{ccc} 1 & 2 & 3 \end{array} \\ \begin{array}{r} A \\ \text{Type machine } B \\ C \end{array} & \begin{bmatrix} 1 & 1 & 1 \\ 1 & 2 & 2 \\ 2 & 1 & 1 \end{bmatrix} \end{array}$$

One day the company sent out the following fleet:

$$\begin{array}{r c} & \text{Number} \\ \begin{array}{r} 1 \\ \text{Truck type } 2 \\ 3 \end{array} & \begin{bmatrix} 4 \\ 2 \\ 3 \end{bmatrix} \end{array}$$

If each truck was fully loaded, how many machines of each type were transported that day?

5. In Example 6.6, in how many ways can a message be sent from point 2 to point 3 using exactly one relay point? Using no more than one relay point?

6. Find the communication matrix for the network of Figure 6–3. Using no more than one relay point, in how many ways can a message be sent from point 1 to point 4? From point 4 to point 3? From point 2 to point 4?

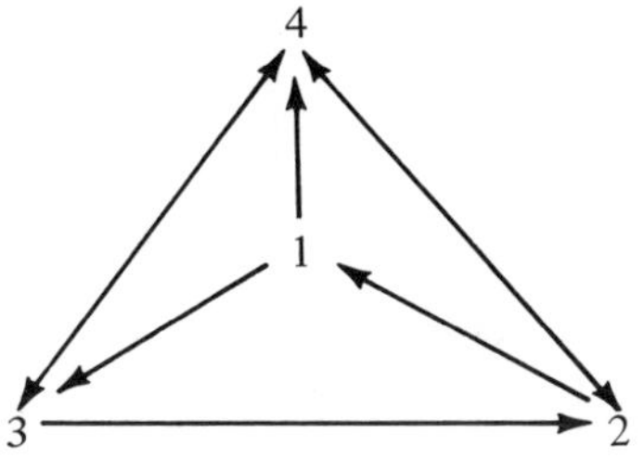

FIGURE 6–3

7. Find the dominance matrix showing that team A beats teams B and C, team B beats team D, team C beats team B, and team D beats teams A and C. What is the power of each team?

8. Matrix multiplication is associative. Make up a problem which illustrates this fact. Make up another problem to illustrate the fact that matrix multiplication distributes over matrix addition.

*6.4 IDENTITY AND INVERSE MATRICES

6.4.1 Identity Matrices

We have seen that the system of matrices displays many of the properties which we studied in Chapter 4. There is an operation called addition. This operation is both commutative and associative. There is, for each matrix, an identity matrix relative to the addition operation. This is the matrix **0**. For every matrix **A** there is an additive identity matrix $-\mathbf{A}$ such that $\mathbf{A} + (-\mathbf{A}) = \mathbf{0}$. There is a second operation called matrix multiplication which is associative but not commutative. Because there is a matrix multiplication, it seems reasonable to pose these two questions:

1. Is there an identity matrix relative to multiplication?
2. If there is an identity relative to multiplication, do any of the matrices have multiplicative inverses?

The answers to these questions are the subject matter of this section.

Consider the set of square matrices which have 1's diagonally from top left to bottom right corners and 0's as all other entries. Three of these matrices are:

$$\begin{bmatrix} 1 & 0 \\ 0 & 1 \end{bmatrix} \quad \begin{bmatrix} 1 & 0 & 0 \\ 0 & 1 & 0 \\ 0 & 0 & 1 \end{bmatrix} \quad \begin{bmatrix} 1 & 0 & 0 & 0 \\ 0 & 1 & 0 & 0 \\ 0 & 0 & 1 & 0 \\ 0 & 0 & 0 & 1 \end{bmatrix}$$

Every matrix in this set is named **I**. As was the case with the zero matrix, it is not necessary to specify the dimensions of a particular **I** since these can easily be determined by context.

EXAMPLE 6.15 If $\mathbf{A} = \begin{bmatrix} 3 & 2 & 1 \\ 1 & 4 & 5 \\ 6 & 2 & 8 \end{bmatrix}$, compute **AI**.

Solution By definition **I** is a square matrix and, if **I** is to be multiplied by **A** on the left, it must have the same number of rows as **A** has columns. Therefore **I** must be 3×3.

$$\mathbf{AI} = \begin{bmatrix} 3 & 2 & 1 \\ 1 & 4 & 5 \\ 6 & 2 & 8 \end{bmatrix} \begin{bmatrix} 1 & 0 & 0 \\ 0 & 1 & 0 \\ 0 & 0 & 1 \end{bmatrix} = \begin{bmatrix} 3 & 2 & 1 \\ 1 & 4 & 5 \\ 6 & 2 & 8 \end{bmatrix}$$

We may show for any square matrix **A** that

$$\mathbf{AI} = \mathbf{A} \qquad \text{and} \qquad \mathbf{IA} = \mathbf{A}$$

For this reason **I** is called an **identity matrix** for multiplication.

The matrix **I** is unique for any given dimensions. That is, there is no 2×2 matrix other than $\begin{bmatrix} 1 & 0 \\ 0 & 1 \end{bmatrix}$ which will act as an identity for matrix multiplication on the set of 2×2 matrices. The only matrix which will serve as a multiplicative identity for 3×3 matrices is

$$\begin{bmatrix} 1 & 0 & 0 \\ 0 & 1 & 0 \\ 0 & 0 & 1 \end{bmatrix} \text{ etc.}$$

6.4.2 Inverses

The existence of an identity matrix suggests that there may be pairs of matrices **A** and **B** such that **AB** = **I** and **BA** = **I**. If so, **A** and **B** are multiplicative inverses: **B** is the inverse of **A**, and **A** is the inverse of **B**. The symbol for "multiplicative inverse of **A**" is $\mathbf{A}^{-1}$.

EXAMPLE 6.16 Consider

$$\mathbf{A} = \begin{bmatrix} 0 & \frac{1}{2} \\ \frac{1}{2} & -\frac{1}{4} \end{bmatrix} \quad \text{and} \quad \mathbf{B} = \begin{bmatrix} 1 & 2 \\ 2 & 0 \end{bmatrix}$$

$$\mathbf{AB} = \begin{bmatrix} 1 & 0 \\ 0 & 1 \end{bmatrix} = \mathbf{I} \quad \text{and} \quad \mathbf{BA} = \mathbf{I}$$

So **A** and **B** are inverses.

One might suppose that all square matrices, except **0**, have inverses. This is not true.

EXAMPLE 6.17 Suppose that $\mathbf{A} = \begin{bmatrix} 3 & 0 \\ 4 & 0 \end{bmatrix}$ has an inverse; then it must have the form $\mathbf{B} = \begin{bmatrix} b_{11} & b_{12} \\ b_{21} & b_{22} \end{bmatrix}$ where b_{11}, b_{12}, b_{21} and b_{22} are real numbers.

Then

$$\mathbf{AB} = \begin{bmatrix} 3 & 0 \\ 4 & 0 \end{bmatrix} \begin{bmatrix} b_{11} & b_{12} \\ b_{21} & b_{22} \end{bmatrix}$$

$$= \begin{bmatrix} 3b_{11} + 0 \cdot b_{21} & 3b_{12} + 0 \cdot b_{22} \\ 4b_{11} + 0 \cdot b_{21} & 4b_{12} + 0 \cdot b_{22} \end{bmatrix}$$

$$= \begin{bmatrix} 3b_{11} & 3b_{12} \\ 4b_{11} & 4b_{12} \end{bmatrix}$$

But if **A** and **B** are inverses, it must be true that

$$\mathbf{AB} = \begin{bmatrix} 1 & 0 \\ 0 & 1 \end{bmatrix}$$

Thus

$$\begin{bmatrix} 3b_{11} & 3b_{12} \\ 4b_{11} & 4b_{12} \end{bmatrix} = \begin{bmatrix} 1 & 0 \\ 0 & 1 \end{bmatrix}$$

and by the definition of equal matrices, it follows that

$$3b_{11} = 1 \qquad \text{and} \qquad 4b_{11} = 0$$

But $3b_{11} = 1$ if and only if $b_{11} = \frac{1}{3}$, and $4b_{11} = 0$ if and only if $b_{11} = 0$. Therefore, if **A** has an inverse, b_{11} must be 0 and it must also be $\frac{1}{3}$. Clearly this is not possible, since $0 \neq \frac{1}{3}$. Therefore we must conclude that it is impossible to find a matrix **B** which is the inverse of matrix **A** in this example.

As a consequence of Examples 6.16 and 6.17 it is clear that some square matrices have inverses but some do not. In the next section we shall consider the problem of finding inverses when they exist. The method we shall use depends upon techniques which are the subject matter of the remainder of the present section.

6.4.3 Elementary Row Operations

A **row operation** is an operation performed on a single row of a matrix. There are two of these which must be considered at this time:

ELEMENTARY ROW OPERATIONS

Operation 1 Multiply each member of any row of the matrix by any nonzero number.

Operation 2 Add a scalar multiple of one row to another row. The latter row is replaced by the sum. (See Example 6.18.)

Because addition and subtraction are inverse operations, Operation 2 allows a multiple of one row to be subtracted from any other row.

EXAMPLE 6.18 Consider

$$\mathbf{A} = \begin{bmatrix} 3 & 1 & 2 \\ 2 & 4 & 2 \\ 5 & 1 & 3 \end{bmatrix} \qquad \mathbf{B} = \begin{bmatrix} 3 & 1 & 2 \\ 1 & 2 & 1 \\ 5 & 1 & 3 \end{bmatrix} \qquad \mathbf{C} = \begin{bmatrix} 3 & 1 & 2 \\ 1 & 2 & 1 \\ -1 & -1 & -1 \end{bmatrix}$$

B is formed from **A** by multiplying each element of **A**'s second row by $\frac{1}{2}$. **C** is formed from **B** by adding -2 times **B**'s first row to its third row.

In practice we usually start with some matrix **A**, change it by means of a row operation to a second matrix **B**, change **B** to **C**, change **C** to **D**, and so on until the final matrix in the sequence has some predetermined form.

EXAMPLE 6.19 By means of a sequence of elementary row operations, convert

$$\begin{bmatrix} 3 & 6 & 2 \\ 1 & 4 & 3 \\ 0 & 1 & 1 \end{bmatrix} \quad \text{to} \quad \begin{bmatrix} 1 & 0 & -2 \\ 1 & 1 & 0 \\ 0 & 1 & 1 \end{bmatrix}$$

Solution There are many sequences which might be used. One of these is

$$\mathbf{A} = \begin{bmatrix} 3 & 6 & 2 \\ 1 & 4 & 3 \\ 0 & 1 & 1 \end{bmatrix} \qquad \mathbf{B} = \begin{bmatrix} 1 & -2 & -4 \\ 1 & 4 & 3 \\ 0 & 1 & 1 \end{bmatrix}$$

$$\mathbf{C} = \begin{bmatrix} 1 & 0 & -2 \\ 1 & 4 & 3 \\ 0 & 1 & 1 \end{bmatrix} \qquad \mathbf{D} = \begin{bmatrix} 1 & 0 & -2 \\ 1 & 1 & 0 \\ 0 & 1 & 1 \end{bmatrix}$$

Matrix **B** is obtained from matrix **A** by taking row 1 − 2 times row 2 in **A** and letting that replace row 1 of **A**. Matrix **C** is obtained from matrix **B** by taking row 1 + 2 times row 3 and letting this replace row 1 of **B**. Matrix **D** is obtained from matrix **C** by taking row 2 − 3 times row 3 of **C** and letting this replace row 2 of **C**.

We have seen that there is a special subset of the matrices which serve as identity matrices for multiplication. These are square matrices with 1's diagonally from their upper left to their lower right corners and 0's in every other position. We have also observed that in order for a matrix to have an inverse, relative to multiplication, it must be a square matrix. However, all square matrices do not have inverses.

In the next section we shall use the elementary row operations of this section to construct inverse matrices when they exist.

Exercise Set 6.4

1. Multiply each pair of matrices to determine whether or not the pairs are inverses.

(a) $\begin{bmatrix} 1 & 2 \\ 2 & 3 \end{bmatrix}$ and $\begin{bmatrix} -3 & 2 \\ 2 & -1 \end{bmatrix}$

(b) $\begin{bmatrix} 2 & -1 \\ -5 & 3 \end{bmatrix}$ and $\begin{bmatrix} 3 & 1 \\ 5 & 2 \end{bmatrix}$

(c) $\begin{bmatrix} -9 & 4 & 2 \\ 11 & -4 & -3 \\ -3 & 1 & 1 \end{bmatrix}$ and $\begin{bmatrix} 1 & 2 & 4 \\ 2 & 3 & 5 \\ 1 & 3 & 8 \end{bmatrix}$

(d) $\begin{bmatrix} 1 & 0 & 0 & 1 \\ 2 & 1 & 0 & 2 \\ 0 & 1 & 1 & 0 \\ 1 & 3 & 1 & 2 \end{bmatrix}$ and $\begin{bmatrix} -2 & 2 & 1 & -1 \\ -2 & 1 & 0 & 0 \\ 2 & -1 & 1 & 0 \\ 3 & -2 & -1 & 1 \end{bmatrix}$

2. The first of each of the following pairs of matrices can be transformed to the second by performing exactly one of the elementary row operations on it. Describe the operation in each case.

(a) $\begin{bmatrix} 2 & 1 \\ 5 & 8 \end{bmatrix}$ to $\begin{bmatrix} 2 & 1 \\ 1 & 6 \end{bmatrix}$

(b) $\begin{bmatrix} 1 & 0 & 2 \\ 1 & 1 & 1 \\ 0 & 3 & 1 \end{bmatrix}$ to $\begin{bmatrix} 1 & 0 & 2 \\ 1 & 1 & 1 \\ -1 & 2 & 0 \end{bmatrix}$

(c) $\begin{bmatrix} 3 & 4 & 2 \\ 4 & 2 & 1 \end{bmatrix}$ to $\begin{bmatrix} -5 & 0 & 0 \\ 4 & 2 & 1 \end{bmatrix}$

(d) $\begin{bmatrix} 8 & 3 \\ 5 & -1 \end{bmatrix}$ to $\begin{bmatrix} 8 & 3 \\ -5 & 1 \end{bmatrix}$

3. The first of each of these pairs of matrices can be transformed to the second by a sequence of elementary row operations. Describe a possible sequence in each case.

(a) $\begin{bmatrix} 1 & 1 \\ 1 & 2 \end{bmatrix}$ to $\begin{bmatrix} 1 & 0 \\ 0 & 1 \end{bmatrix}$

(b) $\begin{bmatrix} 3 & 6 \\ 1 & 4 \end{bmatrix}$ to $\begin{bmatrix} 1 & 2 \\ 0 & 2 \end{bmatrix}$

(c) $\begin{bmatrix} 1 & 3 & 2 \\ 2 & 1 & 4 \\ 1 & 0 & 1 \end{bmatrix}$ to $\begin{bmatrix} 0 & 3 & 1 \\ 0 & 1 & 2 \\ 1 & 0 & 1 \end{bmatrix}$

(d) $\begin{bmatrix} 3 & 4 & 6 \\ 1 & 2 & 4 \\ 5 & 0 & -3 \end{bmatrix}$ to $\begin{bmatrix} 1 & \frac{4}{3} & 2 \\ 0 & \frac{2}{3} & 2 \\ 0 & -\frac{20}{3} & -13 \end{bmatrix}$

4. Describe the series of row operations which were used to change the first matrix to the last one in each case.

(a) $\mathbf{A} = \begin{bmatrix} 2 & 3 \\ 1 & 5 \end{bmatrix}$ to $\mathbf{B} = \begin{bmatrix} 1 & -2 \\ 1 & 5 \end{bmatrix}$ to $\mathbf{C} = \begin{bmatrix} 1 & -2 \\ 0 & 7 \end{bmatrix}$ to $\mathbf{D} = \begin{bmatrix} 1 & -2 \\ 0 & 1 \end{bmatrix}$ to

$\mathbf{E} = \begin{bmatrix} 1 & 0 \\ 0 & 1 \end{bmatrix}$

(b) $\mathbf{A} = \begin{bmatrix} 1 & 4 & 3 \\ -1 & 3 & 4 \\ -2 & 2 & 0 \end{bmatrix}$ to $\mathbf{B} = \begin{bmatrix} 1 & 4 & 3 \\ 0 & 7 & 7 \\ -2 & 2 & 0 \end{bmatrix}$ to $\mathbf{C} = \begin{bmatrix} 1 & 4 & 3 \\ 0 & 7 & 7 \\ 0 & 10 & 6 \end{bmatrix}$ to

$\mathbf{D} = \begin{bmatrix} 1 & 4 & 3 \\ 0 & 1 & 1 \\ 0 & 10 & 6 \end{bmatrix}$ to $\mathbf{E} = \begin{bmatrix} 1 & 0 & -1 \\ 0 & 1 & 1 \\ 0 & 10 & 6 \end{bmatrix}$ to $\mathbf{F} = \begin{bmatrix} 1 & 0 & -1 \\ 0 & 1 & 1 \\ 0 & 0 & -4 \end{bmatrix}$

5. Show that if **I** is 3×3, **I** is its own inverse relative to multiplication.
6. Show that **0** has no multiplicative inverse.
7. Show that if **A** and **B** are square matrices with **AB** = **I**, then **BA** = **I** also. Start with **BA** = **BA**, multiply each member by **B** on the right, and use associativity.

*6.5 *FINDING INVERSES*

The method of determining the inverse of a given matrix is based upon the following theorem.

THEOREM 6.1 If an elementary row operation is performed on **I** and the resulting matrix is multiplied by **A**, the product is the same matrix as that obtained by performing the row operation on **A** to begin with.

EXAMPLE 6.20 Consider

$$\mathbf{A} = \begin{bmatrix} 1 & 1 & 0 \\ -2 & 1 & 2 \\ 1 & 1 & 1 \end{bmatrix} \quad \text{and} \quad \mathbf{I} = \begin{bmatrix} 1 & 0 & 0 \\ 0 & 1 & 0 \\ 0 & 0 & 1 \end{bmatrix}$$

Let **B** be the matrix formed by adding 2 times row 1 of **I** to its second row and replacing the second row.

$$\mathbf{B} = \begin{bmatrix} 1 & 0 & 0 \\ 2 & 1 & 0 \\ 0 & 0 & 1 \end{bmatrix}$$

Then

$$\mathbf{BA} = \begin{bmatrix} 1 & 0 & 0 \\ 2 & 1 & 0 \\ 0 & 0 & 1 \end{bmatrix}\begin{bmatrix} 1 & 1 & 0 \\ -2 & 1 & 2 \\ 1 & 1 & 1 \end{bmatrix} = \begin{bmatrix} 1 & 1 & 0 \\ 0 & 3 & 2 \\ 1 & 1 & 1 \end{bmatrix}$$

This product is precisely the same that results from adding twice **A**'s first row to its second row.

To find the inverse of a given matrix **A**, we observe that

$$\mathbf{A} = \mathbf{IA}$$

Theorem 6.1 tells us that we may perform the same row operation on both **A** and **I** in the above expression and still maintain the relation of equality. Suppose that a single row operation is all that is necessary to transform **A** to **I**, and suppose the matrix resulting from performing that row operation on **I** is called **B**. We then have

$$\mathbf{I} = \mathbf{BA}$$

Then from the result of Problem 7, Exercise Set 6.4, we know

$$\mathbf{I} = \mathbf{AB}$$

also, and **B** is the inverse of A

EXAMPLE 6.21 Find the inverse of

$$\mathbf{A} = \begin{bmatrix} 1 & 0 & 0 \\ 3 & 1 & 0 \\ 0 & 0 & 1 \end{bmatrix}$$

Solution First we write **A** = **IA**:

$$\begin{bmatrix} 1 & 0 & 0 \\ 3 & 1 & 0 \\ 0 & 0 & 1 \end{bmatrix} = \begin{bmatrix} 1 & 0 & 0 \\ 0 & 1 & 0 \\ 0 & 0 & 1 \end{bmatrix} \begin{bmatrix} 1 & 0 & 0 \\ 3 & 1 & 0 \\ 0 & 0 & 1 \end{bmatrix}$$

Then by elementary row operations we convert **A** to **I**. The same row operations performed on **I** changes **I** to $\mathbf{A}^{-1}$.

(*Add −3 times row 1 to row 2.*) $$\begin{bmatrix} 1 & 0 & 0 \\ 0 & 1 & 0 \\ 0 & 0 & 1 \end{bmatrix} = \begin{bmatrix} 1 & 0 & 0 \\ -3 & 1 & 0 \\ 0 & 0 & 1 \end{bmatrix} \begin{bmatrix} 1 & 0 & 0 \\ 3 & 1 & 0 \\ 0 & 0 & 1 \end{bmatrix}$$

The required inverse is

$$\begin{bmatrix} 1 & 0 & 0 \\ -3 & 1 & 0 \\ 0 & 0 & 1 \end{bmatrix}$$

In Example 6.21 it was possible to change **A** to **I** by means of a single elementary row operation. Often, however, several row operations must be used to convert a given matrix to **I**. In such cases we make use of the following theorem.

THEOREM 6.2 If **A** can be changed to **I** by a sequence of elementary row operations, then the same sequence of row operations performed on **I** will produce $\mathbf{A}^{-1}$.

EXAMPLE 6.22 Find $\mathbf{A}^{-1}$ if

$$\mathbf{A} = \begin{bmatrix} 1 & 3 \\ 1 & 2 \end{bmatrix}$$

Solution Set **A** and **I** side by side and, using one row operation at a time, change **A** to **I**. Perform the same sequence of row operations on **I**.

$$\begin{bmatrix} 1 & 3 \\ 1 & 2 \end{bmatrix} \quad \Bigg| \quad \begin{bmatrix} 1 & 0 \\ 0 & 1 \end{bmatrix}$$

(*Subtract row 1 from row 2.*)
$$\begin{bmatrix} 1 & 3 \\ 0 & -1 \end{bmatrix} \quad \Bigg| \quad \begin{bmatrix} 1 & 0 \\ -1 & 1 \end{bmatrix}$$

(*Add 3 times row 2 to row 1.*)
$$\begin{bmatrix} 1 & 0 \\ 0 & -1 \end{bmatrix} \quad \Bigg| \quad \begin{bmatrix} -2 & 3 \\ -1 & 1 \end{bmatrix}$$

(*Multiply row 2 by −1.*)
$$\begin{bmatrix} 1 & 0 \\ 0 & 1 \end{bmatrix} \quad \Bigg| \quad \begin{bmatrix} -2 & 3 \\ 1 & -1 \end{bmatrix}$$

Then

$$\mathbf{A}^{-1} = \begin{bmatrix} -2 & 3 \\ 1 & -1 \end{bmatrix}$$

To check the above computations, verify that the product $\mathbf{A} \cdot \mathbf{A}^{-1} = \mathbf{I}$.

$$\begin{bmatrix} 1 & 3 \\ 1 & 2 \end{bmatrix}\begin{bmatrix} -2 & 3 \\ 1 & -1 \end{bmatrix} = \begin{bmatrix} 1 \times (-2) + 3 \times 1 & 1 \times 3 + 3 \times (-1) \\ 1 \times (-2) + 2 \times 1 & 1 \times 3 + 2 \times (-1) \end{bmatrix}$$

$$= \begin{bmatrix} -2 + 3 & 3 + (-3) \\ -2 + 2 & 3 + (-2) \end{bmatrix}$$

$$= \begin{bmatrix} 1 & 0 \\ 0 & 1 \end{bmatrix}$$

Example 6.23 Find $\mathbf{A}^{-1}$ if

$$\mathbf{A} = \begin{bmatrix} 1 & 2 & 1 \\ 1 & 3 & 1 \\ 0 & 1 & 3 \end{bmatrix}$$

Solution

$$\begin{bmatrix} 1 & 2 & 1 \\ 1 & 3 & 1 \\ 0 & 1 & 3 \end{bmatrix} \left| \begin{bmatrix} 1 & 0 & 0 \\ 0 & 1 & 0 \\ 0 & 0 & 1 \end{bmatrix} \right.$$

(*Subtract row 1 from row 2.*)

$$\begin{bmatrix} 1 & 2 & 1 \\ 0 & 1 & 0 \\ 0 & 1 & 3 \end{bmatrix} \left| \begin{bmatrix} 1 & 0 & 0 \\ -1 & 1 & 0 \\ 0 & 0 & 1 \end{bmatrix} \right.$$

(*Subtract 2 times row 2 from row 1 and Subtract row 2 from row 3.*)

$$\begin{bmatrix} 1 & 0 & 1 \\ 0 & 1 & 0 \\ 0 & 0 & 3 \end{bmatrix} \left| \begin{bmatrix} 3 & -2 & 0 \\ -1 & 1 & 0 \\ 1 & -1 & 1 \end{bmatrix} \right.$$

(*Multiply row 3 by* $\frac{1}{3}$.)

$$\begin{bmatrix} 1 & 0 & 1 \\ 0 & 1 & 0 \\ 0 & 0 & 1 \end{bmatrix} \left| \begin{bmatrix} 3 & -2 & 0 \\ -1 & 1 & 0 \\ \frac{1}{3} & -\frac{1}{3} & \frac{1}{3} \end{bmatrix} \right.$$

(*Subtract row 3 from row 1.*)

$$\begin{bmatrix} 1 & 0 & 0 \\ 0 & 1 & 0 \\ 0 & 0 & 1 \end{bmatrix} \left| \begin{bmatrix} \frac{8}{3} & -\frac{5}{3} & -\frac{1}{3} \\ -1 & 1 & 0 \\ \frac{1}{3} & -\frac{1}{3} & \frac{1}{3} \end{bmatrix} \right.$$

Then

$$\mathbf{A}^{-1} = \frac{1}{3} \begin{bmatrix} 8 & -5 & -1 \\ -3 & 3 & 0 \\ 1 & -1 & 1 \end{bmatrix}$$

Check:

$$\mathbf{A}^{-1}\mathbf{A} = \frac{1}{3} \begin{bmatrix} 8 & -5 & -1 \\ -3 & 3 & 0 \\ 1 & -1 & 1 \end{bmatrix} \begin{bmatrix} 1 & 2 & 1 \\ 1 & 3 & 1 \\ 0 & 1 & 3 \end{bmatrix}$$

$$= \frac{1}{3} \begin{bmatrix} 8 - 5 + 0 & 16 - 15 - 1 & 8 - 5 - 3 \\ -3 + 3 + 0 & -6 + 9 + 0 & -3 + 3 + 0 \\ 1 - 1 + 0 & 2 - 3 + 1 & 1 - 1 + 3 \end{bmatrix}$$

$$= \frac{1}{3}\begin{bmatrix} 3 & 0 & 0 \\ 0 & 3 & 0 \\ 0 & 0 & 3 \end{bmatrix}$$

$$= \begin{bmatrix} 1 & 0 & 0 \\ 0 & 1 & 0 \\ 0 & 0 & 1 \end{bmatrix}$$

EXAMPLE 6.24 Find $\mathbf{A}^{-1}$ if

$$\mathbf{A} = \begin{bmatrix} 1 & 0 & 1 \\ 1 & 1 & 1 \\ 1 & 2 & 1 \end{bmatrix}$$

Solution

$$\left[\begin{array}{ccc|ccc} 1 & 0 & 1 & 1 & 0 & 0 \\ 1 & 1 & 1 & 0 & 1 & 0 \\ 1 & 2 & 1 & 0 & 0 & 1 \end{array}\right]$$

(*Subtract row 1 from rows 2 and 3.*)

$$\left[\begin{array}{ccc|ccc} 1 & 0 & 1 & 1 & 0 & 0 \\ 0 & 1 & 0 & -1 & 1 & 0 \\ 0 & 2 & 0 & -1 & 0 & 1 \end{array}\right]$$

(*Subtract 2 times row 2 from row 3.*)

$$\left[\begin{array}{ccc|ccc} 1 & 0 & 1 & 1 & 0 & 1 \\ 0 & 1 & 0 & -1 & 1 & 0 \\ 0 & 0 & 0 & 1 & -2 & 1 \end{array}\right]$$

But now there is no way to eliminate the 1 in the first row, third column of the left-hand matrix. This means that **A** has no inverse.

Exercise Set 6.5

Find the inverse of each of these matrices when possible.

1. $\begin{bmatrix} 2 & -1 \\ 4 & 3 \end{bmatrix}$

2. $\begin{bmatrix} 4 & 2 \\ 7 & 2 \end{bmatrix}$

3. $\begin{bmatrix} 4 & -2 \\ 2 & 1 \end{bmatrix}$

4. $\begin{bmatrix} 1 & 0 & 1 \\ 3 & 1 & 0 \\ 1 & 0 & 0 \end{bmatrix}$

5. $\begin{bmatrix} 4 & 1 & 4 \\ 0 & 1 & 0 \\ 2 & 0 & 3 \end{bmatrix}$

8. $\begin{bmatrix} 0 & 0 & 1 \\ 0 & 1 & 0 \\ 1 & 0 & 0 \end{bmatrix}$

6. $\begin{bmatrix} 2 & 0 & 3 \\ -1 & 0 & 2 \\ 0 & 1 & 1 \end{bmatrix}$

9. $\begin{bmatrix} 1 & 4 & 0 \\ 2 & -1 & -1 \\ 3 & 2 & 1 \end{bmatrix}$

7. $\begin{bmatrix} 1 & 2 & 0 \\ 1 & 3 & 1 \\ 2 & 4 & 1 \end{bmatrix}$

10. $\begin{bmatrix} 1 & 0 & 0 & 0 \\ 1 & 1 & 0 & 0 \\ 1 & 1 & 1 & 0 \\ 1 & 1 & 1 & 1 \end{bmatrix}$

*6.6 SYSTEMS OF LINEAR EQUATIONS

A **system of equations** is a set of equations in which at least one of the equations of the set contains two or more variables. A **solution** of a system is a set of replacements, one for each variable in the system, which makes every equation in the system true.

As an example,

$$\begin{aligned} x + y + z &= 6 \\ 2x \quad\quad + z &= 7 \\ x - y + z &= 2 \end{aligned}$$

is a system of three equations in the three variables x, y, and z. The solution of this system is $x = 3$, $y = 2$, and $z = 1$.

If a system contains only linear equations, those in which the variables occur only in the first power and no term is the product of variables, we may use matrix methods in finding the solution if there is one.

EXAMPLE 6.25 Solve the system:

$$\begin{aligned} x + 2y \quad\quad &= 4 \\ x + 3y + z &= 2 \\ 2x + 4y + z &= 0 \end{aligned}$$

Solution The system can be considered to be two equal column vectors:

$$\begin{bmatrix} 1x + 2y + 0z \\ 1x + 3y + 1z \\ 2x + 4y + 1z \end{bmatrix} = \begin{bmatrix} 4 \\ 2 \\ 0 \end{bmatrix}$$

The vector on the left is the product

$$\begin{bmatrix} 1 & 2 & 0 \\ 1 & 3 & 1 \\ 2 & 4 & 1 \end{bmatrix} \begin{bmatrix} x \\ y \\ z \end{bmatrix}$$

Therefore the system can be rewritten in matrix form as

$$\begin{bmatrix} 1 & 2 & 0 \\ 1 & 3 & 1 \\ 2 & 4 & 1 \end{bmatrix} \begin{bmatrix} x \\ y \\ z \end{bmatrix} = \begin{bmatrix} 4 \\ 2 \\ 0 \end{bmatrix}$$

The inverse of the 3 × 3 matrix on the left (from Problem 7, Exercise Set 6.5) is

$$\begin{bmatrix} -1 & -2 & 2 \\ 1 & 1 & -1 \\ -2 & 0 & 1 \end{bmatrix}$$

Multiply both members of the matrix equation, from the left, by this inverse,

$$\begin{bmatrix} -1 & -2 & 2 \\ 1 & 1 & -1 \\ -2 & 0 & 1 \end{bmatrix} \left(\begin{bmatrix} 1 & 2 & 0 \\ 1 & 3 & 1 \\ 2 & 4 & 1 \end{bmatrix} \begin{bmatrix} x \\ y \\ z \end{bmatrix} \right) = \begin{bmatrix} -1 & -2 & 2 \\ 1 & 1 & -1 \\ -2 & 0 & 1 \end{bmatrix} \begin{bmatrix} 4 \\ 2 \\ 0 \end{bmatrix}$$

and since matrix multiplication is associative,

$$\left(\begin{bmatrix} -1 & -2 & 2 \\ 1 & 1 & -1 \\ -2 & 0 & +1 \end{bmatrix} \begin{bmatrix} 1 & 2 & 0 \\ 1 & 3 & 1 \\ 2 & 4 & 1 \end{bmatrix} \right) \begin{bmatrix} x \\ y \\ z \end{bmatrix} = \begin{bmatrix} -1 & -2 & 2 \\ 1 & 1 & -1 \\ -2 & 0 & 1 \end{bmatrix} \begin{bmatrix} 4 \\ 2 \\ 0 \end{bmatrix}$$

Thus

$$\begin{bmatrix} 1 & 0 & 0 \\ 0 & 1 & 0 \\ 0 & 0 & 1 \end{bmatrix} \begin{bmatrix} x \\ y \\ z \end{bmatrix} = \begin{bmatrix} -8 \\ 6 \\ -8 \end{bmatrix}$$

or

$$\begin{bmatrix} x \\ y \\ z \end{bmatrix} = \begin{bmatrix} -8 \\ 6 \\ -8 \end{bmatrix}$$

That is, $x = -8$, $y = 6$, and $z = -8$ is the solution of the system.

The 3×3 matrix of Example 6.25, whose components are the coefficients of x, y, and z, is called the **coefficient matrix** of the system. Whenever the coefficient matrix of a system has an inverse, that system has a unique solution. This is shown by the following theorem.

THEOREM 6.3 Suppose a system of linear equations is expressed in matrix form as

$$\mathbf{Ax} = \mathbf{c}$$

where $\mathbf{A}$ is the coefficient matrix, $\mathbf{x}$ is the column vector of variables, and $\mathbf{c}$ is the column vector of constants from the system. Let $\mathbf{A}$ have an inverse, $\mathbf{A}^{-1}$. Then $\mathbf{A}^{-1}\mathbf{c}$ is the matrix of solutions for the system.

Proof

$$\begin{aligned} \mathbf{Ax} &= \mathbf{c} \\ \mathbf{A}^{-1}(\mathbf{Ax}) &= \mathbf{A}^{-1}\mathbf{c} \\ (\mathbf{A}^{-1}\mathbf{A})\mathbf{x} &= \mathbf{A}^{-1}\mathbf{c} \\ \mathbf{Ix} &= \mathbf{A}^{-1}\mathbf{c} \\ \mathbf{x} &= \mathbf{A}^{-1}\mathbf{c} \end{aligned}$$

In theory, to solve a system of equations, we should determine the inverse of the coefficient matrix, then multiply the inverse, on the right, by the vector of constants from the system. In practice the method can be abbreviated somewhat.

Suppose the coefficient matrix is $\mathbf{A}$, the column vector of variables is $\mathbf{x}$, and the vector of constants is $\mathbf{c}$. Then

$$\mathbf{Ax} = \mathbf{c}$$

or

$$\mathbf{Ax} = \mathbf{Ic}.$$

We wish to perform elementary row operations on **A** converting it to **I** and the same operations on **I** converting it to $\mathbf{A}^{-1}$. But Theorem 6.1 tells us that performing the operations directly on **c** will have the same result as first performing them on **I** and then multiplying by **c**, on the right. Accordingly, we convert **A** to **I** and with the same row operations convert **c** to the solution vector.

EXAMPLE 6.26 Solve:

$$\begin{aligned} x + y + z \quad &= 1 \\ x + 2y + z + w &= 0 \\ y - z + w &= 0 \\ x \quad - z + w &= 2 \end{aligned}$$

Solution

$$\begin{bmatrix} 1 & 1 & 1 & 0 \\ 1 & 2 & 1 & 1 \\ 0 & 1 & -1 & 1 \\ 1 & 0 & -1 & 1 \end{bmatrix} \begin{bmatrix} x \\ y \\ z \\ w \end{bmatrix} = \begin{bmatrix} 1 \\ 0 \\ 0 \\ 2 \end{bmatrix}$$

Convert the coefficient matrix to **I** using a sequence of elementary row operations. Perform the same sequence of operations on the vector,

$$\begin{bmatrix} 1 \\ 0 \\ 0 \\ 2 \end{bmatrix}$$

(*Subtract row 1 from rows 2 and 4.*)

$$\begin{bmatrix} 1 & 1 & 1 & 0 \\ 0 & 1 & 0 & 1 \\ 0 & 1 & -1 & 1 \\ 0 & -1 & -2 & 1 \end{bmatrix} \begin{bmatrix} x \\ y \\ z \\ w \end{bmatrix} = \begin{bmatrix} 1 \\ -1 \\ 0 \\ 1 \end{bmatrix}$$

(*Subtract row 2 from rows 1 and 3. Add row 2 to row 4.*)

$$\begin{bmatrix} 1 & 0 & 1 & -1 \\ 0 & 1 & 0 & 1 \\ 0 & 0 & -1 & 0 \\ 0 & 0 & -2 & 2 \end{bmatrix} \begin{bmatrix} x \\ y \\ z \\ w \end{bmatrix} = \begin{bmatrix} 2 \\ -1 \\ 1 \\ 0 \end{bmatrix}$$

(*Add row 3 to row 1. Subtract 2 times row 3 from row 4.*)

$$\begin{bmatrix} 1 & 0 & 0 & -1 \\ 0 & 1 & 0 & 1 \\ 0 & 0 & -1 & 0 \\ 0 & 0 & 0 & 2 \end{bmatrix} \begin{bmatrix} x \\ y \\ z \\ w \end{bmatrix} = \begin{bmatrix} 3 \\ -1 \\ 1 \\ -2 \end{bmatrix}$$

(*Multiply row 3 by −1. Multiply row 4 by $\frac{1}{2}$.*)

$$\begin{bmatrix} 1 & 0 & 0 & -1 \\ 0 & 1 & 0 & 1 \\ 0 & 0 & 1 & 0 \\ 0 & 0 & 0 & 1 \end{bmatrix} \begin{bmatrix} x \\ y \\ z \\ w \end{bmatrix} = \begin{bmatrix} 3 \\ -1 \\ -1 \\ -1 \end{bmatrix}$$

(*Add row 4 to row 1. Subtract row 4 from row 2.*)

$$\begin{bmatrix} 1 & 0 & 0 & 0 \\ 0 & 1 & 0 & 0 \\ 0 & 0 & 1 & 0 \\ 0 & 0 & 0 & 1 \end{bmatrix} \begin{bmatrix} x \\ y \\ z \\ w \end{bmatrix} = \begin{bmatrix} 2 \\ 0 \\ -1 \\ -1 \end{bmatrix}$$

$$\begin{bmatrix} x \\ y \\ z \\ w \end{bmatrix} = \begin{bmatrix} 2 \\ 0 \\ -1 \\ -1 \end{bmatrix}$$

In summary, to solve a system of linear equations, write the system in matrix form

$$\mathbf{Ax} = \mathbf{c}$$

Then perform elementary row operations on **A** to change it to **I**. The identical row operations performed on **c** will convert it to the vector of solutions of the system. If **A** cannot be converted to **I** by elementary row operations, then the given system either has no solution or the solution is not unique.

Exercise Set 6.6

1. Solve each system by the method of Example 6.25. The necessary inverses were computed in Exercise Set 6.5. The problems to which you should refer are indicated in parentheses.

   (a) $2x - y = 3$ (Problem 1)
   $4x + 3y = 7$

   (b) $4x + 2y = 3$ (Problem 2)
   $7x + 2y = 5$

(c) $$\begin{aligned} 4x + y + 4z &= 8 \\ y &= 10 \\ 2x + 3z &= 1 \end{aligned}$$ (Problem 5)

(d) $$\begin{aligned} x + 4y &= 0 \\ 2x - y - z &= 1 \\ 3x + 2y + z &= 4 \end{aligned}$$ (Problem 9)

2. Solve each of these systems by the method of Example 6.26.

(a) $$\begin{aligned} x + y &= 3 \\ 2x - y &= 4 \end{aligned}$$

(b) $$\begin{aligned} 3x + 2y &= 4 \\ 5x + 2y &= 0 \end{aligned}$$

(c) $$\begin{aligned} 6x + 3y + 2z &= 1 \\ 3x - 4z &= 4 \\ 5x - y &= 14 \end{aligned}$$

(d) $$\begin{aligned} x - z &= 2 \\ y + 3z &= 1 \\ x - 2y &= 7 \end{aligned}$$

3. To control a certain crop disease it is necessary to use 5 units of Chemical P, 13 units of Q, and 9 units of R. Commercial spray can be purchased under three brand names. One barrel of each brand contains the following number of units of each chemical.

Brand	Chemical P	Chemical Q	Chemical R
A	2	3	4
B	1	3	5
C	2	6	1

How many barrels of each brand should be used to spread the exact amount of chemicals?

SUMMARY

Terms you should know:

1. Matrix
2. Dimensions
3. Component
4. Row vector
5. Column vector
6. Scalar product

Facts you should know:

1. Matrix addition is closed, commutative, associative.
2. There is an identity matrix **0**.

3. For every matrix **A** there is a matrix $-\mathbf{A}$ such that $\mathbf{A} + (-\mathbf{A}) = \mathbf{0}$.
4. Matrix multiplication is associative.
5. There is an identity matrix **I**.
6. For some square matrices **A** there are inverse matrices $\mathbf{A}^{-1}$ such that $\mathbf{A}\mathbf{A}^{-1} = \mathbf{A}^{-1}\mathbf{A} = \mathbf{I}$.
7. Matrix multiplication distributes over matrix addition.

REVIEW EXERCISES

1. If **A** is 2×3, **B** is 3×2, **C** is 3×2, and **D** is 1×2, what are the dimensions of:

   (a) $\mathbf{B} + \mathbf{C}$ (c) $\mathbf{AB}$ (e) $\mathbf{BC}$

   (b) $\mathbf{B} + \mathbf{A}$ (d) $\mathbf{BA}$ (f) $\mathbf{D(AB)}$

2. Find values for the variables, if possible, which will make the equalities true:

   (a) $\begin{bmatrix} x & 3 \\ 2 & 1 \end{bmatrix} = \begin{bmatrix} 1 & 3 \\ 2 & 1 \end{bmatrix}$

   (b) $\begin{bmatrix} y & 1 \\ 3 & 2 \end{bmatrix} = \frac{1}{2}\begin{bmatrix} 7 & 2 \\ 6 & 4 \end{bmatrix}$

   (c) $\begin{bmatrix} x & y \\ z & w \end{bmatrix} = \mathbf{I}$

   (d) $\begin{bmatrix} x & y \\ z & w \end{bmatrix} = \begin{bmatrix} 3 & 2 \\ 1 & 5 \end{bmatrix} + \begin{bmatrix} 1 & 8 \\ 4 & 3 \end{bmatrix}$

   (e) $[x \quad y] = \begin{bmatrix} 1 & 3 & 4 \\ 4 & 0 & 1 \end{bmatrix}\begin{bmatrix} 2 \\ 1 \\ 1 \end{bmatrix}$

3. Perform the following computations if possible:

   (a) $\begin{bmatrix} 2 \\ 1 \\ 1 \end{bmatrix} + \begin{bmatrix} 3 \\ 2 \\ 1 \end{bmatrix}$

(b) $3\begin{bmatrix} 1 & 2 \\ 3 & -1 \\ 4 & 0 \end{bmatrix}$

(c) $[1 \quad 3 \quad 1]\begin{bmatrix} 1 & 2 \\ 2 & 0 \\ 1 & 1 \end{bmatrix}$

(d) $\begin{bmatrix} 1 & 1 \\ 1 & 2 \end{bmatrix}\begin{bmatrix} 3 & 1 \\ 1 & 4 \end{bmatrix} + \begin{bmatrix} 1 & 0 \\ 2 & 1 \end{bmatrix}$

(e) $\begin{bmatrix} 1 & 2 \\ 0 & 0 \\ 6 & 3 \\ 2 & 1 \\ 1 & 0 \end{bmatrix}\begin{bmatrix} 2 & 1 & 5 & 3 & 1 \\ 1 & 0 & 2 & 1 & 1 \end{bmatrix}$

(f) $\begin{bmatrix} 0 & 0 & 0 \\ 1 & 0 & 0 \\ 2 & 3 & 0 \end{bmatrix}^3$

4. A communication network is shown in Figure 6–4.

   (a) Show this network as a 5×5 matrix, $\mathbf{A}$.

   (b) Compute $\mathbf{A}^2$.

   (c) Write the matrix which shows the total number of communication lines open between two stations using no more than two relay stations.

   (d) In how many ways can station 1 send a message back to itself using no more than two relay stations?

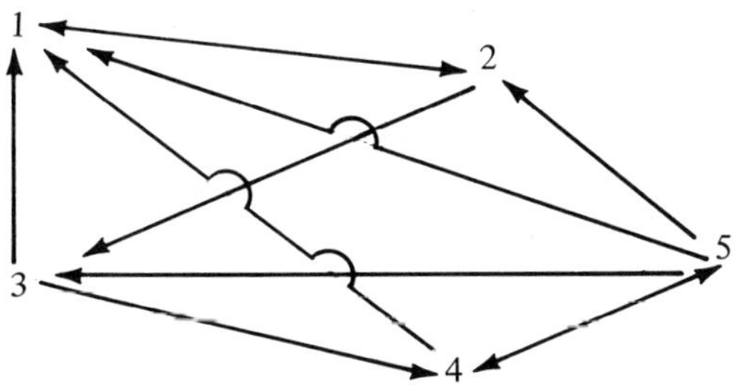

FIGURE 6–4

5. The following table shows the prices of stocks of three companies for the months of June, July, and August.

Company	June	July	August
Fledgling Airways	$45\frac{1}{4}$	$40\frac{1}{2}$	38
Rustmore Metals	$27\frac{3}{4}$	26	$24\frac{1}{2}$
Played Out Mines	$39\frac{1}{2}$	38	36

(a) Write a matrix showing the information in the table.

(b) An investor bought each month 100 shares of Fledgling Airways, 200 shares of Rustmore Metals, and 80 shares of Played Out Mines. Write a matrix to represent these facts.

(c) Use matrix multiplication to find a matrix showing the total amount the investor spent on stock each month.

6. Write the inverse of:

$$\text{(a)} \begin{bmatrix} 1 & 0 & 0 \\ 1 & 1 & 1 \\ 0 & 1 & 1 \end{bmatrix} \qquad \text{(b)} \begin{bmatrix} 1 & 2 & 1 \\ 2 & 1 & 5 \\ 0 & 0 & 1 \end{bmatrix} \qquad \text{(c)} \begin{bmatrix} 2 & 1 & 1 & 1 \\ 3 & 2 & 1 & 0 \\ 0 & 0 & 2 & 1 \\ 0 & 1 & 0 & 0 \end{bmatrix}$$

7. If three types of trucks are each equipped to haul three different types of heavy machines per load,

$$\begin{array}{rc} & \text{Type of truck} \\ & \begin{array}{ccc} 1 & 2 & 3 \end{array} \\ \text{Type of machine} \begin{array}{c} A \\ B \\ C \end{array} & \begin{bmatrix} 1 & 1 & 2 \\ 1 & 2 & 1 \\ 2 & 1 & 0 \end{bmatrix} \end{array}$$

how many trucks should be used to haul 6 A's, 8 B's, and 5 C's?

7 PROBABILITY

There are many situations in which, although we cannot be certain of the outcomes, we can make predictions and feel reasonably sure that the predictions will come true. Examples of predictions of this type are "College graduates have a greater earning power than nongraduates"; "Women have a longer life expectancy than men"; and "When a coin is tossed, there is an even chance that it will land heads up."

The mathematical science of describing the uncertain but likely outcomes of an experiment or happening is called *probability theory.* The rudiments of this theory were first developed in the seventeenth century in an attempt to discover the laws governing games of chance. The first practical applications were to insurance, and it is the insurance industry which must be given credit for keeping interest in probability alive through the eighteenth and nineteenth centuries.

In this century there has been a rebirth of interest in probability. Today its applications are so diverse that it is difficult to find an area that is not dependent, to some extent, upon this science.

The topics in this chapter are selected to provide a basic understanding of the applications of probability to real life situations. Many of the problems considered are coin, card, and dice problems. There are two reasons for using examples from these areas. One reason is that most of you have, at one time or another, tossed a coin, played a card game, and rolled dice. Therefore you are already familiar with the outcomes which are possible when these implements are used. The second reason for using gaming problems is that doing so allows the arithmetic computations to be kept at a fairly simple level.

7.1 THE PROBABILITY MEASURE

7.1.1 Vocabulary

Suppose that a regular deck of playing cards is shuffled and a single card is drawn from the deck at random. The color of the card, its suit, and its denomination are three things which may be considered as outcomes. The fundamental problem of probability theory is to define the possible outcomes and assign to each outcome some numerical value that will indicate the relative frequency with which that outcome can be expected to occur. This relative frequency can be found approximately through experimentation. Experiments are always subject to random fluctuations. It is usually assumed that over the long run these fluctuations tend to cancel each other, but this need not be so. In the game of matching pennies, for instance, if one player wins 10 times in a row, his opponent will probably suspect that something more than luck has entered the game. If, on the other hand, 100,000 pairs of people match pennies, one can expect that in more than 90 of the pairs one player will win the first 10 consecutive plays.

Another difficulty of using experimentation to determine probability lies in the fact that in many cases the situations cannot be repeated. What is the probability that it will rain tomorrow? What is the probability that Candidate X will be elected in the next Presidential race? In neither of these cases can the probability be discovered by experimentation; some other method must be used.

In a great many situations, experience, logic, and experimentation suggest the rule that the outcomes are equally likely. For example, when a die is rolled, it is assumed that in any given number of rolls each of the six faces of the die will show with the same frequency. If 1000 slips of paper, each containing the name of a different person, are placed in a hat and one slip is drawn, the assumption is made that each of the persons named has the same chance that his name will be on that slip.

SAMPLE SPACE We shall call any situation which has one or more possible outcomes a **trial**. Any sequence of one or more trials will be called an **experiment**. The set of all possible outcomes of an experiment is its **sample space**.

EXAMPLE 7.1 Suppose an experiment consists of two trials. In each trial a single coin is tossed. What is the sample space for the experiment?

Solution For this experiment we may use a set of ordered pairs to show the sample space:

$$\{(H, H), (H, T), (T, H), (T, T)\}$$

The first element of each pair represents the face of the coin which was showing on the first toss, and the second element in the pair represents the face of the coin which was showing on the second toss.

EVENT Any subset of the sample space of an experiment is called an **event.**

In Example 7.1, the set $\{(H, T), (T, H)\}$ is the event "On one toss the coin shows a head, and on the other toss it shows a tail."

7.1.2 Computing Probability

PROBABILITY MEASURE If every outcome of an experiment is equally likely to occur, then the probability that an event E will occur is the ratio of the number of outcomes in E to the number of outcomes in the sample space of the experiment. That is,

$$P(E) = \frac{N(E)}{N(S)}$$

in which $P(E)$ stands for the probability that E occurs, $N(E)$ is the number of outcomes in E, and $N(S)$ is the number of elements in S, the sample space of the experiment.

EXAMPLE 7.2 A coin is tossed twice. What is the probability that both tosses are heads?

Solution The sample space was determined in Example 7.1.

$$S = \{(H, H), (H, T), (T, H), (T, T)\}$$

The event "Both tosses are heads" is the set

$$E = \{(H, H)\}$$

If the coin is fair, each of the outcomes in S are equally likely; thus the probability that both tosses are heads is

$$P(E) = \frac{N(E)}{N(S)} = \frac{1}{4}$$

EXAMPLE 7.3 A single die is rolled. What is the probability that the die stops with either a 4 spot or a 6 spot on the upper face?

Solution The sample space is

$$S = \{1, 2, 3, 4, 5, 6\}$$

The desired event is

$$E = \{4, 6\}$$

Therefore

$$P(E) = \frac{N(E)}{N(S)} = \frac{2}{6} = \frac{1}{3}$$

EXAMPLE 7.4 A pair of dice are rolled. What is the probability that the sum of the spots showing is 7?

Solution Assume that it is possible to distinguish between the two dice (perhaps one is red and the other is white). Then the first die will show one of six faces identified by the number of spots on that face. The second die will also show one face, and that face will be identified by the number of spots on it. Thus the sample space for the experiment may be shown as the ordered pairs in the cross product set $\{1, 2, 3, 4, 5, 6\} \times \{1, 2, 3, 4, 5, 6\}$. Thus

$$S = \left\{\begin{matrix} (1, 1) & (1, 2) & (1, 3) & (1, 4) & (1, 5) & (1, 6) \\ (2, 1) & (2, 2) & (2, 3) & (2, 4) & (2, 5) & (2, 6) \\ (3, 1) & (3, 2) & (3, 3) & (3, 4) & (3, 5) & (3, 6) \\ (4, 1) & (4, 2) & (4, 3) & (4, 4) & (4, 5) & (4, 6) \\ (5, 1) & (5, 2) & (5, 3) & (5, 4) & (5, 5) & (5, 6) \\ (6, 1) & (6, 2) & (6, 3) & (6, 4) & (6, 5) & (6, 6) \end{matrix}\right\}$$

The event "The sum of the spots is 7" is the set

$$E = \{(1, 6), (2,5), (3,4), (4, 3), (5, 2), (6, 1)\}.$$

There are 6 pairs in this set and 36 pairs in the sample space; therefore the probability that the sum of the spots is 7 is

$$P(\text{sum is } 7) = \frac{N(E)}{N(S)} = \frac{6}{36} = \frac{1}{6}$$

It is very important that only equally likely events be considered when the probability formula is applied. In Example 7.4, one might reason that when two dice are rolled only eleven sums are possible, 2, 3, 4, 5, 6, 7, 8, 9, 10, 11, and 12, and therefore the probability of a 7 is $\frac{1}{11}$. While it is true that each of the eleven sums is an outcome, these outcomes are not equally likely in this experiment since, for example, there is only one way for the two dice to land so that the sum of the spots showing is 12, but there are several ways in which the dice can land such that the sum of the spots is 8.

EXAMPLE 7.5 Assume that girl and boy babies are equally likely to be born to a family. What is the probability that in a family of three children exactly two will be girls?

Solution The sample space for this experiment can be expressed as the set of ordered triples

$$S = \{(G, G, G), (G, G, B), (G, B, G), (G, B, B), \\ (B, G, G), (B, G, B), (B, B, G)\ (B, B, B)\}$$

The event "Exactly two girls in the family" is

$$E = \{(G, G, B), (G, B, G), (B, G, G)\}$$

Thus

$$P(E) = \frac{N(E)}{N(S)} = \frac{3}{8}$$

Exercise Set 7.1

1. Using the sample space of Example 7.4, find the probability that the sum of the spots on a single throw of two dice is:

   (a) A 6
   (b) An 11
   (c) A 7 or an 11
   (d) A sum greater than 7
   (e) A 1
   (f) A 13

2. A single card is drawn from an ordinary deck of 52 cards. What is the probability that it is:

   (a) Red
   (b) A king
   (c) A king or an ace
   (d) The ace of spades
   (e) A black jack
   (f) A red 7 or a black 9
   (g) A heart
   (h) A heart or a spade

3. Find the sample space for an experiment that consists of a single toss of three coins.

4. Using the sample space of Problem 3, find the probability that when three coins are tossed the outcome is:

   (a) Three heads
   (b) Exactly two heads
   (c) At least one head
   (d) Either two heads and a tail or two tails and a head

5. There are four children in a family. Show a sample space that indicates both the sex of the children and the order of birth.

6. Using the sample space of Problem 5, find the probability that:

   (a) There are three boys in the family
   (b) There are two boys and two girls
   (c) The eldest child is a boy and the youngest child is a girl
   (d) There are more girls than boys in the family

7. There are exactly four coins in my pocket, a penny, a nickel, a dime, and a quarter. If I withdraw two coins at random, what is the probability that the value of the coins withdrawn is greater than 20¢?

8. A commuter train runs through towns A, B, C, and D in that order. One evening three people board the train at town A. Make a sample space showing the ways they may disembark if none of them go past town D. What is the probability that at least one of them gets off at town C?

7.2 PROPERTIES OF PROBABILITY

With the ideas of the previous section fresh in mind, we turn to some basic properties of probability measure. The use of these properties will enable us to handle more complicated experiments than those considered thus far.

Two events E_1 and E_2 are **mutually exclusive** if $E_1 \cap E_2 = \varnothing$, that is, if they cannot both happen at the same time. Consider an experiment which consists of drawing a single card from a deck of 52 cards. The events {the card is an ace} and {the card is a king} are mutually exclusive. The events {the card is an ace} and {the card is a spade} are not mutually exclusive since the card could be the ace of spades.

If E_1 and E_2 are mutually exclusive events, then we know from Chapter 2 that

$$N(E_1 \cup E_2) = N(E_1) + N(E_2)$$

If the sample space for the experiment is S,

$$P(E_1 \cup E_2) = \frac{N(E_1 \cup E_2)}{N(S)}$$

Thus

$$\begin{aligned} P(E_1 \cup E_2) &= \frac{N(E_1) + N(E_2)}{N(S)} \\ &= \frac{N(E_1)}{N(S)} + \frac{N(E_2)}{N(S)} \\ &= P(E_1) + P(E_2) \end{aligned}$$

We have proved the following:

THEOREM 7.1 If E_1 and E_2 are mutually exclusive events, the probability that one or the other occurs is the sum of their individual probabilities. That is,

$$P(E_1 \cup E_2) = P(E_1) + P(E_2)$$

EXAMPLE 7.6 What is the probability that a single roll of a pair of dice results in a sum that is 7 or 11?

Solution Of the 36 possible outcomes in the sample space of this experiment the event {7} = {(1, 6), (2, 5), (3, 4), (4, 3), (5, 2), (6, 1)}. Thus

$$P(7) = \tfrac{6}{36}$$

The event {11} = {(6, 5), (5, 6)}; therefore

$$P(11) = \tfrac{2}{36}$$

{7} and {11} are mutually exclusive, and so

$$P(7 \text{ or } 11) = \tfrac{6}{36} + \tfrac{2}{36} = \tfrac{8}{36} = \tfrac{2}{9}$$

Theorem 7.1 can be restated for a finite number of events:

THEOREM 7.1 (*General Form*) If $E_1, E_2, E_3, \ldots, E_k$ are mutually exclusive events,

$$P(E_1 \cup E_2 \cup E_3 \cup \cdots \cup E_k) = P(E_1) + P(E_2) + P(E_3) + \cdots + P(E_k)$$

Remember, Theorem 7.1 can only be used when $E_1, E_2, E_3, \ldots, E_k$ are disjoint sets.

Suppose that $E_1, E_2, E_3, \cdots, E_k$ are mutually exclusive events with

$$S = E_1 \cup E_2 \cup E_3 \cup \cdots \cup E_k$$

Then

$$\begin{aligned} P(E_1) + P(E_2) + P(E_3) + \cdots + P(E_k) &= P(E_1 \cup E_2 \cup E_3 \cup \cdots \cup E_k) \\ &= P(S) \\ &= \frac{N(S)}{N(S)} \\ &= 1 \end{aligned}$$

since the sample space S cannot be empty.

When the probability of an event is 1, we say that the event is **certain.** From the above it follows that an event which is certain contains every element of the sample space of its experiment. On the other hand, an event which can never occur is represented by the empty set. Since $P(\varnothing) = N(\varnothing)/N(S) = 0/N(S) = 0$, an event which cannot happen has probability 0. If E is any non-empty proper subset of sample space S, $N(E)$ is less than $N(S)$, and since

$$P(E) = \frac{N(E)}{N(S)}$$

$P(E)$ is less than 1.

The above discussion can be summarized as follows:

If E is any event, then $0 \leq P(E) \leq 1$. If E cannot occur, $P(E) = 0$, and if E is sure to occur, $P(E) = 1$.

Suppose E_1 and E_2 have elements in common. Then from Chapter 2 we know that

$$N(E_1 \cup E_2) = N(E_1) + N(E_2) - N(E_1 \cap E_2)$$

thus

$$\begin{aligned} P(E_1 \cup E_2) &= \frac{N(E_1 \cup E_2)}{N(S)} \\ &= \frac{N(E_1) + N(E_2) - N(E_1 \cap E_2)}{N(S)} \\ &= \frac{N(E_1)}{N(S)} + \frac{N(E_2)}{N(S)} - \frac{N(E_1 \cap E_2)}{N(S)} \\ &= P(E_1) + P(E_2) - P(E_1 \cap E_2) \end{aligned}$$

EXAMPLE 7.7 What is the probability that one card drawn from a deck of 52 cards is either a spade or an ace?

Solution The sample space for this experiment is the set of all the cards in the deck. It will not be necessary to display this set since it is clear that it contains 52 elements. There are 13 spades in the deck; thus the probability that the card is a spade is

$$P(\text{spade}) = \tfrac{13}{52}$$

There are 4 aces in the deck; thus

$$P(\text{ace}) = \tfrac{4}{52}$$

Finally, because the events {the card is a spade} and {the card is an ace} are not mutually exclusive, we must compute the probability that the card is the ace of spades. This is

$$P(\text{ace of spades}) = \tfrac{1}{52}$$

Then

$$\begin{aligned} P(\text{spade or ace}) &= P(\text{spade}) + P(\text{ace}) - P(\text{ace of spades}) \\ &= \tfrac{13}{52} + \tfrac{4}{52} - \tfrac{1}{52} = \tfrac{4}{13} \end{aligned}$$

EXAMPLE 7.8 What is the probability that a single card drawn from a deck of 52 cards is neither an ace nor a spade?

Solution Let E_1 be the event "The card is not an ace and it is not a spade." If S is the set of the 52 cards of the deck, then E_1 is represented by the shaded region in Figure 7–1.

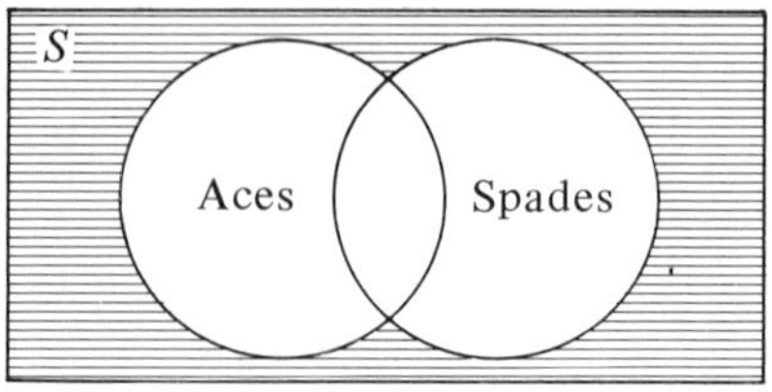

FIGURE 7–1

Clearly E_1 is the complement of E_2 = {the card is an ace or a spade} and $E_1 \cup E_2 = S$. Therefore, since $E_1 \cap E_2 = \varnothing$, $P(E_1 \cup E_2) = P(S)$ and

$$P(E_1) + P(E_2) = 1$$

or

$$P(E_1) = 1 - P(E_2)$$

From Example 7.7, we know $P(E_2) = \frac{4}{13}$. Therefore

$$P(\text{the card is neither an ace nor a spade}) = 1 - \tfrac{4}{13} = \tfrac{9}{13}$$

Example 7.8 illustrates the final property to be considered in this section. That property is

$$P(\tilde{E}) = 1 - P(E)$$

where $\tilde{E}$ represents the complement of E, that is, all the outcomes of the sample space except those which are in E.

In the following exercise set, first decide if the given events are mutually exclusive. If they are, then the probability that one or the other of them occurs is the sum of their individual probabilities. If the events are not mutually exclusive, then the probability that one or the other of them occurs is the sum of their individual probabilities diminished by the probability that both of them occur.

Exercise Set 7.2

1. A pair of dice are rolled once. Consider these events:

E_1 = the same number of spots show on each die
E_2 = the sum is a 6
E_3 = the sum is an odd number
E_4 = one die shows a 5
E_5 = the sum is 11

Which of these events are mutually exclusive?

(a) E_1 and E_2 (e) E_2 and E_3
(b) E_1 and E_3 (f) E_2 and E_4
(c) E_1 and E_4 (g) E_3 and E_5
(d) E_1 and E_5 (h) E_3, E_4, and E_5

2. Using the events of Problem 1, what is the probability of:

(a) $E_1 \cup E_2$ (e) $E_2 \cup E_3$
(b) $E_1 \cup E_3$ (f) $E_2 \cup E_4$
(c) $E_1 \cup E_4$ (g) $E_3 \cup E_5$
(d) $E_1 \cup E_5$ (h) $E_3 \cup E_4 \cup E_5$

3. A card is drawn from an ordinary deck of 52 playing cards. What is the probability that it is:

(a) A picture card (king, queen, or jack)
(b) Not a picture card
(c) A black ace or a red queen
(d) A picture card or a heart
(e) An ace, a king, or a diamond
(f) An ace, a king, or a 9
(g) A red card or a black card
(h) Not an ace nor a jack nor a red card

4. A student needs an A on the last hour exam or a C on the final in order to pass the course. He estimates that his chance of getting an A on the hour exam is $\frac{1}{10}$, that of getting a C on the final is $\frac{3}{5}$, and that his probability of getting an A on the hour exam and a C on the final is $\frac{1}{100}$. What is the probability that he passes the course?

5. There are three children in a family. Assuming that there is an equal probability of a child being of either sex, find the probability that:

(a) There are two boys or two girls in the family.
(b) The children are all the same sex.
(c) The eldest child is a boy or a girl.

6. What is the probability that the next person you meet:

(a) Has the same birthday as you
(b) Was born in March
(c) Was born on Tuesday
(d) Was not born in April

7. Three marbles are placed in a bag. One of them is red, one is green, and the third one is blue. Two marbles are drawn from the bag. What is the probability that:

    (a) One is red
    (b) Neither is green
    (c) Both are red
    (d) At least one is blue
    (e) One of the marbles is red or green or blue

8. In a group of 50 students, 20 are undergraduate women, 10 are undergraduate men, 12 are graduate men, and the rest are graduate women. If one member of the group is chosen at random, what is the probability that the student is:

    (a) A man
    (b) An undergraduate
    (c) Either a man or an undergraduate
    (d) Either a man or an undergraduate woman

7.3 CONDITIONAL PROBABILITY

Suppose we have some additional information about one of the outcomes of an experiment. Will this have an effect on the probability of an event which contains that outcome? It seems reasonable to expect that it will, and this expectation is verified in the following example.

EXAMPLE 7.9 Two dice are rolled once. If one of the dice shows a 5, what is the probability that the sum of the spots on the two dice is 7?

Solution There are 36 outcomes in the sample space (see Example 7.4). If E_1 is the event "One of the dice shows a 5" and E_2 is the event "The sum is 7":

$$E_1 = \{(1, 5), (2, 5), (3, 5), (4, 5), (5, 5), (6, 5), (5, 1), (5, 2), (5, 3), (5, 4), (5, 6)\}$$
$$E_2 = \{(1, 6), (2, 5), (3, 4), (4, 3), (5, 2), (6, 1)\}$$

Because it is known that one die shows a 5, E_1 may be used as the sample space for the experiment (see Figure 7–2). If this is done, the event "The sum is 7 given that one die shows a 5" is

$$E_2^* = \{(2, 5), (5, 2)\}$$

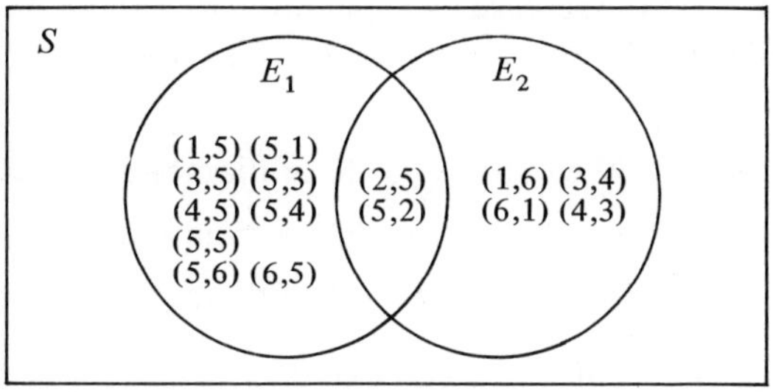

S = possible outcomes when two dice are rolled

FIGURE 7–2

since those are the only two elements of E_2 which are also in E_1. Then

$$P(E_2^*) = \frac{N(E_2^*)}{N(E_1)} = \frac{2}{11}$$

The probability of a 7 from Example 7.4 is $\frac{1}{6}$. However, the probability is changed when it is known in advance what number will come up on one of the dice.

Probabilities of the type illustrated above are called **conditional probabilities** since they are obtained by putting a condition on one or more of the outcomes. We shall denote a conditional probability as follows:

CONDITIONAL PROBABILITY $P(E_2/E_1)$ means the probability that event E_2 occurs given that event E_1 has occurred; that is, E_1 is used as the sample space. A formula for conditional probability is

$$P(E_2/E_1) = \frac{P(E_2 \cap E_1)}{P(E_1)}$$

where $P(E_2 \cap E_1)$ and $P(E_1)$ are computed using the sample space S.

For the problem of Example 7.9, P(one die shows a 5) = $\frac{11}{36}$ and P(one die shows a 5 and the sum of the spots is 7) = $\frac{2}{36}$. Thus by the formula above:

$$P(\text{sum is 7/one die shows a 5}) = \frac{\frac{2}{36}}{\frac{11}{36}} = \frac{2}{11}$$

EXAMPLE 7.10 A card is drawn at random from a deck of 52 playing cards. If it is a picture card, what is the probability that it is a king?

Solution We know that the card is a picture card; therefore we may use the set of picture cards as the sample space for this experiment. There are 12 of these cards in the deck:

$$\begin{Bmatrix} \text{K of spades} & \text{K of hearts} & \text{K of diamonds} & \text{K of clubs} \\ \text{Q of spades} & \text{Q of hearts} & \text{Q of diamonds} & \text{Q of clubs} \\ \text{J of spades} & \text{J of hearts} & \text{J of diamonds} & \text{J of clubs} \end{Bmatrix}$$

Four of the picture cards are kings. Therefore

$$P(\text{K/picture card}) = \frac{N(\text{picture cards which are also kings})}{N(\text{picture cards})}$$

$$= \frac{4}{12} = \frac{1}{3}$$

EXAMPLE 7.11 There are 39 families in a small town; of these, 23 are Republican and 16 are Democratic. There are 7 families headed by a woman, and of these 5 are Democratic. A poll taker chooses one family at random. Find the probability that the family is Democratic given that it is headed by a woman.

Solution We are asked to find $P(D/W)$. A Venn diagram showing the distribution of families is shown in Figure 7–3. The numbers in the various regions were determined using the method of Chapter 2. For example, 5 of the Democratic families are headed by a woman; therefore a 5 is written in the region $D \cap W$. There are 16 Democratic families in all; thus there are 11 families that are Democratic but not headed by a woman. Thus an 11 is entered in the region that shows the Democratic families which are not headed by a woman. Because it is known that the family in question is headed by a woman, we need to consider as a sample space only the set W. Therefore

$$P(D/W) = \frac{N(D \cap W)}{N(W)} = \frac{5}{7}$$

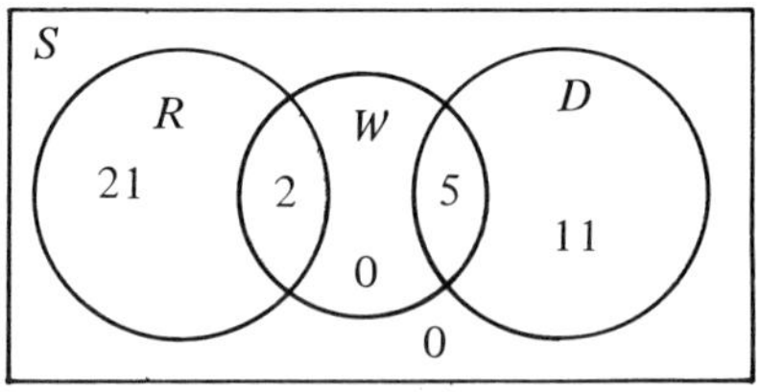

S = Families in a small town
R = Republican families
D = Democratic families
W = Families headed by a woman

FIGURE 7–3

Alternate Solution From Figure 7.3, we see that $N(W) = 7$ and $N(D \cap W) = 5$. There are 39 families; therefore $P(W) = \frac{7}{39}$ and $P(D \cap W) = \frac{5}{39}$. Then by formula

$$P(D/W) = \frac{P(D \cap W)}{P(W)} = \frac{\frac{5}{39}}{\frac{7}{39}} = \frac{5}{7}$$

In general, to calculate the probability that event E_2 occurs if it is known that E_1 occurs, use E_1 as the sample space and then

$$P(E_2/E_1) = \frac{N(E_1 \cap E_2)}{N(E_1)}$$

An alternative way to solve the same problem is to compute both $P(E_1 \cap E_2)$ and $P(E_1)$ and then use the formula

$$P(E_2/E_1) = \frac{P(E_1 \cap E_2)}{P(E_1)}$$

Exercise Set 7.3

1. Two dice are rolled. What is the probability that:
   (a) A sum of 11 is rolled if one of the dice shows a 5
   (b) The sum is 8 if a 4 shows on one die
   (c) The roll is a pair if the sum is even
   (d) One die shows a 4 if the other shows a 6
   (e) One die shows an odd number if the sum is 8
2. The enrollment of a community college is 1000; 600 students are men and 400 are women. Two hundred of the men and 100 of the women ride bikes to school.
   (a) What is the probability that a student chosen at random will ride a bike to school?
   (b) What is the probability that a woman student chosen at random rides a bike?
   (c) What is the probability that a man student chosen at random is a bike rider?
   (d) Sandy Simms is one of the students who rides a bike. What is the probability that Sandy is a man?

3. Suppose that 50 of 100 students take math, 40 take physics, and 20 take both math and physics. What is the probability that a student selected at random from the group of 100:

   (a) Takes neither math nor physics
   (b) Takes math if he takes physics
   (c) Takes physics if he takes math
   (d) Takes math if he doesn't take physics

4. A card is drawn at random from a deck of playing cards. What is the probability that the card is:

   (a) A king, given that it is an honor (an honor is an ace, king, queen, jack, or 10)
   (b) An honor, given that it is red
   (c) A 5, given that it is between 2 and 7 inclusive

5. A student takes a five-question true-false exam. If he guesses on every answer, what is the probability that he gets a perfect paper:

   (a) If he knows that the instructor always starts true-false tests with a question which is "true"
   (b) If he knows there are more "true" than "false" questions
   (c) If he knows that the instructor never puts three questions in a row with the same answer
   (d) If he knows that the first and last questions must have opposite answers

7.4 EXPERIMENTS WITH SEVERAL TRIALS

7.4.1 Conditional Probability

Most of the experiments that have been considered thus far have consisted of a single trial or, at the most, two or three trials. In this section we present a general method for use in computing probability when the experiment consists of many trials. The dealing of a five-card poker hand is one such experiment; five cards are dealt one at a time, and thus the deal may be thought of as five separate trials. Another, much more complicated, example of an experiment with many trials is the game called craps. In craps, the player throws a pair of dice and if he does not win by throwing a 7 or an 11 or lose by throwing a 2 or a 12 on the first roll, he is allowed to throw the dice again and again until either he wins by repeating the sum he threw on the first roll or he loses by rolling a 7 or an 11.

From the formula for conditional probability,

$$P(E_2/E_1) = \frac{P(E_2 \cap E_1)}{P(E_1)}$$

we obtain the rule

$$P(E_2 \cap E_1) = P(E_2/E_1) \cdot P(E_1)$$

In words, this formula says the probability that both of two events will occur is equal to the probability that the first event will occur multiplied by the conditional probability that the second event will occur when it is known that the first event will definitely occur.

In many experiments there is a definite time order for the individual trials. In other cases this time order is missing, but because of the nature of the experiment it is possible to arbitrarily assign an order to the trials. In the experiment in which three coins are thrown, for instance, we may assume that this experiment is the same as one in which a single coin is tossed three times. In the case of a five-card poker hand the order in which the cards are dealt has no bearing on the strength of the hand, and for this reason any one of them may be assumed to be the first dealt, any of the remaining four may be assumed to be the second one, and so on.

EXAMPLE 7.12 Two cards are dealt from an ordinary deck. What is the probability that they are both aces?

Solution Let E_1 be the event "The first card is an ace" and let E_2 be the event "The second card is an ace." Because, in the beginning, there are 4 aces in the deck of 52 cards,

$$P(E_1) = \tfrac{4}{52}$$

To find $P(E_2/E_1)$, it is necessary to consider only those experimental outcomes that exist after E_1 has occurred. If the first card drawn is an ace, 51 cards remain in the deck and only 3 of these are aces. Therefore

$$P(E_2/E_1) = \frac{3}{51}$$

Then

$$\begin{aligned} P(E_2 \cap E_1) &= P(E_2/E_1) \cdot P(E_1) \\ &= \frac{3}{51} \cdot \frac{4}{52} \\ &= \frac{1}{221} \end{aligned}$$

EXAMPLE 7.13 Two identical boxes are labeled A and B on their bottoms where you cannot see the labels. Box A contains one \$5 bill and two \$1 bills. Box B contains three \$1 bills and two \$5 bills. You select a box at random and draw one bill from that box. What is the probability that the bill is a \$5 bill?

Solution Because the boxes are identical, the probability that you select box A is $\frac{1}{2}$, and that is also the probability that box B is selected. The conditional probability of drawing a \$5 bill given that box A is chosen is $\frac{1}{3}$ (since box A contains three bills only one of which is a \$5 bill). Similarly, the conditional probability of drawing a \$5 bill given that B is the box selected is $\frac{2}{5}$ (B contains two \$5 and three \$1 bills). Therefore

$$\begin{aligned} P(A \text{ and } \$5) &= P)\$5/A) \cdot P(A) \\ &= \tfrac{1}{3} \cdot \tfrac{1}{2} \\ &= \tfrac{1}{6} \end{aligned}$$

and

$$\begin{aligned} P(B \text{ and } \$5) &= P(\$5/B) \cdot P(B) \\ &= \tfrac{2}{5} \cdot \tfrac{1}{2} \\ &= \tfrac{1}{5} \end{aligned}$$

You may make only one selection in this experiment: either box A or box B, but not both boxes. Therefore E_1, "Box A is chosen and the bill is \$5," and E_2, "Box B is chosen and the bill is \$5," are mutually exclusive events. Thus

$$\begin{aligned} P(E_1 \text{ or } E_2) &= P(E_1) + P(E_2) \\ &= \tfrac{1}{6} + \tfrac{1}{5} \\ &= \tfrac{11}{30} \end{aligned}$$

That is, the probability that you draw a \$5 bill is $\frac{11}{30}$.

7.4.2 Independent Events

Two events are thought of as being independent if neither has any influence on the other. The easiest way to determine independence of two events is to answer this question, "Does the trial which has the first event as a possible outcome in any way alter the probability of the second event?" If the answer is "yes" then the events are dependent, but if the answer is "no" the events are independent. As an example consider the experiment in which a coin is tossed twice. On the first toss there are two possible outcomes, heads or tails.

No matter which way the coin lands on the first toss, the possible outcomes on the second toss are the same, heads or tails. Thus the outcomes from repeated tosses of a coin are independent. The outcomes from drawing two cards from a deck of cards are also independent *provided that the first card is replaced before the second card is drawn.* For if the card is replaced, the deck is exactly the same immediately before the second card is drawn as it was before the first card was drawn. If two cards are drawn from a deck *without replacement*, however, the deck contains one less card before the second draw than it did before the first draw. In this case the outcome of the second draw is dependent upon the outcome of the first draw.

We may define independence mathematically in the following way:

INDEPENDENCE Events E_1 and E_2 are independent if and only if

$$P(E_2/E_1) = P(E_2)$$

From the above definition we may prove the following:

THEOREM 7.2 Events E_1 and E_2 are independent if and only if

$$P(E_1 \cap E_2) = P(E_1) \cdot P(E_2)$$

Proof

$$P(E_2/E_1) = \frac{P(E_2 \cap E_1)}{P(E_1)}$$

Therefore

$$P(E_2 \cap E_1) = P(E_1) \cdot P(E_2/E_1)$$

By the definition of independence,

$$P(E_2/E_1) = P(E_2)$$

if and only if E_1 and E_2 are independent. Then

$$P(E_2 \cap E_1) = P(E_1) \cdot P(E_2)$$

if and only if E_1 and E_2 are independent.

EXAMPLE 7.14 A card is drawn from a deck of playing cards, then replaced, and the deck is shuffled and a second drawing made. What is the probability that both of the cards were aces?

Solution Since the first card was replaced before the second card was drawn, the events "The first card is an ace" and "The second card is an ace" are independent events, each with probability $\frac{4}{52} = \frac{1}{13}$. Therefore the probability that both cards are aces is the product

$$\frac{1}{13} \cdot \frac{1}{13} = \frac{1}{169}$$

Three or more events are independent if no one of them is dependent on any of the others. In that case we have:

GENERAL PROPERTY OF INDEPENDENT EVENTS If events E_1, E_2, $E_3, \ldots, E_k$ are independent,

$$P(E_1 \cap E_2 \cap E_3 \cap \cdots \cap E_k) = P(E_1) \cdot P(E_2) \cdot P(E_3) \cdots P(E_k)$$

EXAMPLE 7.15 What is the probability that when five coins are tossed every coin will land "heads"?

Solution We may consider this to be the same as the experiment of tossing one coin five times. Then on each toss the probability of a head is $\frac{1}{2}$, and the five outcomes are independent since no toss changes the sample space for any other toss. Therefore

$$\begin{aligned} P(\text{five heads}) &= (\tfrac{1}{2})(\tfrac{1}{2})(\tfrac{1}{2})(\tfrac{1}{2})(\tfrac{1}{2}) \\ &= \tfrac{1}{32} \end{aligned}$$

EXAMPLE 7.16 Suppose that the probability that a newborn child is a girl is $\frac{58}{100}$. What is the probability that in a family of three children there are exactly two girls?

Solution Since the probability that a child is a girl is $\frac{58}{100}$, the probability that a child is a boy is $1 - \frac{58}{100}$ or $\frac{42}{100}$. If we consider the order of birth, the family may be (B, G, G), (G, B, G), or (G, G, B). We may assume the sex of two successive births to be independent events; therefore

$$\begin{aligned} P(B, G, G) &= (\tfrac{42}{100})(\tfrac{58}{100})(\tfrac{58}{100}) \\ &= \tfrac{141288}{1000000} \\ P(G, B, G) &= (\tfrac{58}{100})(\tfrac{42}{100})(\tfrac{58}{100}) \\ &= \tfrac{141288}{1000000} \\ P(G, G, B) &= (\tfrac{58}{100})(\tfrac{58}{100})(\tfrac{42}{100}) \\ &= \tfrac{141288}{1000000} \end{aligned}$$

In a family of three children, two of whom are girls, if the sexes of the children are ordered by birthdates, then only one of (B, G, G), (G, B, G), or (G, G, B) can occur; that is, these three events are mutually exclusive, and

$$\begin{aligned} P(\text{two girls}) &= \tfrac{141288}{1000000} + \tfrac{141288}{1000000} + \tfrac{141288}{1000000} \\ &= \tfrac{423864}{1000000} \end{aligned}$$

The numerator and denominator of this fraction are too large to be really meaningful to most of us. Let us divide both the numerator and the denominator by 10,000. We then have

$$P(\text{two girls}) = \frac{42.3864}{100}$$

That is, the probability is that in 42 out of 100 families with three children exactly two of the children will be girls.

EXAMPLE 7.17 What is the probability that in a room containing 45 people, no two will share the same birthday?

Solution We assume each day is equally likely to be a birthday, and to make things a bit more simple we shall assume that every year contains 365 days. Choose one person to be the first person. Then his birthday is one of the 365 days of the year. A second person may have a different birthday on any one of 364 days; thus the probability that his birthday is different from that of the first person is $\frac{364}{365}$. A third person may have a different birthday than the first two in 363 ways; therefore the probability that his birthday is different from the birthdays of the first two persons is $\frac{363}{365}$. We continue eliminating birthdays in this manner until we find that the probability of the 45th person having a different birthday than the other 44 is $\frac{321}{365}$. We may assume birthdays are independent events. Therefore

$$\begin{aligned} P(\text{the 45 people have different birthdays}) &= \left(\tfrac{364}{365}\right)\left(\tfrac{363}{365}\right)\left(\tfrac{362}{365}\right)\left(\tfrac{361}{365}\right)\cdots\left(\tfrac{321}{365}\right) \\ &= 0.059 \text{ approximately} \end{aligned}$$

In the following exercise set you are asked to find the probabilities of sequences of trials. Some of the trials have independent outcomes and some do not. The basic formula in either case is

$$P(E_2 \cap E_1) = P(E_2/E_1)\cdot P(E_1)$$

If E_2 and E_1 are independent events,

$$P(E_2/E_1) = P(E_2)$$

which leads to the simpler formula:

$$P(E_2 \cap E_1) = P(E_2) \cdot P(E_1)$$

Exercise Set 7.4

1. A bag contains three red and four blue marbles. Two marbles are drawn from the bag one at a time without replacement. Find the probability that:
   (a) Both of the marbles are red
   (b) The first marble is red and the second is blue
   (c) The first marble is blue and the second is red
   (d) The marbles are different colors
   (e) Both marbles are the same color
2. Repeat Problem 1, but this time replace the first marble before drawing the second one.
3. Three identical bags are labeled *A*, *B*, and *C*. Bag *A* contains three red and one black marbles; bag *B* contains two red and two black marbles; and bag *C* contains one red and three black marbles. A bag is chosen at random, and a marble is drawn from it. What is the probability that:
   (a) The marble is red
   (b) The marble is black
4. Two people are selected at random from a group of 30.
   (a) What is the probability that both were born on Wednesday?
   (b) What is the probability that neither was born on Wednesday?
   (c) What is the probability that only one of them was born on a Wednesday?
   (d) What is the probability that at least one of them was born on a Wednesday?
5. A die is rolled. If a 1 or a 6 appears, a marble is drawn from a bag that contains four white marbles and one red marble. If any other number except a 1 or a 6 appears, a marble is drawn from a bag that contains one white marble and four red marbles. What is the probability that a white marble is drawn?
6. Suppose that the probability that a newborn child is a girl is $\frac{58}{100}$. What is the probability that in a family of three children:
   (a) All are girls
   (b) There is exactly one girl
   (c) There are no girls

7. A company has three machines. The hourly production for each machine is

   A makes 8 units, 1 of which is defective
   B makes 10 units, 3 of which are defective
   C makes 12 units, 2 of which are defective

   Every hour an inspector selects a machine at random and also selects, at random, a unit from those produced by the machine during the preceding hour. What is the probability that the unit selected is defective?

8. A bag contains six blue marbles, four white marbles, and eight red marbles. Three marbles are drawn from the bag without replacement. What is the probability that:

   (a) All three are red
   (b) Two are red and one is white
   (c) All three of the marbles are the same color
   (d) The three marbles are different colors

7.5 EXPECTED VALUE

7.5.1 Odds

"The odds are 3 to 2 in favor of the Giants." "What odds will you give that the fight will go four rounds?" Statements such as these are heard whenever two or more sports fans get together. In this section we investigate the meaning of the term **odds** and also the meaning of the related term **expected value**.

Let us suppose that a game is organized between two players in this way: Player A agrees to pay player B \$1 for each time that B rolls a 1 or a 6 with one die. What should B pay A each time he does not roll a 1 or a 6, in order to make the game a fair one? Since a 1 or a 6 occurs with probability $\frac{1}{3}$, if B plays enough times he can expect to win about $\frac{1}{3}$ of the time. Therefore he can expect to win \$1 for every three rolls. Since B expects to win in one out of three rolls, he expects to lose in two out of three rolls. Therefore, he should pay A \$0.50 every time he loses.

The amount to be paid if the event does not occur, divided by the payoff if the event does occur, is called the odds *in favor of the event*. In the above game, the odds in favor of rolling a 1 or a 6 with a single die are $\frac{1}{2}$. In general,

$$\text{The odds in favor of event } E = \frac{P(E)}{P(\tilde{E})}$$

EXAMPLE 7.18 Find the odds in favor of throwing three heads in a row with a single coin.

Solution We know

$$P(H, H, H) = \tfrac{1}{8}$$

therefore

$$P(\widetilde{H, H, H}) = 1 - \tfrac{1}{8} = \tfrac{7}{8}$$

Thus

$$\text{Odds in favor of } (H, H, H) = \frac{\frac{1}{8}}{\frac{7}{8}} = \frac{1}{7}$$

7.5.2 Expected Value

If the possible outcomes of an experiment occur with probabilities $p_1, p_2, p_3, \ldots, p_n$ and the payoffs on the event are, respectively, $a_1, a_2, a_3, \ldots, a_n$, then the **expected value** v of the experiment is

$$v = a_1p_1 + a_2p_2 + a_3p_3 + \cdots + a_np_n$$

It is important to remember that the expected value is not the value which will necessarily occur in a single trial, but it is the most probable average value per trial if the experiment is repeated a large number of times. If the expected value of a game of chance is greater than the expected loss, the game is said to be favorable; if the expected value is equal to the expected loss, the game is fair; and if the expected value is less than the expected loss, the game is unfavorable. Few games of chance are favorable or even fair, but this fact does not seem to inhibit the players.

EXAMPLE 7.19 A game is played in which a single die is rolled and the person rolling the die is paid \$1 for each spot on the face which lands up. What is the value of the game to that person?

Solution Each face of the die comes up with the same probability, $\frac{1}{6}$. Therefore the expected value is

$$\begin{aligned} v &= \$1(\tfrac{1}{6}) + \$2(\tfrac{1}{6}) + \$3(\tfrac{1}{6}) + \$4(\tfrac{1}{6}) + \$5(\tfrac{1}{6}) + \$6(\tfrac{1}{6}) \\ &= \$\tfrac{21}{6} \\ &= \$3.50 \end{aligned}$$

This means that if the game is played long enough the player may expect to win an average of \$3.50 for each play. Therefore he should be willing to pay \$3.50 per roll of the die to make the game fair.

EXAMPLE 7.20 A box contains two \$1 bills, a \$5 bill, and a \$10 bill. You may draw one bill from the box and keep it. What is the value of the game to you?

Solution There are four bills in the box, two \$1's, one \$5, and one \$10. Thus

$$P(\$1) = \tfrac{1}{2} \qquad P(\$5) = \tfrac{1}{4} \qquad P(\$10) = \tfrac{1}{4}$$

Then

$$\begin{aligned} v &= \$1(\tfrac{1}{2}) + \$5(\tfrac{1}{4}) + \$10(\tfrac{1}{4}) \\ &= \$0.50 + \$1.25 + \$2.25 \\ &= \$4.00 \end{aligned}$$

EXAMPLE 7.21 A game is played as follows. A bag contains three red, two blue, and one white marbles. A player draws one marble from the bag. If it is red he receives \$1, if it is white he receives \$4.80, but if it is blue he must pay \$4.50. What is the value of the game to the player?

Solution There are 6 marbles in the bag. Three are red, two are blue, and one is white. Therefore

$$\begin{aligned} P(\text{red}) &= \tfrac{3}{6} \\ P(\text{blue}) &= \tfrac{2}{6} \\ P(\text{white}) &= \tfrac{1}{6} \end{aligned}$$

The player wins \$1 if he draws a red marble, and he wins \$4.80 if he draws a white marble. Therefore he can expect his winnings to be

$$\begin{aligned} \$1(\tfrac{3}{6}) + \$4.80(\tfrac{1}{6}) &= \$0.50 + \$0.80 \\ &= \$1.30 \end{aligned}$$

On the other hand, he loses \$4.50 if he draws a blue marble. His expected losses are

$$\$4.50(\tfrac{2}{6}) = \$1.50$$

Since the expected loss is greater than the expected gain on each draw, the player may expect to lose \$0.20 (\$1.50 − 1.30) on each play of the game.

Exercise Set 7.5

1. What are the odds in favor of:

   (a) Three tails on a single toss of three coins

   (b) A 7 or an 11 on a roll of two dice

   (c) Drawing an ace from a deck of 52 cards

   (d) Having a flush (five cards of the same suit) dealt from a deck of 52 cards?

2. What are the odds *against*:

   (a) A person being born in January

   (b) Guessing the correct answer on a multiple choice question with five answers from which to choose

   (c) Drawing an ace or a king in one draw from an ordinary deck

3. Pairs of dancing partners (a man and a woman) are chosen by lot from a group of seven men and five women. What are the odds in favor of Sam and Sally (two members of the group) being partners?

4. One hundred tickets are sold for a raffle. There are three prizes. First prize is \$20, second prize is \$15, and third prize is \$10. What is the expected value of the raffle to a person who holds one ticket?

5. A box contains four \$1 bills and two \$5 bills. Two bills are drawn from the box and kept by the one doing the drawing. What is a fair price to pay for the privilege of being the player who withdraws the bills?

6. A box contains two \$1 bills, one \$5 bill, and one \$10 bill. For the price of \$5, you may draw out one bill which you may keep. Should you play the game? Explain.

7. The probability that a man aged 20 will live for five years is $\frac{97}{100}$. How much should he pay for a five-year paid-up life insurance policy worth \$10,000?

8. Suppose that of all children born in a certain country the probability of a girl child is $\frac{6}{10}$. If a newly married couple from that country plan a family of four children, how many girls may they expect to have?

9. The cost of making a certain TV tube is \$3.80, and the tube sells for \$6.00. If 2% of the tubes manufactured by a company are defective, what profit can the company expect to make on one of this particular kind of tube?

10. The Acme Construction Company estimates that its probability of making a successful bid on a contract is 0.30. If it will cost Acme \$1000 to prepare a bid, and the profit on the contract will be \$50,000, what is the expected value for Acme on this type of job?

SUMMARY

Terms you should know:

1. Trial
2. Experiment
3. Sample space
4. Event
5. Mutually exclusive events
6. Conditional probability
7. Independent events
8. Odds
9. Expected value

Formulas you should understand:

1. $P(E) = \dfrac{N(E)}{N(S)}$
2. $P(E_1 \cup E_2) = P(E_1) + P(E_2) - P(E_1 \cap E_2)$
3. $P(\tilde{E}) = 1 - P(E)$
4. $P(E_2/E_1) = \dfrac{P(E_2 \cap E_1)}{P(E_1)}$
5. $P(E_2 \cap E_1) = P(E_2/E_1) \cdot P(E_1)$
6. Odds in favor of $E = \dfrac{P(E)}{P(\tilde{E})}$
7. $v = a_1p_1 + a_2p_2 + \cdots + a_np_n$

REVIEW EXERCISES

1. A spinner of the type found in many popular games is made in such a way that the needle is equally likely to point to any one of the numerals 1, 2, 3, 4. Show the sample space for an experiment which consists of two successive spins.
   (a) What is the probability that the needle stops on a 3 on two successive spins?
   (b) What is the probability that the needle stops on the same number on two successive spins?
   (c) What is the probability that the needle stops on a 4 exactly once in two spins?
   (d) What is the probability that the needle stops on a 4 at least once in two successive spins?

2. Show a sample space for an experiment which consists of throwing a die and then drawing a marble from a bag which contains one red, one white, and one blue marble.

   (a) What is the probability that the die shows a 6 and the marble is red?

   (b) What is the probability that the die shows an even number and the marble is red?

   (c) What is the probability that the die shows a 6 or the marble is red?

   (d) What is the probability that the die shows an even number or the marble is red?

   (e) What is the probability that the die shows a 6 or a 5 and the marble is red or white?

3. If the probability of rain on Saturday is $\frac{1}{10}$, what is the probability that it will not rain on Saturday?

4. What is the probability that you will win the door prize at a party attended by 150 people including yourself? What is the probability that you will not win the door prize?

5. What is the probability of rolling a 2 or a 4 or a 6 when rolling a die?

6. A bag contains five white marbles, four blue marbles, and three black marbles. What is the probability that a marble drawn is:

   (a) White

   (b) White or black

   (c) Not a black marble

   (d) A black marble if it is not a white marble

7. Let S be composed of the equally likely outcomes a, b, c, d, e, and f. Let $E_1 = \{a, b\}$, $E_2 = \{a, b, c\}$, and $E_3 = \{c, d, e, f\}$. Find:

   (a) $P(E_1)$ (e) $P(E_1 \text{ and } E_3)$

   (b) $P(E_2)$ (f) $P(E_1 \text{ or } E_2)$

   (c) $P(E_3)$ (g) $P(E_1 \text{ and } E_2)$

   (d) $P(E_1 \text{ or } E_3)$ (h) $P(E_1/E_2)$

8. Suppose a red box holds five white marbles and five black marbles and a green box holds six white marbles and three black marbles. What is the probability that a blindfolded person draws a black marble from the green box?

9. The probability that a man will live 10 more years is $\frac{1}{4}$. The probability that his wife will live 10 more years is $\frac{1}{2}$. What is the probability that both will be alive in 10 years?

10. Two coins are tossed. What is the probability that two heads will appear? What is the probability that one head will appear? What is the probability that zero heads will appear? What is the expected value of this game to a person who will get $3.00 for each head that appears?

11. To encourage Jim to study harder this semester, his father has offered to pay him $5 for each A he receives, $3 for each B he receives, and $1 for each C he receives. However, Jim must pay his father $4 for each D he receives and $5 for each F he receives. Jim estimates the following probabilities for obtaining the given grades in his five classes.

	Grade				
Subject	A	B	C	D	F
Math	0.10	0.20	0.30	0.20	0.20
Chem	0.10	0.10	0.50	0.20	0.10
History	0.10	0.50	0.30	0.10	0.00
English	0.00	0.30	0.40	0.20	0.10
P.E.	0.50	0.40	0.10	0.00	0.00

How much money should Jim expect to make (or lose) on this agreement?

12. The Bang 'Em Up Aircraft Corporation must choose between building four different models. The table below gives the cost of each model, the selling price of each model, and the probability that the company can put each of the models into the air. What is the expected value for each model? Which model should the Bang 'Em Up Aircraft Corporation choose to build (if any)?

Model Type	Cost	Selling Price	Probability That the Model Will Fly
I	$10,000	$15,000	0.90
II	$18,000	$24,000	0.80
III	$20,000	$30,000	0.75
IV	$30,000	$60,000	0.65

8
STATISTICAL ANALYSIS

It is very common, in this modern age, for large bodies of information to be collected relative to some specific situation. For example, every 10 years the Federal Government conducts a National Census, the FBI is continuously amassing information relating to criminal activities, TV polls are conducted in an attempt to determine the desires of the viewers, and department stores keep credit and sales records.

Once the facts have been collected, they must be organized and analyzed so that patterns may be recognized and decisions made. The task of organizing, analyzing, and predicting is the lot of the **statistician**, and his work is called **statistical analysis**. Statistical analysis may be directed toward describing and interpreting only the information collected; this is referred to as **descriptive statistics**. On the other hand, generalizations may be made which go beyond the facts collected; in such cases the analysis is called **statistical inference**. For example, after the National Census takers have made their rounds, the information they have collected is summarized by means of tables and graphs so that comprehension of the information is made easier. The statistical methods used in this phase of the work belong to descriptive statistics. After the census information has been summarized, predictions are made from it concerning the entire population of the United States. The methods used in making these predictions are those of statistical inference.

In the first two sections of this chapter we shall consider methods used in descriptive statistics; the remaining sections of the chapter will be devoted to some of the methods of inference.

8.1 VOCABULARY—CENTRAL TENDENCY

8.1.1 Basic Terminology

Statistical analysis, like most specialized fields, has its own technical vocabulary, some of which must be learned by anyone who wishes to understand what it is that the statistician does.

DATA The facts gathered to be used in a statistical analysis are called **data**. Each fact is called a **data point**. If data are expressed as numbers, it is common practice to call a data point a **score**. Often a data point is referred to as a **statistic**; thus the word **statistics** has a double meaning—it may mean the set of data being studied or it may mean the method by which the study is conducted.

In a study of wages paid to factory workers, the data (or statistics) of the study are the wages paid to the workers. The wages of an individual worker constitute a data point (or statistic). The methods by which the data are collected, analyzed, and interpreted are also called statistics.

POPULATION The set of things or people to which the results of a statistical analysis apply is called the **population**.

A population may be quantitative, such as the set of scores on a certain test, or qualitative, such as the answers "favor" or "opposed" given to the questions on an opinion poll. A population may be concrete or conceptual. The set of automobiles which pass through a certain intersection on a certain day is a concrete population. The set of all problems which can be solved using a computer is a conceptual population, for we can imagine such a set but it would be impossible to list all the elements of the set.

SAMPLE Because it is often impractical or impossible to collect data from each member of a population, the statistician obtains a subset of the population from which to collect facts. This subset is called a **sample**.

The sample contains pertinent information about the population. We should not expect the same information about the population from every sample. Of all the possible samples imaginable, a few might possess all the characteristics of a population; however, most samples fall short by varying degrees. Therefore predictions based on any one sample are subject to error.

To minimize the degree of uncertainty caused by sampling, it is desirable to select a sample in such a way that we may validly apply the rules of probability to calculate the chances of errors. We do this by random sampling.

RANDOM SAMPLE A sample of n elements is a **random sample** if it is selected from a population by any procedure that gives every sample of n population elements the same likelihood of being selected.

Several techniques are used to ensure randomness; a common one is the lottery in which the name of each member of the population is placed in a container and after all the names have been thoroughly mixed, a certain number are withdrawn to make up the sample.

Data are collected in many forms, but often they are in the form of numerical quantities that measure a specific characteristic of points of the sample. Data for medical studies include height, weight, blood pressure, and sugar content of the blood, to name but a few of the characteristics on which data are collected. Included in the information collected by the public school systems are age, sex, number of children in the family, IQ scores, aptitude test scores, achievement test scores, as well as grades, attendance records, and deportment records.

EXAMPLE 8.1 A mathematics test was given to a class of 25 students. Each test paper was scored with the number of problems correctly answered. The scores were

16	14	18	14	12	15	10	15	17	16
16	15	17	14	16	20	12	16	12	18
13	11	11	17	15					

The sample in this instance is not a random sample, but it is the set of all the students in the class. The data are the set of scores.

8.1.2 Displaying Data

Data in the form shown in Example 8.1 do not reveal much about the class. However, if they are arranged in order from highest to lowest and if tally marks are used to record the occurrence of each datum, then the frequency with which each data point occurs can be determined. Table 8–1 is a **frequency distribution** for the data of Example 8.1.

TABLE 8–1 Frequency Distribution

Score	Tally	Frequency
20	|	1
19		0
18	||	2
17	|||	3
16	~~||||~~	5
15	||||	4
14	|||	3
13	|	1
12	|||	3
11	||	2
10	|	1
	Total	25

Often data is more meaningful if it is displayed in a graphical form. We shall consider only one of the many graphical forms commonly used, the **histogram**.

A histogram is a series of rectangles each having as its base a convenient width which represents the difference between successive data points. The frequency with which each data point occurs is represented by the height of its rectangle. The histogram for the test scores of Example 8.1 is shown in Figure 8–1.

Observe that the sum of the areas of the rectangles of the histogram of Figure 8–1 is 25 square units, and that 25 is also the number of test scores in the

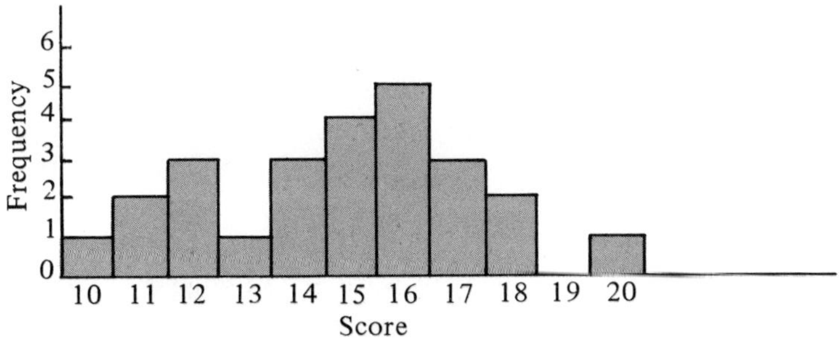

FIGURE 8–1 A Histogram.

distribution which was pictured by the histogram. In a well-constructed histogram this will always be so: *The sum of the areas in the rectangles in the histogram is numerically equal to the number of the items of data being graphed.* Since the fraction of the total population (sample space) that corresponds to an event is the probability of that event, in a histogram the fraction of the area that represents the event is equal to the probability of that event. Thus the probability that one of the students, picked at random from the class of Example 8.1, scored 12 on the test is

$$\frac{\text{Area of third rectangle from left}}{\text{Total area}} = \frac{3}{25}$$

and the probability that he scored between 15 and 17 inclusive is

$$\tfrac{4}{25} + \tfrac{5}{25} + \tfrac{3}{25} = \tfrac{12}{25}$$

8.1.3 Measures of Central Tendency

When the scores of Example 8.1 have been organized in one or more of the ways discussed above, it begins to take on meaning. One of the first things which we notice is that most of the scores bunch up in the mid-region of the distribution with relatively few scores at either the low or the high end. This is a rather common characteristic of many collections of data, and it has caused the statistician to search for the single score that can be considered the most typical. This most typical score is called a **measure of central tendency**. Three measures of central tendency in common use are the **mode**, the **median**, and the **mean**.

MODE The **mode** of a set of scores is the score which occurs with the greatest frequency. In Example 8.1 the mode is 16.

MEDIAN If the number of data points in a set is odd, the **median** is the middle data point when all the data points are ranked in order of size. In Example 8.1 there are 25 scores; therefore the median is the 15, the thirteenth score when the scores are arranged:

10, 11, 11, 12, 12, 12, 13, 14, 14, 14, 15, 15, (15,) 15,
16, 16, 16, 16, 16, 17, 17, 17, 18, 18, 20

If there are an even number of data points in a set, the median is half-way between the two middle data points when the data points are ranked in order of size. Thus in the set

4, 4, 5, 6, 6, 7, 7, 7, 8, 8

the two middle scores are 6 and 7; so the median is 6.5.

MEAN The **mean** is found by dividing the sum of the scores by the number of scores. In the set

$$4, \quad 4, \quad 5, \quad 6, \quad 6, \quad 7, \quad 7, \quad 7, \quad 8, \quad 8$$

the mean is

$$\frac{4 + 4 + 5 + 6 + 6 + 7 + 7 + 7 + 8 + 8}{10} = 6.2$$

We shall use the capital letter X to stand for the set of data in a distribution. A lowercase x will be used to represent an element of set X, in which it is understood that x is a variable quantity which can assume, one at a time, the identities of some or all of the elements of X. Thus if $X = \{5, 3, 7, 9, 4\}$, then x may at one time be 3 and at another time be 5 or 7 or 9 or 4.

We shall use the Greek capital letter sigma ($\sum$) to mean "the sum of." Thus if a distribution designated by X consists of the five numbers

$$X = \{3, 4, 6, 7, 8\}$$

then

$$\sum x = 3 + 4 + 6 + 7 + 8$$

$$= 28$$

Similarly $\sum (x - 2)$ means "subtract 2 from each of the numbers in set X and sum the differences." For the set

$$X = \{3, 4, 6, 7, 8\}$$

$$\sum (x - 2) = 1 + 2 + 4 + 5 + 6$$

$$= 18$$

and

$$\sum (x - 2)^2 = 1^2 + 2^2 + 4^2 + 5^2 + 6^2$$

$$= 1 + 4 + 16 + 25 + 36$$

$$= 82$$

The symbol we shall use for the mean of a sample is $\bar{x}$.† Using this symbol and the one for *sum* described above, the formula for the mean of the n scores contained in X is

$$\bar{x} = \frac{\sum x}{n}$$

The three measures of central tendency, the mode, the median, and the mean, are widely used in statistics. Each has its advantages and each has its disadvantages.

The mode requires no computation, and consequently it is the easiest to find. A second advantage of the mode is that it is not affected by extremely high or extremely low values in the data. On the other hand, there may be no mode, or there may be two or more modes in the same set of data.

The median is relatively easy to find, and like the mode it is not affected by extreme values of the data. The biggest disadvantage of the median is that it does not lend itself well to computations involving several sets of data.

The mean is the most difficult of the three measures to compute. It is affected by extreme values of the data. However, it does have some advantages which make up for its shortcomings:

1. If the mean $\bar{x}$ and the total number of elements in the set of data n are known, then the sum $\sum x$ can be found by multiplying $\bar{x}$ by n, that is,

   $$\sum x = \bar{x} \cdot n$$

   As an example consider a company which employs 30 men at a mean annual salary of \$7000. Then the annual payroll for the company is

   $$(\$7000)(30) = \$210{,}000$$

2. The mean varies less from sample to sample for a large population.
3. The mean is easiest to use in arithmetic computations in statistical methods. For instance, the means of several sets of data can be combined into an overall mean.
4. Another name for mean is **average**, and this is a term which is familiar to most people, whereas few people know the meaning of either median or mode.

The question of which measure of central tendency to use in a particular situation is not an easy one to answer. In fact, it is not uncommon that the

† It is common practice to denote the mean of a sample by $\bar{x}$ and the mean of a population by the lowercase Greek mu (μ). We shall use this convention in this text.

measure which is used is the one that best reflects the viewpoint of the person who makes the analysis. Thus if the football team's mean yardage gain is 4 yards and the mode of the yardage gained is 25 yards, the coach will very likely use the mean as a whip to make the offensive team perform better; but he may choose to quote the mode when he is the guest speaker at the Kiwanis Club luncheon.

EXAMPLE 8.2 The weights (in pounds) of 15 seven-year-old children are

50, 55, 56, 57, 57, 58, 59, 60, 61, 62,
63, 65, 67, 68

Find the mode, median and mean. Which measure would it be best to use in summarizing the weights of the children?

Solution The weights 57 and 58 each occur twice in the distribution, and no other weight occurs as frequently. Therefore both 57 and 58 are modes. The median weight is 59, the eighth weight in the listing. The mean is

$$\bar{x} = \frac{50+55+56+57+57+58+58+59+60+61+62+63+65+67+68}{15}$$

$$= \frac{896}{15} = 59.7$$

Since there are two modes, the mode is of little value as a representative of the weights. The mean and the median, however, are nearly the same and either one can be used to summarize the weights.

Exercise Set 8.1

1. For each of the following sets of data find the mode, the median, and the mean. Which measure is most representative of the typical data point in each case?

   (a) 9, 8, 7, 8, 5, 7, 8, 6, 9, 8

   (b) 2, 5, 8, 7, 3, 8, 6, 6, 4, 8, 7, 2, 6, 5, 8

   (c) 1, 8, 7, 4, 8, 8, 10, 9, 8, 2, 6, 8, 9

2. Prepare a frequency table and a histogram for each of the sets of data in Problem 1.

3. Given set $X = \{3, 4, 5, 8, 2\}$, find:

   (a) $\sum x$ (d) $\sum x^2$

   (b) $\sum (x - 1)$ (e) $(\sum x)^2$

   (c) $\sum (x - 1)^2$

4. The names of every male student in a certain school were written on slips of paper and put into a hat. Twenty slips were drawn from the hat, and the heights in inches of the boys selected were:

71	71	69	67	65	76	71	68	72	75
69	74	70	73	72	72	72	65	69	68

   Make a frequency table and histogram for this data. Find the mode, median, and mean. Which of these best characterizes the heights of the male students in the school?

5. A test was administered to a number of people. The results of the test are pictured in Figure 8–2. Use the histogram to answer the following questions.

   (a) What score occurred with the greatest frequency?

   (b) How many people took the test?

   (c) What was the median score?

   (d) What percentage of the people who took the test scored 82?

   (e) What percentage of the people scored above 81?

   (f) What is the probability that a student, selected at random, scored above 80? Between 80 and 82 inclusive?

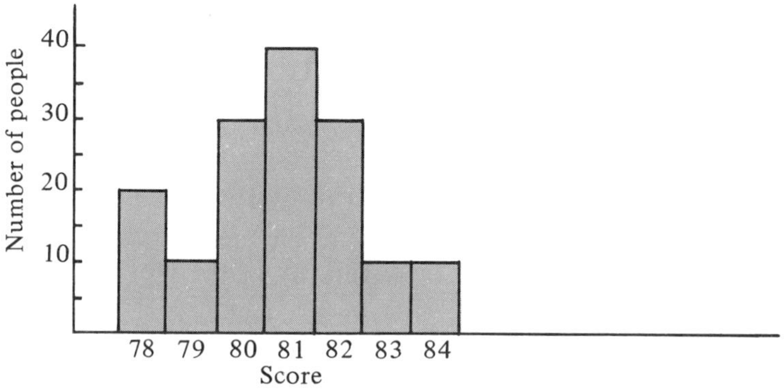

FIGURE 8–2

6. A company employs 20 workers. The foreman earns \$400 a week, his assistant earns \$320, two workers earn \$200 each, five earn \$150 each, ten earn \$100 each, and one earns \$90. Find the mode, median, and mean.

Which of these do you think the management is likely to use in reports to the stockholders? Which is most likely to be used by the workers when they meet with management to discuss labor contracts for next year?

7. The mean score on a set of 20 tests was 75. What is the sum of the scores on the tests?

8. Jim has an average of 85 from the four math tests he has taken this semester. If each of the five tests counts 100, what score must he receive on the final test in order to have an average score of 90 from the five tests?

*8.2 DISPERSION TENDENCIES

8.2.1 Range

Let us consider two sets of data taken from samples of the same population.

$$A = \{6, 6, 7, 9, 10, 10, 11, 11, 11\}$$

$$B = \{2, 3, 4, 5, 9, 11, 11, 12, 13, 20\}$$

The mode for set A is 11, its median is 10, and its mean is 9. Set B has a mode of 11, a median of 10, and a mean of 9. Here are two sets of data which Figure 8–3 shows are quite dissimilar, and yet they have identical modes, medians, and

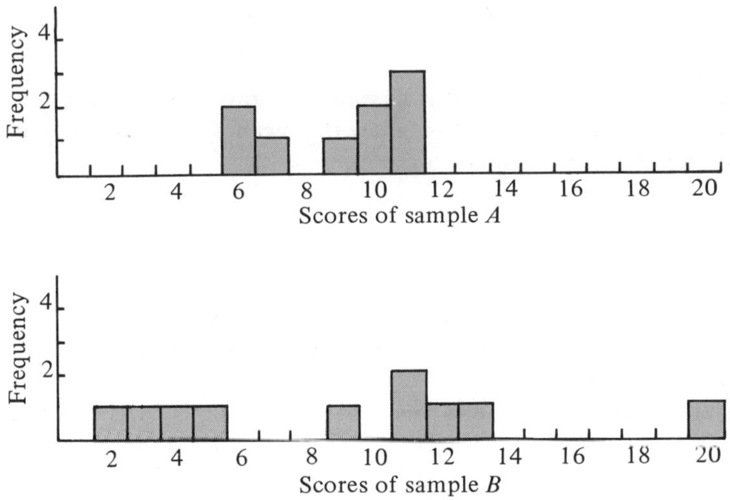

FIGURE 8–3

means. Clearly, central tendency, by itself, is not sufficient to describe a distribution of data. At least one other measure is needed, a measure to describe the spread or **dispersion** of the data.

One measure of dispersion is the **range**, the difference between the largest and the smallest of the scores in the distribution. The range of set A above is 5, and the range of set B is 18. The range is easy to compute and it does give some notion of the spread of the data, but it does not give any information about the behavior of the scores between the extreme values.

Since the dispersion of a set of data is *small* if the data are bunched closely about their mean and it is *large* if they are spread over considerable distances away from their mean, it seems reasonable to define variation in terms of the distances by which the data points depart from their mean. If $X = \{x_1, x_2, \ldots, x_n\}$ is a set of numbers whose mean is $\bar{x}$, the quantities $(x_1 - \bar{x})$, $(x_2 - \bar{x}), \ldots, (x_n - \bar{x})$ are called the **deviations from the mean**. For set $A = \{6, 6, 7, 9, 10, 10, 11, 11, 11\}$ with mean $\bar{x} = 9$, the deviations are shown in the following table.

Score	Deviation
6	$(6 - 9) = -3$
6	$(6 - 9) = -3$
7	$(7 - 9) = -2$
9	$(9 - 9) = 0$
10	$(10 - 9) = +1$
10	$(10 - 9) = +1$
11	$(11 - 9) = +2$
11	$(11 - 9) = +2$
11	$(11 - 9) = +2$

A positive sign on a deviation tells us that the score is above the mean, whereas a negative sign indicates that it is below the mean. A zero deviation indicates that a particular score is equal to the mean.

8.2.2 Standard Deviation

We wish to find some measure that is representative of the deviations. We cannot use the mean of the deviations because the sum of the deviations is 0 and thus the mean of the deviations is also 0. To get around this difficulty, the deviations are squared and the mean of the squares is computed. The deviations for set A were $-3, -3, -2, 0, 1, 1, 2, 2, 2$, and the squares of the deviations are

$$9, \quad 9, \quad 4, \quad 0, \quad 1, \quad 1, \quad 4, \quad 4, \quad 4$$

The mean of these numbers is

$$\frac{9 + 9 + 4 + 0 + 1 + 1 + 4 + 4 + 4}{9} = \frac{36}{9} = 4$$

This number is called the **variance** of the scores, and it could be used as a measure of dispersion. We recall to obtain the variance, we squared the deviations. To compensate for this, we take the square root of the variance and call this number the **standard deviation**. For the case under consideration, this is $\sqrt{4} = 2$. Since the deviations of set A are between -3 and $+2$, 2 seems a reasonable representative of the deviations.

We shall denote the standard deviation for a sample by a lowercase s.†

EXAMPLE 8.3 Compute the standard deviation of the set

$$B = \{2, 3, 4, 5, 9, 11, 11, 12, 13, 20\}$$

Solution Recall that the mean for this distribution is 9. Then:

Score x	Deviation from Mean $x - \bar{x}$	Deviation² $(x - \bar{x})^2$
2	−7	49
3	−6	36
4	−5	25
5	−4	16
9	0	0
11	2	4
11	2	4
12	3	9
13	4	16
20	11	121
		$\sum (x - \bar{x})^2 = 280$

$$s = \sqrt{\frac{\sum (x - \bar{x})^2}{n}}$$

$$= \sqrt{\frac{280}{10}}$$

$$= \sqrt{28}$$

† Following common practice, we shall denote the standard deviation of a sample by s and that of a population by a lowercase Greek sigma (σ).

We may approximate $\sqrt{28}$ by means of Table I in Appendix B. To use the table, locate 28 in the column headed n. Opposite the 28, in the column headed $\sqrt{n}$, find 5.29. Thus

$$s = 5.29 \ (approx.)$$

The formula $s = \sqrt{(x - \bar{x})^2/n}$ was easy to use in the preceding examples due to the fact that the x's and their means, and therefore the deviations from the means, were all integers. Had this not been the case, it might have been advantageous to use the following short-cut formula for s:

$$s = \sqrt{\frac{x^2}{n} - (\bar{x})^2}$$

This formula can be derived from the other formula for s. The advantage of this short-cut formula is that we do not have to go through the process of actually finding the deviations from the mean; instead we calculate $\sum x^2$, the sum of the squares of the x's, and substitute directly into the formula. Thus for set B above,

$$\sum x^2 = 2^2 + 3^2 + 4^2 + 5^2 + 9^2 + 11^2 + 11^2 + 12^2 + 13^2 + 20^2$$
$$= 4 + 9 + 16 + 25 + 81 + 121 + 121 + 144 + 169 + 400$$
$$= 1090$$

and

$$s = \sqrt{\tfrac{1090}{10} - (9)^2}$$
$$= \sqrt{109 - 81}$$
$$= \sqrt{28}$$
$$= 5.29$$

The standard deviations, 2 for set A and 5.29 for set B, show that the scores of set B are more widely dispersed than those of set A. The exact meaning of this difference in dispersions of scores will be investigated in Section 8.4. For the present it is sufficient to remember that the smaller the standard deviation, the closer to the mean are the scores.

The standard deviation may be computed about any of the measures of central tendency; however, it is common practice to compute it for the mean only.

Exercise Set 8.2

1. Find s for each of the sets of data in Problem 1 of Exercise Set 8.1. Use the short-cut formula.
2. Find the standard deviation of the heights in Problem 4 of Exercise Set 8.1.
3. The following were the scores of an English class on a test.

   98, 96, 95, 93, 90, 89, 87, 86, 85, 83, 82, 79,
   77, 76, 75

   Compute the mean and the standard deviation.
4. The temperatures at noon for seven days were 74°, 72°, 76°, 81°, 73°, 76°, 73°. What was the mean noontime temperature for that week? What was the standard deviation of the temperatures?
5. Eight light bulbs were drawn at random from a shipment. The number of hours of life for the bulbs were

   100, 230, 120, 160, 110, 180, 40, 280

   What is the mean lifetime of a bulb from the shipment? What is the standard deviation?
6. After a math test you go to your teacher's office to find out your grade. He is not there, but on his desk you see the following results:

   Class A: Mean 75, standard deviation 5
   Class B: Mean 75, standard deviation 10

   Which class would you rather be in? Why?

*8.3 THE BINOMIAL DISTRIBUTION

8.3.1 Binomial Experiments

There are many problems involving experiments with only two outcomes, one of which may be called "success" and the other "failure." Examples of this kind of experiment include coin tossing, answers to true-false tests, outcomes of a dice game in which success is rolling a certain number and failure is not rolling that number, and a cloud seeding experiment in which precipitation from a seeded cloud is success and failure is no precipitation from the cloud.

Experiments such as those mentioned above are called binomial experiments because they have only two possible outcomes.

In many binomial experiments we are interested in the probability of x successes in n trials. In the problems we shall consider in this book it will always be assumed that:

1. There is a fixed number n of trials.
2. The probability of a success is the same for each trial.
3. The trials are all independent.

This means that the theory we shall discuss will not apply to finding the probability that a particular team wins the World Series since in the World Series the number of trials (games) is not fixed. Nor will the theory apply to the drawing of a card from a deck of cards unless the card is replaced after each draw so that the probability of drawing any particular card remains the same in each ensuing draw. Moreover the theory cannot be used to find the probability that Joe Cool dates three of the four Homecoming Princesses since, to some extent, Joe's charm is enhanced (or diminished) after he has dated one of the girls.

To handle problems that do meet the above specifications, we use a special kind of distribution called the **binomial distribution**. The binomial distribution is the set of x's along with the set of probabilities associated with a binomial experiment and which are in the form $P(x$ successes in n trials).

EXAMPLE 8.4 A modified craps game is played in which the player may roll a pair of dice three times. Each time he rolls a 7 he wins, and each time he does not roll a 7 he loses. Find the probability that he:

(a) Wins on all three rolls
(b) Wins on two rolls out of the three
(c) Wins on one roll out of the three
(d) Loses on all three rolls

Solution Normally we must consider 36 outcomes for a single roll of two dice, and in three consecutive rolls there would be (36)(36)(36) = 46,656 outcomes. In this case, however, we are interested only in whether an outcome is a 7 (a win) or some other number (a loss). Therefore there are only eight outcomes for consideration: WWW, WWL, WLW, WLL, LWW, LWL, LLW, and LLL. Three of these, WWL, WLW, and LWW, are different ways in which there can be two wins in three rolls. Furthermore, these three outcomes are mutually exclusive; therefore the probability that one or the other of them occurs is the sum of their individual probabilities. In Chapter 7 we learned that the probability of rolling a 7 on one roll of a pair of dice is $\frac{1}{6}$, and the probability of not rolling a 7 is $1 - P(7) = \frac{5}{6}$. Therefore the probability of a win on each roll of the dice in this experiment is $\frac{1}{6}$, and the probability of a loss is $\frac{5}{6}$. Successive throws of a pair of dice are independent events; thus $P(WWL) =$

TABLE 8–2 Binomial distribution for an experiment having three trials with the probability of success in one trial equal to $\frac{1}{6}$.

Number of Wins x	Ways the Wins Can Occur	Number of Different Ways the Wins Can Occur	P(x wins in n trials)
3	*WWW*	1	$1\cdot\frac{1}{6}\cdot\frac{1}{6}\cdot\frac{1}{6} = \frac{1}{216}$
2	*WWL, WLW, LWW*	3	$3\cdot\frac{1}{6}\cdot\frac{1}{6}\cdot\frac{5}{6} = \frac{15}{216}$
1	*WLL, LWL, LLW*	3	$3\cdot\frac{1}{6}\cdot\frac{5}{6}\cdot\frac{5}{6} = \frac{75}{216}$
0	*LLL*	1	$1\cdot\frac{5}{6}\cdot\frac{5}{6}\cdot\frac{5}{6} = \frac{125}{216}$
			Total $\frac{216}{216} = 1$

$\frac{1}{6}\cdot\frac{1}{6}\cdot\frac{5}{6} = \frac{5}{216}$. But this is the same as $P(WLW)$ and $P(LWW)$ since each of these is the product of two $\frac{1}{6}$ factors and one $\frac{5}{6}$ factor. Thus P(two wins in three rolls) $= \frac{5}{216} + \frac{5}{216} + \frac{5}{216}$. This sum is the same as the product $3\cdot\frac{5}{216} = \frac{15}{216}$. In a similar manner we find that three of the outcomes, *WLL*, *LWL*, and *LLW*, are different ways in which there is one win in three rolls. The probability of any one of these occurring is $\frac{1}{6}\cdot\frac{5}{6}\cdot\frac{5}{6} = \frac{25}{216}$; therefore

$$P(\text{one win in three rolls}) = 3\cdot\frac{25}{216} = \frac{75}{216}$$

There is just one way that all three rolls can be 7's and the probability that this occurs is $\frac{1}{6}\cdot\frac{1}{6}\cdot\frac{1}{6} = \frac{1}{216}$. Also there is only one way that none of the rolls can be a 7, and the probability that this happens is $\frac{5}{6}\cdot\frac{5}{6}\cdot\frac{5}{6} = \frac{125}{216}$. The binomial distribution for this problem is summarized in Table 8–2. The distribution is shown graphically in Figure 8–4. Note that the graph is similar to the histograms

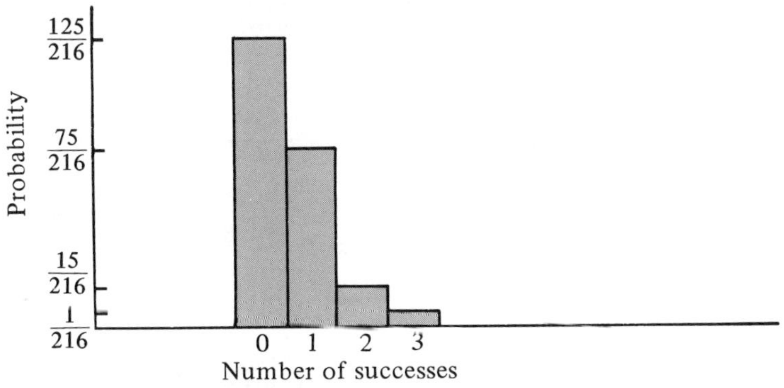

FIGURE 8–4 Graph showing binomial distribution for three trials with the probability of success on a single trial equal to $\frac{1}{6}$.

studied earlier, except that the lengths of the bars represent probabilities rather than frequencies.

8.3.2 A Formula for Binomial Probability

Before attempting another binomial problem, let us summarize the steps in computing $P(x$ successes in n trials).

1. Determine the different ways in which x successes and $n - x$ failures can occur.
2. If the probability of success in one trial is p, then the probability of x successes and $n - x$ failures is computed by multiplying the number of ways x successes can occur by p for each success and by $1 - p$ for each failure, that is,

$$P(x \text{ successes}) = (\text{number of ways } x \text{ successes can occur}) \times (p^x)\cdot(1 - p)^{n-x}$$

EXAMPLE 8.5 A bag contains three red marbles and two white marbles. A marble is drawn from the bag and then replaced. Find the binomial distribution for the experiment of drawing two marbles from the bag if success is the drawing of a white marble.

Solution The outcomes which concern us are: two successes, one success, and no successes. That these outcomes are not equally likely can be seen from Table 8–3.

TABLE 8–3

x Successes	Ways x Can Occur	Number of Different Ways x Can Occur
2	WW	1
1	WR, RW	2
0	RR	1

Since there are five marbles in the bag and two are white,

$$P(\text{Success on any one trial}) = \tfrac{2}{5}$$

and

$$P(\text{Failure on any one trial}) = \tfrac{3}{5}$$

The probabilities are computed in Table 8–4.

TABLE 8–4 Binomial distribution table for $p = \frac{2}{5}$.

x	Different Ways x Can Occur	$P(x$ successes in n trials)
2	1	$1 \cdot \frac{2}{5} \cdot \frac{2}{5} = \frac{4}{25}$
1	2	$2 \cdot \frac{2}{5} \cdot \frac{3}{5} = \frac{12}{25}$
0	1	$1 \cdot \frac{3}{5} \cdot \frac{3}{5} = \frac{9}{25}$
		Total $\frac{25}{25} = 1$

We can read the desired probabilities from the table.

The probability of two successes is $\frac{4}{25}$.
The probability of one success is $\frac{12}{25}$.
The probability of no success is $\frac{9}{25}$.

The graph for these probabilities is shown in Figure 8–5.

In the preceding examples we found the number of different ways x successes can occur by listing all the ways x can occur and then counting these ways. There are ways to obtain these numbers without counting. One of these ways is attributed to the French mathematician Blaise Pascal (1623–1662). He constructed the following triangle of numbers.

$$
\begin{array}{ccccccccccccc}
 & & & & & 1 & & 1 & & & & & \\
 & & & & 1 & & 2 & & 1 & & & & \\
 & & & 1 & & 3 & & 3 & & 1 & & & \\
 & & 1 & & 4 & & 6 & & 4 & & 1 & & \\
 & 1 & & 5 & & 10 & & 10 & & 5 & & 1 & \\
1 & & 6 & & 15 & & 20 & & 15 & & 6 & & 1
\end{array}
$$

Each row of this triangle gives the number of possible arrangements of x successes in n trials. For instance, the numbers in the third row from the top are 1, 3, 3, 1 and these are the numbers we computed in Example 8.3 as the number of ways of getting 3 successes in 3 trials, 2 successes in 3 trials, 1 success in 3 trials, and no successes in 3 trials. To find the number of ways of getting x successes in 6 trials, count down 6 rows from the top of the table and in the

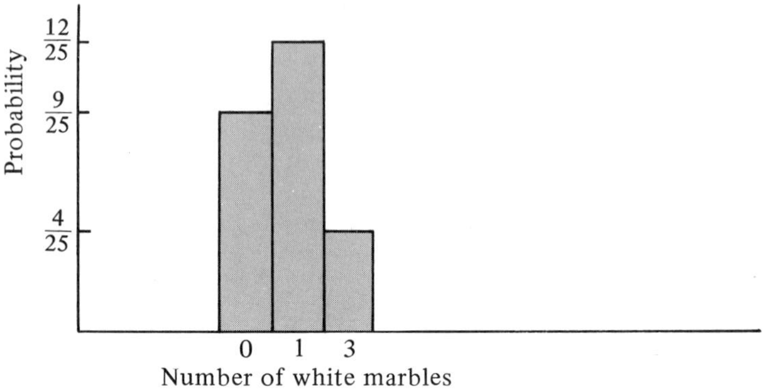

FIGURE 8–5 Graph showing binomial distribution for two trials with probability of success on one trial equal to $\frac{2}{5}$.

sixth row find the numbers 1, 6, 15, 20, 15, 6, 1. Then there must be 1 way of getting 6 successes in 6 trials, 6 ways of getting 5 successes, 15 ways of getting 4 successes, 20 ways of getting 3 successes, etc.

EXAMPLE 8.6 Five cards are selected from a regular deck one at a time, each card drawn replaced, and the deck thoroughly shuffled before the next card is drawn. A draw is a success if the card drawn is a spade. Make a table showing the probability distribution for this experiment and from the table find:

(a) The probability that three spades are drawn
(b) The probability that at least three spades are drawn

Solution On each trial the deck contains 52 cards, 13 of which are spades. Therefore $P(\text{success on one trial}) = \frac{13}{52} = \frac{1}{4}$ and $P(\text{failure on one trial}) = \frac{3}{4}$. From Pascal's triangle the number of ways of obtaining 5, 4, 3, 2, 1, and 0 successes in 5 trials are 1, 5, 10, 10, 5, and 1, respectively. The computation of the probabilities is shown in Table 8–5.

(a) The probability that three spades are drawn is $\frac{90}{1024}$.
(b) The probability that at least three spades are drawn is

$$\begin{aligned} P(3 \text{ or } 4 \text{ or } 5) &= P(3) + P(4) + P(5) \\ &= \frac{90}{1024} + \frac{15}{1024} + \frac{1}{1024} \\ &= \frac{106}{1024} \end{aligned}$$

It is an easy matter to enlarge Pascal's triangle. Note that each row begins and ends with a 1. Each of the other numbers is the sum of the two numbers to the left and right of it in the preceding row. To write the numbers in the seventh row, begin with 1 and then write 7, the sum of the 1 and 6 in the

TABLE 8–5 Distribution of probabilities in a binomial experiment of five trials with the probability of success on any one trial equal to $\frac{1}{4}$.

x	Number of Ways x Can Occur	$P(x$ successes in 5 trials)
5	1	$1 \cdot ppppp = 1 \cdot \frac{1}{4} \cdot \frac{1}{4} \cdot \frac{1}{4} \cdot \frac{1}{4} \cdot \frac{1}{4} = \frac{1}{1024}$
4	5	$5 \cdot pppp(1-p) = 5 \cdot \frac{1}{4} \cdot \frac{1}{4} \cdot \frac{1}{4} \cdot \frac{1}{4} \cdot \frac{3}{4} = \frac{15}{1024}$
3	10	$10 \cdot ppp(1-p)(1-p) = 10 \cdot \frac{1}{4} \cdot \frac{1}{4} \cdot \frac{1}{4} \cdot \frac{3}{4} \cdot \frac{3}{4} = \frac{90}{1024}$
2	10	$10 \cdot pp(1-p)(1-p)(1-p)$ $= 10 \cdot \frac{1}{4} \cdot \frac{1}{4} \cdot \frac{3}{4} \cdot \frac{3}{4} \cdot \frac{3}{4} = \frac{270}{1204}$
1	5	$5 \cdot (1-p)(1-p)(1-p)(1-p)$ $= 5 \cdot \frac{1}{4} \cdot \frac{3}{4} \cdot \frac{3}{4} \cdot \frac{3}{4} \cdot \frac{3}{4} = \frac{405}{1024}$
0	1	$1 \cdot (1-p)(1-p)(1-p)(1-p)(1-p)$ $= 1 \cdot \frac{3}{4} \cdot \frac{3}{4} \cdot \frac{3}{4} \cdot \frac{3}{4} \cdot \frac{3}{4} = \frac{243}{1024}$
		Total $\frac{1024}{1024} = 1$

first two positions of the sixth row. The third number in the seventh row is $6 + 15 = 21$, the fourth number is $15 + 20 = 35$, etc.

8.3.3 μ and σ for Binomial Distributions

Statisticians have proven that the mean μ and the standard deviation σ of a binomial distribution are given by the formulas $\mu = np$ and $\sigma = \sqrt{np(1-p)}$ in which n is the number of trials and p is the probability of a success on any trial.

EXAMPLE 8.7 Find μ and σ for the binomial distribution of Example 8.4.

Solution The experiment of Example 8.4 consisted of three rolls of a pair of dice with success being the occurrence of a 7. Thus $p = \frac{1}{6}$ and $n = 3$. Therefore

$$\begin{aligned}
\mu &= np \\
&= 3 \cdot \tfrac{1}{6} \\
&= \tfrac{1}{2} \\
\sigma &= \sqrt{np(1-p)} \\
&= \sqrt{3(\tfrac{1}{6})(\tfrac{5}{6})} \\
&= \sqrt{\tfrac{15}{36}}
\end{aligned}$$

Using the fact that the square root of a common fraction is equal to the square root of the numerator divided by the square root of the denominator, we have

$$\begin{aligned}\sigma &= \sqrt{\tfrac{15}{36}} \\ &= \frac{\sqrt{15}}{6} \\ &= \frac{3.87}{6} \qquad \textit{(from Table I in Appendix B)} \\ &= 0.63\end{aligned}$$

If we compute the expected number of 7's in three rolls of a pair of dice, using the method of Chapter 7, we have

$$\begin{aligned}v &= 3\cdot\tfrac{1}{216} + 2\cdot\tfrac{15}{216} + 1\cdot\tfrac{75}{216} + 0\cdot\tfrac{125}{216} \\ &= \tfrac{3}{216} + \tfrac{30}{216} + \tfrac{75}{216} + \tfrac{0}{216} \\ &= \tfrac{108}{216} \\ &= \tfrac{1}{2}\end{aligned}$$

Obviously this does not mean that in three rolls of a pair of dice we can expect one-half of a 7 to come up. What it does mean is that if the experiment is repeated many times, we can expect, *on the average*, one 7 in every two trials. This is, of course, the identical interpretation which we should give to the mean $\bar{x}$ of the distribution. That the expected value in this experiment is the same as the mean of the distribution is no accident. *In any binomial distribution, the mean of the distribution is the same as the expected number of successes of the experiment.*

We have considered binomial experiments in this section, and we have found that there is a distribution of probabilities associated with each binomial experiment. This distribution can be computed using the formula

$$P(x \text{ successes}) = (\text{number of ways } x \text{ successes can occur})\cdot(p^x)(1 - p)^{n-x}$$

where n is the number of trials in the experiment, p is the probability of success on a single trial, and the number of ways x successes can occur can be found from Pascal's triangle.

The binomial probability distribution is used in statistical inference; some examples of this use were shown above. Other uses of the binomial distribution for the purposes of inference will be discussed in the following section.

Exercise Set 8.3

1. Which of the following are binomial experiments?
   (a) Answers to yes-no questions
   (b) Answers in a five-question multiple choice test in terms of right answers and wrong answers
   (c) Grades in a pass-fail system
   (d) The make-up of a five-man committee if the members of the committee are chosen by lot from a group of 25 people.
2. Write the numbers in the seventh and eighth rows of Pascal's triangle.
3. A fair coin is tossed four times.
   (a) Complete the binomial distribution table below.

Number of Heads x	Number of Ways x Can Occur (from Pascal's triangle)	$P(x)$
4		
3		
2		
1		
0		

   (b) Draw the graph showing these results.
   (c) What is the probability that less than two of the tosses will be heads?
   (d) What is the probability that two or three of the tosses will be heads?
4. A true-false quiz is given with five questions. A passing score is at least three questions answered correctly.
   (a) What is the probability that a student who guesses on every question passes the test?
   (b) How many of the questions can a student who guesses on all of them expect to answer correctly?
5. If the probability that a newborn child is a girl is $\frac{1}{2}$, what is the probability that a couple who plans a family of five children will have three or four girls?
6. Both Steve and Bill have asked Judy for a date on each of the three evenings of next weekend, but she cannot make up her mind whom she should date. She decides to leave the decision to chance. She will spin the spinner

shown in Figure 8–6 once for each day. If it stops on red or yellow she will date Bill that evening, but if it stops on green she will date Steve. What is the probability that Judy:

(a) Dates Bill on two of the three evenings

(b) Dates the same boy on all three evenings

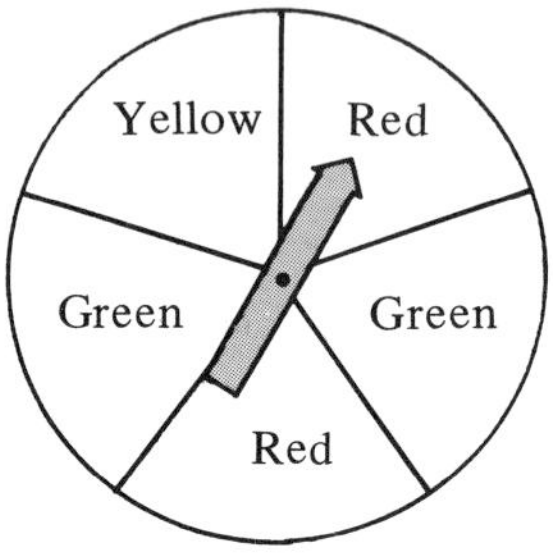

FIGURE 8–6

7. Among 25 income tax forms filed with the Internal Revenue Service, 6 contain errors favoring the taxpayer. What is the probability that among 5 of these forms selected for audit, exactly 2 contain such an error?

8. Certain machine parts are shipped in lots of 50. Three parts are selected from each lot, and the lot is accepted if none of the parts is defective. What is the probability that a lot will be accepted if it contains five defectives?

9. Find the mean and standard deviation of each of the following distributions:

   (a) The number of heads obtained in 100 tosses of a fair coin

   (b) The number of 7's obtained in 300 rolls of a pair of fair dice

   (c) The number of seeds that will germinate among those contained in a package of 200 if the probability that any one of the seeds will germinate is 0.60

   (d) The number of people in a group of 40 at an office party who should not drive home after the party if 40 percent of the people at office parties can be expected to drink enough to make them dangerous drivers

10. The manager of a restaurant figures that 8 out of 10 of his customers have cocktails before dinner.

   (a) Find the probabilities that in a party of 4 diners at the restaurant at least 1 will have pre-dinner cocktails.

   (b) How many customers in a party of 6 can be expected to have cocktails before dinner?

*8.4 THE NORMAL CURVE

8.4.1 Probability Curves

Consider a binomial experiment with five trials in which the probability of success on any one trial is 0.4. The graph for this experiment is shown in Figure 8–7.

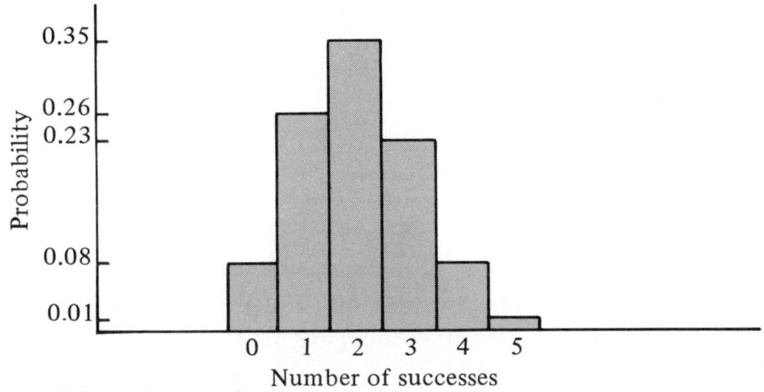

FIGURE 8–7 Binomial distribution: 5 trials with p = 0.4.

If we were interested in a similar experiment but with $n = 10$ or $n = 20$, we would get the graphs shown in Figures 8–8 and 8–9.

In the binomial experiments we have considered thus far, we have assumed that the number of successes could be integral only. It is often convenient

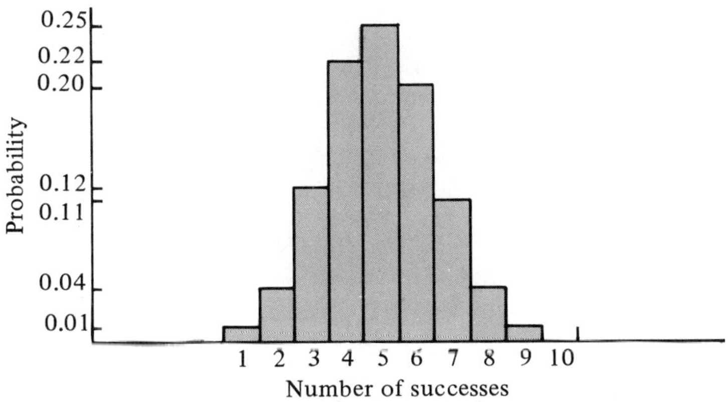

FIGURE 8–8 Binomial distribution: 10 trials with p = 0.4.

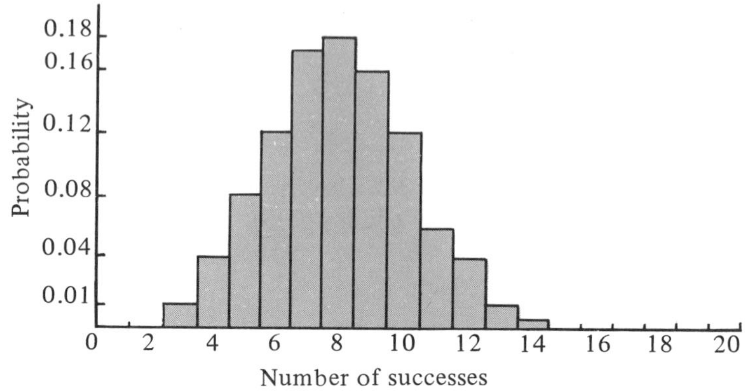

FIGURE 8–9 Binomial distribution: 20 trials with $p = 0.4$.

to assume that the variable x in $P(x$ successes) is **continuous**, that is, that x can take on all values fractional as well as integral. For example, in a binomial experiment of five trials with $p = 0.4$, x might be any number between 0 and 5 inclusive. If so, we may no longer represent the distribution of probabilities by a series of rectangles as we did for the case in which the variable was integral. Instead we draw a smooth curve through the points which would have been the midpoints of the tops of the rectangles in the integral case. Figure 8–7 shows the distribution when x is integral; Figure 8–10 is the graph for the continuous case. Graphs representing the continuous cases of the experiments depicted by Figures 8–8 and 8–9 are shown in Figures 8–11 and 8–12. The graphs of Figures 8–10, 8–11, and 8–12 are examples of **probability curves**.

In drawing the graph of a probability curve, the corners of some of the rectangles of the graph of the integral case are cut off by the curve (see Figures 8–10 to 8–12), and thus some of the area of the rectangles is not included in the

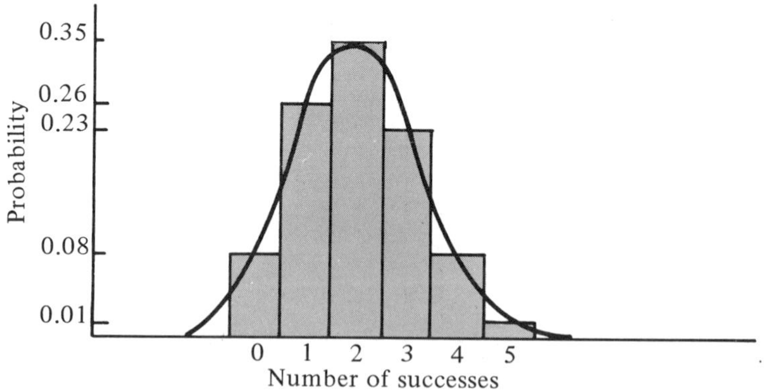

FIGURE 8–10 Probability curve for binomial experiment with continuous variable $n = 5$, $p = 0.4$.

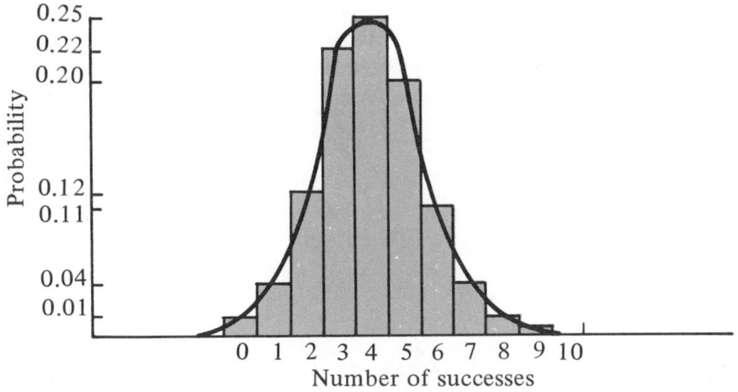

FIGURE 8–11 Probability curve for binomial experiment with continuous variable, $n = 10$, $p = 0.4$.

area beneath the curve. On the other hand, there is some area under the curve which was not inside the rectangles. One might surmise that the area of the rectangles which was left out from under the curve is nearly compensated for by the area under the curve which was not in any of the rectangles. Such a surmise would be correct. Recall that the area of the graph of a binomial probability experiment in which there are only integral numbers of successes has a total area of 1. Therefore the area under a probability curve which is drawn by connecting the midpoints of the tops of the rectangles of the integral case is approximately 1 also.

When we first discussed histograms, we pointed out that the frequencies associated with the various events are given by the areas of the rectangles and that the probability of an event is the fraction of the total area which represents the event. The same thing is true in the case of a probability curve:

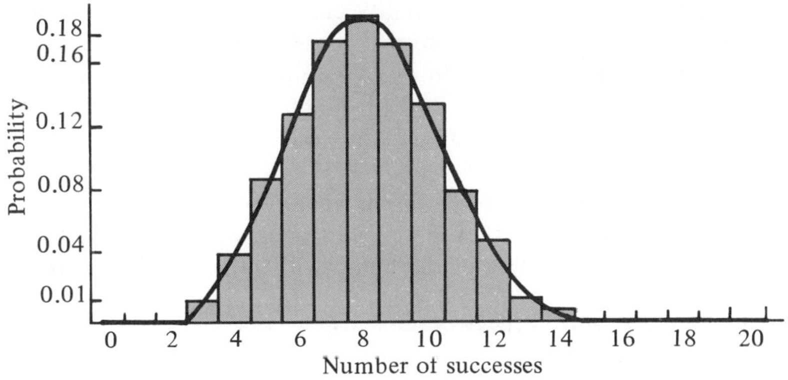

FIGURE 8–12 Probability curve for binomial experiment with continuous variable, $n = 20$, $p = 0.4$.

Suppose x is a continuous variable and $P(x$ successes) is represented graphically by a probability curve. Then the probability that x is a number between x_1 and x_2 is given to the area bounded by the curve, the horizontal axis, and vertical lines through x_1 and x_2.

As an example, suppose that the graph in Figure 8–13 is that of a probability curve and that the shaded area is 0.6. Then the probability that there are more than 1 but less than 3 successes in this experiment is 0.6.

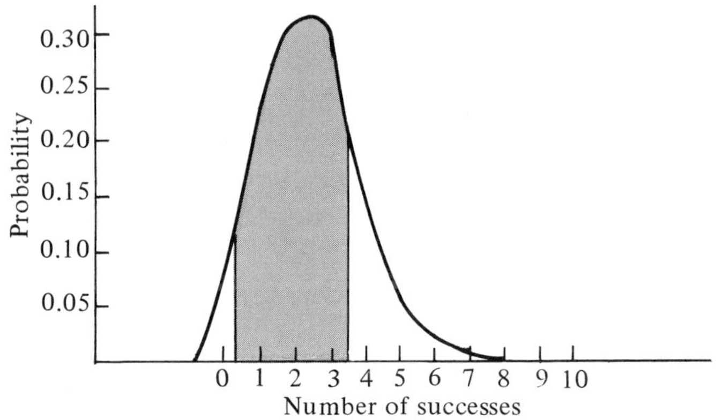

FIGURE 8–13 Shaded region is 60% of total area or 0.60.

8.4.2 The Normal Distribution

Note that the probability curves of Figures 8–10 to 8–13 all display a typical "bell shape." Not all probability curves have this shape, but several kinds do have it. One of these, the **normal distribution curve**, is especially important because it has been found to fit observed data very well in many different fields. A typical normal curve is shown in Figure 8–14. Notice that the curve extends indefinitely in both directions coming nearer and nearer to the horizontal axis without ever reaching it.

A very important feature of the normal curve is that it is symmetrical about its mean. This means that if we fold the drawing of Figure 8–14 along the dashed line, the two halves of the figure will coincide. The statistical significance of this symmetry will be discussed on the following pages.

The normal distribution and its graph, the normal curve, have the following important properties:

1. Approximately 68 percent of the area under the normal curve is between $\mu - \sigma$ and $\mu + \sigma$ where μ is the mean and σ is the standard deviation (see

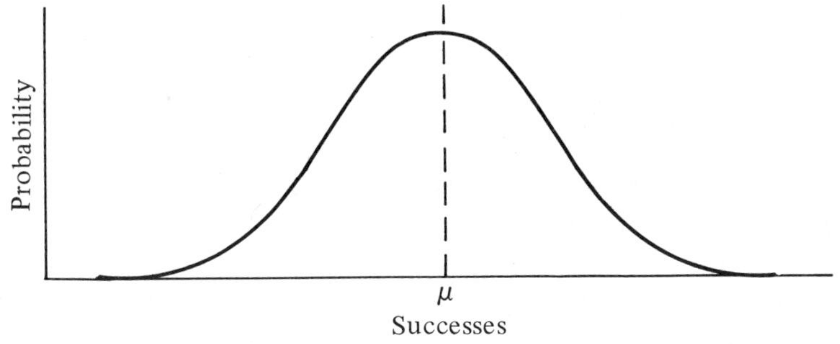

FIGURE 8–14 Normal curve.

Figure 8–15). Therefore 68 percent of the scores in a normal distribution fall between one standard deviation below the mean and one standard deviation above the mean.

2. Approximately 95 percent of the area under the normal curve is between $\mu - 2\sigma$ and $\mu + 2\sigma$. Therefore 95 percent of the scores fall between two standard deviations below the mean and two standard deviations above the mean.
3. The area under the normal curve between $\mu - 3\sigma$ and $\mu + 3\sigma$ is 99.7 percent (to the nearest 0.1 percent) of the total area. Therefore 99.7 percent of the scores in a normal distribution fall between three standard deviations below the mean and three standard deviations above the mean. Thus only 0.3 percent of the scores fall more than three standard deviations from the mean. Because of the symmetry of the normal distribution, half of this 0.3 percent of the scores are located more than three standard deviations above the mean and the other half are located more than three standard deviations below the mean.

It should be noted that the three properties listed above are characteristics of a *normal distribution* and that few, if any, real-world situations have

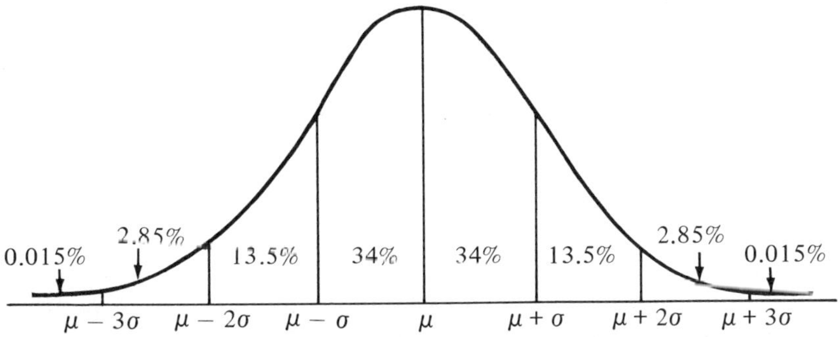

FIGURE 8–15 Proportion of normal curve areas for some intervals.

exact normal distributions. However, there are many distributions which are nearly normal, and for these we may use the properties of the normal distribution. Some of these "nearly normal" distributions will be considered in this chapter.

8.4.3 The Standard Normal Curve

The location and shape of a normal curve are completely determined by its mean μ and its standard deviation σ. Figure 8–16 shows two normal curves, one with $\mu = 3$ and $\sigma = 2$ and one with $\mu = 5$ and $\sigma = 1$.

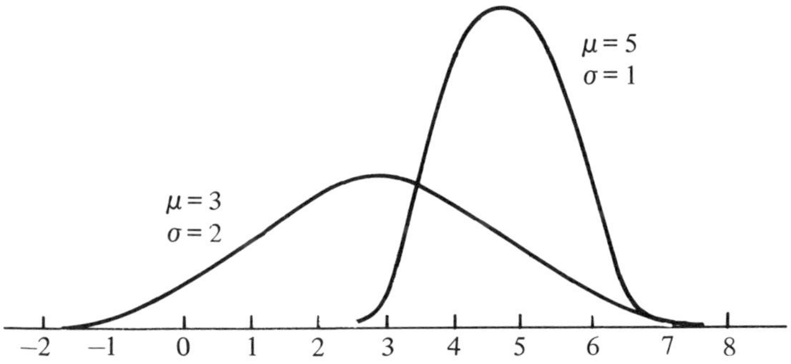

FIGURE 8–16 Comparison of normal curves with different means and standard deviations.

One type of question which we can answer using the normal curve is:

Suppose a set of scores is known to be normally distributed and that the mean and standard deviation are known. What is the probability that a score is between the two values x_1 and x_2?

To answer such a question, it will be necessary that we find the fraction of the area under the curve of the distribution between vertical lines through x_1 and x_2 on the horizontal axis (see Figure 8–17). It is apparent from Figure 8–16 that

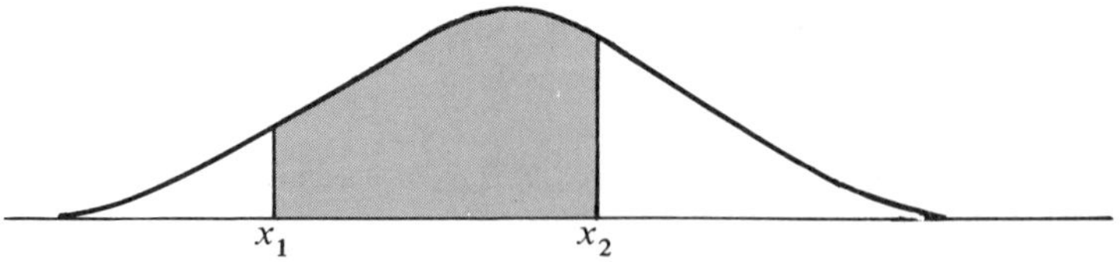

FIGURE 8–17 Shaded region represents probability that a data point x is between x_1 and x_2.

for fixed values of x_1 and x_2 this area will differ with the values of μ and σ. This difficulty is resolved by always working with the **standard normal distribution**, that is, with the normal distribution which has $\mu = 0$ and $\sigma = 1$. Areas under any other normal curve can then be obtained by renumbering the scales on the horizontal axis. For example, to change the normal curve with $\mu = 3$ and $\sigma = 2$ to a standard normal curve, we change the scale by putting a 0 at 3 on the original scale and we let 1 unit of the new scale be 2 units (one standard deviation) of the old scale. The normal curve with both scales is shown in Figure 8–18.

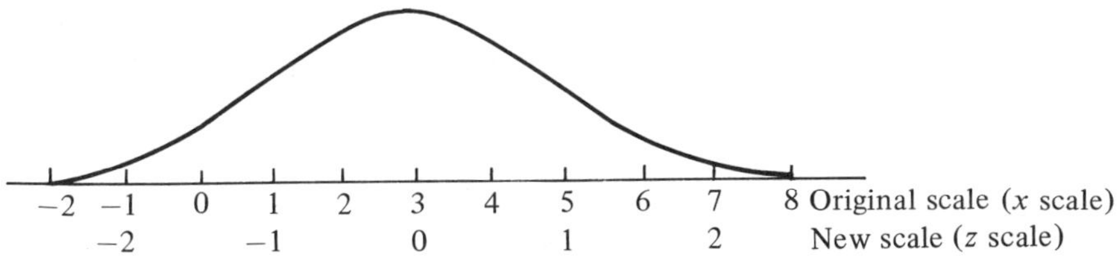

FIGURE 8–18 Normal curve with scale on horizontal axis changed to standard scale.

In practice we do not bother to actually change the scale on the axis of a nonstandard normal curve. Instead we convert the scores in which we have an interest to z scores. A z **score** corresponding to a particular score is the number of standard deviations of that score from the mean of the distribution. If x is a score in a nonstandard normal distribution, then its z score is given by the formula

$$z = \frac{x - \mu}{\sigma}$$

Thus in Figure 8–18, the z score corresponding to the data point 5 is $z = (5 - 3)/2 = 1$. This means that a score of 5 in the nonstandard distribution becomes a score of 1 when the distribution has been standardized. Note that this is exactly what we found to be the case when we changed the scale on the axis of the graph.

Table II in Appendix B gives a list of z scores from $z = 0$ to $z = 3$. For each of the z scores, the table also gives the proportion of the area under a standard normal curve from the mean, 0, to z.

EXAMPLE 8.8 Find the part of the area under the standard normal curve from $z = 1$ to $z = 1.5$.

Solution The area is the shaded part of Figure 8–19. We must subtract the area from $z = 0$ to $z = 1$ from the area from $z = 0$ to $z = 1.5$. To find the area from $z = 0$ to $z = 1.5$, we read down the column of z scores in Table II until we reach the score 1.5. Next to this value in the area column we find 0.433. This means that 43.3 percent of the area under the standard normal curve occurs between the mean and the score which is 1.5 standard deviations above the mean. In a similar manner, we find the area between $z = 0$ and $z = 1$ to be 34.1 percent of the total area under the curve. Thus the area from $z = 1$ to $z = 1.5$ is $43.3 - 34.1 = 9.2$ percent of the total area. This means that 9.2 percent of the scores in a normal distribution will fall between 1 and 1.5 standard deviations above the mean.

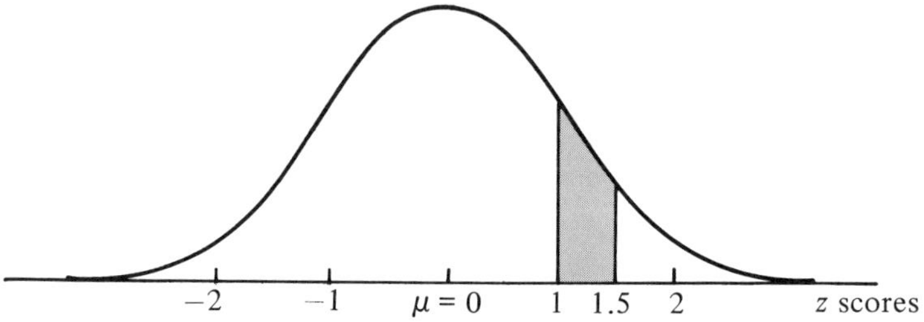

FIGURE 8–19 Normal curve whose shaded region represents probability of a data point from 1 to 1.5 standard units above the mean.

Example 8.9 What percentage of the scores in a normal distribution can be expected to fall within 0.7 standard deviation of the mean?

Solution We are asked to find the area shown in Figure 8–20. Because the scores in a normal distribution are distributed symmetrically about the mean, the area from $z = -0.7$ to $z = 0$ is equal to the area from $z = 0$ to $z = 0.7$. Therefore the required area is twice this last area. From Table II, we find the area between $z = 0$ and $z = 0.7$ to be 0.258 of the total area under the curve. Therefore the required area is $2(0.258) = 0.516$ of the total area. We may expect about 51.6% of the scores of a normal distribution to be within 0.7 standard deviation of the mean.

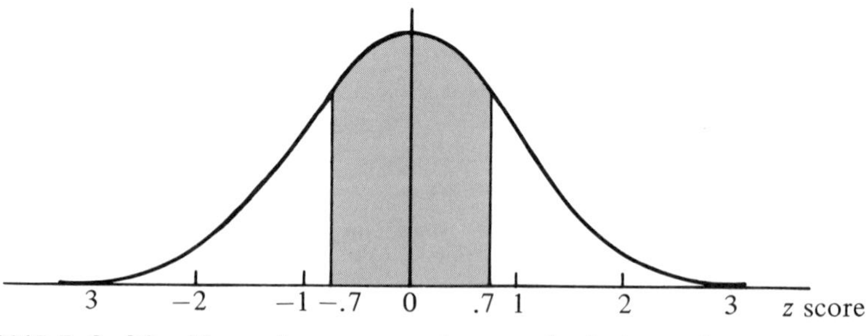

FIGURE 8–20 Normal curve whose shaded region represents probability of a data point from −0.7 to 0.7 standard units from the mean.

EXAMPLE 8.10 Assume that the height of college men is normally distributed with a mean of 69 inches and a standard deviation of 2 inches. What is the probability that a college man selected at random will be between 66 and 70 inches tall?

Solution We first compute the z scores for 66 and 70. Since $\mu = 69$ and since $z = (x - \mu)/\sigma$, the z score corresponding to 66 is

$$\begin{aligned} z_1 &= \frac{66 - 69}{2} \\ &= -1.5 \end{aligned}$$

and the z score corresponding to the score of 70 is

$$\begin{aligned} z_2 &= \frac{70 - 69}{2} \\ &= 0.5 \end{aligned}$$

We are looking for the area under the standard normal curve from $z = -1.5$ to $z = 0.5$. Because of the symmetric property of a normal distribution, the area from $z = -1.5$ to $z = 0$ is the same as the area from $z = 0$ to $z = 1.5$. We have

$$\begin{aligned} \text{(Required area)} &= (\text{area from } z = -1.5 \text{ to } z = 0.5) \\ &= (\text{area from } z = -1.5 \text{ to } z = 0) \\ &\quad + (\text{area from } z = 0 \text{ to } z = 0.5) \\ &= (\text{area from 0 to 1.5}) + (\text{area from 0 to 0.5}) \\ &= 0.433 + 0.191 \qquad (\textit{from Table II}) \\ &= 0.624 \end{aligned}$$

Therefore the probability that a college man is between 66 and 70 inches is 0.624, given the μ and σ in the problem.

EXAMPLE 8.11 What is the probability that one of the college men of Example 8.10 is taller than 6 feet?

Solution The z score for 6 feet (72 inches) is

$$\begin{aligned} z &= \frac{72 - 69}{2} \\ &= 1.5 \end{aligned}$$

The probability that one of the men is taller than 6 feet is the same as that of the area under the standard normal curve which is above $z = 1.5$. We observe that half of the area (50 percent) is below the mean, and Table II tells us that 43.3 percent of the area is above the mean but below $z = 1.5$. Therefore (50 + 43.3 percent) or 93.3 percent of the area is below $z = 1.5$, and this leaves 6.7 percent of the area above $z = 1.5$. The total area is 1; therefore 6.7 percent of the area is 0.067, and we conclude that the probability that a college man is taller than 6 feet is 0.067.

8.4.4 Approximating the Binomial Distribution

We started this section with a consideration of the histograms for three binomial distributions and showed how these histograms could be represented by smooth curves. We then observed that these curves were similar to one particular bell-shaped curve called a normal curve. We conclude the section with a question. Can the normal curve be used to approximate a binomial distribution? If it can, our work will certainly be simplified because converting scores to z scores and looking up normal curve areas in Table II is much easier than computing binomial probabilities as we did in the previous section.

The answer to this question is not a simple one. From Figures 8–10 to 8–12 it would appear that the more trials in the experiment, the more closely the binomial distribution is to a normal distribution. On the other hand, Figure 8–21 shows a histogram for a binomial experiment of 20 trials with $P(\text{success}) = \frac{9}{10}$, and this distribution could not be represented by a normal curve, as can be seen by an inspection of Figure 8–22. A good rule to follow is this one:

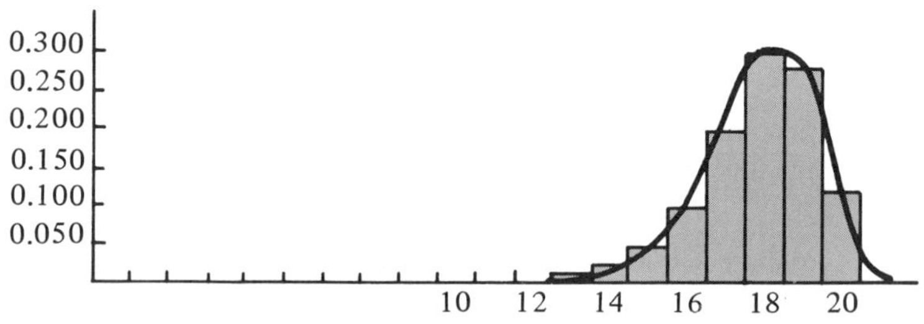

FIGURE 8–21 Binomial probability curve, $n = 20$, $p = 0.9$.

If the expected number of successes np and the expected number of failures $n(1 - p)$ are both greater than 5, then the normal curve is a good approximation of the binomial histogram. If np and $n(1 - p)$ are not both greater than 5, use Pascal's triangle.

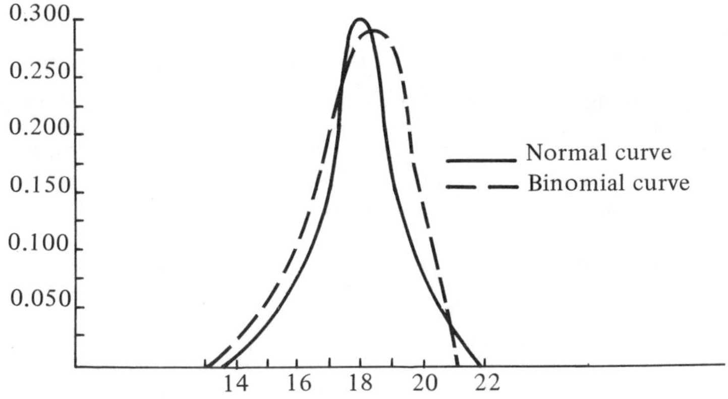

FIGURE 8–22 Comparison of binomial and normal curves both having $\mu = 18$ and $\sigma = (3\sqrt{5})/5$.

Exercise Set 8.4

1. Using Table II of Appendix B, find the percentage of the area under the standard normal curve from:

   (a) $z = 0$ to $z = 0.4$

   (b) $z = 2$ to $z = 2.5$

   (c) $z = -0.5$ to $z = 0.8$

   (d) $z = -1.8$ to $z = -0.4$

2. A set of data is known to be normally distributed with $\mu = 70$ and $\sigma = 5$. Find the percentage of the scores:

   (a) Between 70 and 82

   (b) Between 65 and 75

   (c) Between 68 and 80

   (d) Over 78

   (e) Less than 50

3. Assume that men's shoe sizes are normally distributed with a mean of 10 and a standard deviation of $\frac{1}{2}$. If a shoe company makes 20,000 pairs of men's shoes per year, how many size 11 shoes should the company make? (Men's shoes are made in $\frac{1}{2}$ sizes; thus a man wears a size 11 shoe if his foot measures between $10\frac{3}{4}$ and $11\frac{1}{4}$.)

4. At a certain hospital the weights of newborn infants are normally distributed with $\mu = 7$ pounds and $\sigma = 10$ ounces. Find:

(a) The probability that a newborn child weighs between 6 lb 3 oz and 7 lb 7 oz
(b) The probability that a newborn child is weighed between 6 and 8 pounds
(c) The percentage of babies who weigh more than 8 pounds at birth

5. According to a recent study, college students think about sex on an average of 3.3 hours per day with a standard deviation of 0.6 hour. If the results are normally distributed, what percentage of the students think about sex:

   (a) Less than 3 hours daily
   (b) More than 1.5 hours daily
   (c) More than 5 hours daily

6. Suppose there are 100 true-false questions on a test and you guess on every one of them. What is the probability that you will pass the test if you must get at least 60 questions correct to pass?

7. The probability that a 40-year-old man will have died before age 60 is 0.15. An insurance company insures 1200 men of age 40. What is the probability that fewer than 149 of these men will have died before age 60?

8.5 HYPOTHESIS TESTING

8.5.1 Hypotheses

A large part of the work of the statistician is concerned with testing hypotheses. A **hypothesis** is a claim or assumption.

EXAMPLE 8.12 Clara Voyant claims to have the ability to read minds. As a test of her claim, we select one card from an ordinary deck and note its color. Clara, who cannot see the card, attempts to call the correct color. This experiment is repeated 20 times, the card being returned to the deck and the deck thoroughly shuffled before each new trial. Clara makes 14 correct identifications. Can we agree to her claim?

Solution Whenever a statistician is presented with a statistical claim there are two hypotheses for him to consider:

1. The claim is correct.
2. The claim is not correct.

Usually he has some reason to favor one of these hypotheses; that one is called the **motivated hypothesis**. The other hypothesis is called the **null hypothesis**.

We may feel that Clara, in Example 8.12, can be expected to guess the correct color about half of the time since there are the same number of red cards as black cards in the deck. The fact that she has correctly identified 14 colors may cause us to select as the motivated hypothesis:

Clara has ESP.

Then the null hypothesis is

Clara does not have ESP

or in an equivalent but more useable form,

Clara is guessing.

8.5.2 Testing Hypotheses

The process by which the statistician determines whether the evidence is strong enough to support his motivated hypothesis is called a **hypothesis test**. There are many kinds of hypothesis tests, whose use depends upon the size of the sample, the kind of distribution, whether or not the data are numerical, and whether the data are from one or more samples. Here we shall consider only large single samples with binomial distributions which may be approximated by the standard normal curve.

To do a hypothesis test, the statistician always tests the null hypothesis. The reason for this will become clear when we have completed a few tests.

EXAMPLE 8.13 Test the hypothesis that Clara Voyant's claim (Example 8.12) is false.

Solution Clara correctly named the colors of 14 cards in a sample of 20. Let us concede that she has shown some evidence of an ability to read minds if the probability of her *guessing* and getting 14 out of 20 colors is less than 0.05. (That is, if Clara has ESP, we expect her to be able to do as well as she has done more than 95 out of 100 times that she repeats the experiment.) This number, 0.05 or 5 percent is called the **significance level** of the test. The significance level is always decided upon before the test. It may be any number between 0 and 1. However, it represents the probability of one kind of error we may make in our findings; therefore we usually choose either 0.05 or 0.01.

We must compute the probability that Clara guessed. If she did, then the probability that she gets a color right on each draw is

$$p = \tfrac{1}{2}$$

This is a binomial experiment with $n = 20$; therefore

$$\begin{aligned}\mu &= np \\ &= (20)(\tfrac{1}{2}) \\ &= 10\end{aligned}$$

and

$$\begin{aligned}\sigma &= \sqrt{np(1-p)} \\ &= \sqrt{(20)(\tfrac{1}{2})(\tfrac{1}{2})} \\ &= \sqrt{5} \\ &= 2.24\end{aligned}$$

We wish to find the probability that Clara could guess 14 or more correct colors. The z score corresponding to 14 is

$$\begin{aligned}z &= \frac{x - \mu}{\sigma} \\ &= \frac{14 - 10}{2.24} \\ &= 1.8 \text{ approx.}\end{aligned}$$

From Table II, we find the probability that a score is between the mean and $z = 1.8$ is 0.464. Since the probability is 0.500 that a score is greater than the mean, the probability that Clara guesses 14 or more colors correctly is $0.500 - 0.464 = 0.046$. This means that if Clara is guessing she will do as well as she did in this experiment only 4.6 percent of the time. Since 4.6 percent is less than the 5 percent significance level we decided on at the start, we reject the null hypothesis that Clara has no ESP. That is, we agree that Clara has shown statistical evidence that she can read minds.

8.5.3 Errors in Hypothesis Testing

Note that there are always two possible types of errors that we can make in hypothesis testing:

Type I Error We reject the null hypothesis even though it is true.

Type II Error We retain the null hypothesis even though it is false.

In Example 8.13, the probability of a type I error is 0.046 since if we reject the hypothesis that Clara is guessing, we will be wrong only 4.6 percent of the time.

The possibility of a type II error can be minimized, but there is no way that both types of errors can be minimized in the same problem. If the experimenter wishes to reduce the type II error probability, he can increase either the significance level or the sample size. It is generally true that for a fixed sample size the probability of a type II error decreases as the significance level increases.

EXAMPLE 8.14 You suspect that a certain coin is biased in favor of heads. You decide to toss the coin 30 times and record the results. You reason that heads should occur about 15 times in 30 tosses if the coin is fair. Therefore if too many more than 15 heads occur, you will say that the coin is biased. How many heads must occur if the test is to be conducted at the 5 percent level?

Solution This is a binomial experiment with $n = 30$, $p = \frac{1}{2}$, $\mu = 15$, and

$$\begin{aligned}\sigma &= \sqrt{(30)(\tfrac{1}{2})(\tfrac{1}{2})}\\ &= \sqrt{\tfrac{30}{4}}\\ &= 2.74\end{aligned}$$

We wish to find, in this distribution, a particular score such that the probability of any score being greater than that one is less than 0.05. From Table II we find that approximately 5 percent of the area under the standard normal curve is beyond $z = 1.6$. Thus we wish to find a data point x corresponding to a z score of 1.6. From the formula

$$z = \frac{x - \mu}{\sigma}$$

it follows that

$$\begin{aligned}x &= \mu + z\sigma\\ &= 15 + (1.6)(2.74)\\ &= 15 + 4.4\\ &= 19.4\end{aligned}$$

Therefore if 20 or more heads occur, you should reject the null hypothesis in favor of your motivated hypothesis that the coin is biased; but if the number of heads which occur is between 15 and 19.4, you should retain the null hypothesis and admit that there is no statistical evidence that the coin is not fair.

To summarize the procedure of hypothesis testing discussed above, we may consider the following steps:

1. Based on some idea, the experimenter is motivated to make a claim. This claim is his motivated hypothesis.
2. For statistical purposes he tests the opposite hypothesis. This is his null hypothesis.
3. He selects a significance level (probability of a type I error). That is, he states how large a risk he is willing to take in making the error of claiming his motivated hypothesis is true when it is really false. He will probably select a significance level of either 5 or 1 percent; however, he is free to choose any other he wishes.
4. He decides on the size of the sample he will use.
5. Based on his motivated hypothesis and choice of significance level he calculates the statistics for the decision rule he will use. An example of a decision rule is the one used in the previous example, "Reject the null hypothesis if more than 19 heads come up."
6. He selects a sample and obtains the data.
7. Based on the experimental outcome and the decision rule, he makes one of two decisions:
   (a) Reject the null hypothesis and claim that the motivated hypothesis is correct.
   (b) Retain the null hypothesis and claim that there is no statistical evidence to support the motivated hypothesis.

Exercise Set 8.5

1. In Example 8.12, how many colors would Clara have to call correctly if her claim of being able to read minds is to be accepted at the 1 percent level?
2. You believe that a die is loaded in favor of the side with 6 spots. You toss it 60 times and get fourteen 6's. Using the 5 percent level of significance, do these results cause you to reject the null hypothesis that the die is fair?
3. Madame X says that she can tell by taste whether the tea has been added to boiling water or boiling water has been added to the tea. She sips 10 pairs of cups of tea, one cup made each way, and correctly identifies 9 of the pairs. If her claim is rejected, at what level of significance was the test conducted?
4. A patient suffering from chronic headaches has had 60 percent of a large number of headaches relieved by aspirin without a buffering agent. He tries aspirin with a buffering agent, and 17 of his next 20 headaches are relieved. Would you reject at the 5 percent level the null hypothesis that the buffering agent is not effective in combating his headaches?

5. Joe the bartender at Bennie's Bar claims that only one-fourth of his customers can distinguish between Smooth Suds and Belly Wash Beer. Bill Boozer bets that the number is greater. They gather a random sample of 24 customers. Each customer is given a glass of each brew and asked to identify it. At the 5 percent level, how many customers would have to be correct for Bill to win the bet?

6. The committee for the Reelection of Congressman Windblower is meeting ahead of election day to discuss campaign strategy. Windblower, as the incumbent, is felt to have an advantage over his opponent, but the committee wants some more information about this advantage as a guide to deciding whether to conduct an expensive all-out campaign. The committee decides that they will raise funds for an intensive campaign if Windblower has the support of 60 percent or less of the voters in his district. If Windblower has the support of more than 60 percent of the voters, the committee will conduct a cheaper campaign. In a random sample of 100 voters, they find that 66 will vote for the congressman. Should the committee accept or reject the null hypothesis that the congressman has the support of no more than 60 percent of the voters? Assume a 5 percent significance level.

8.6 CORRELATION

8.6.1 Pearson Correlation Coefficient

Thus far in our discussion of statistics, we have been concerned with ways of describing distributions of sets of data and with some of the methods of statistical inference. We shall conclude the chapter with a discussion of one of the workhorses of statistics, **correlation**.

All of us are familiar with situations in which there are variables that are somehow related. We know, for instance, that the amount of taxes an individual pays is related to the income he earns, that the amount of postage on a letter is a function of the weight of the letter, and that the height of the liquid in a thermometer varies with the surrounding temperature. These are examples of variables which are related because a change in one of them causes a change in the other. There are other cases in which two variables are related but neither of them influences the other. The Dow-Jones averages provide an example of this type of relationship: the changes in the values of the Dow-Jones stocks give an indication of the state of the stock market; in general, however, they do not *cause* changes in the values of other stocks.

We shall confine our discussion of correlation to one type of problem, that of describing the strength of a relationship between two sets of data.

CORRELATION COEFFICIENT A number that measures the strength of a relationship between two sets of data is called a **correlation coefficient.**

There are several kinds of correlation coefficients for measuring various types of relationships between different kinds of measurements. One of these, the **Pearson correlation coefficient**, is widely used, and it is this one that we shall discuss to illustrate the basic ideas of correlation.

EXAMPLE 8.15 Suppose that in a study of the relationship between education and income, a sample of eight cities in a certain geographical region yields the information in Table 8–6. Let us first draw a graph called a **scattergram** to investigate any possible relationship (see Figure 8–23). To draw this graph, we

TABLE 8–6

City	Percent of College Grads	Median Income (thousands of dollars)
A	18	6.5
B	7	5.5
C	16	7.0
D	11	6.0
E	20	7.5
F	24	8.0
G	14	6.5
H	10	6.0

first draw a horizontal axis for the percentages of college graduates and a vertical axis for the median incomes. (This choice is arbitrary; we could have put the median incomes on the horizontal axis and the percentages on the vertical axis.) The percentages of graduates range from 7 to 24 percent, and so the percentage axis must be labeled in such a way that all those percentages can be recorded. Let us choose to make a series of equally spaced marks labeled from 7 to 24, counting 1 at a time. Similarly, because the median incomes range from \$4000 to \$8000, and are given in units of \$500, let us choose to mark the income axis with equally spaced marks labeled from 4 to 8, counting $\frac{1}{2}$ at a time. Note that it is not necessary to start either of the scales at 0, nor is it necessary that the unit of length be the same on both scales.

After the axes have been drawn and labeled, we use one dot for each city. The dot is placed so that it is directly over the city's percentage score and directly to the right of the city's income score. For example, the dot for city *F* is over the 24 on the horizontal axis and to the right of the 8 on the vertical axis.

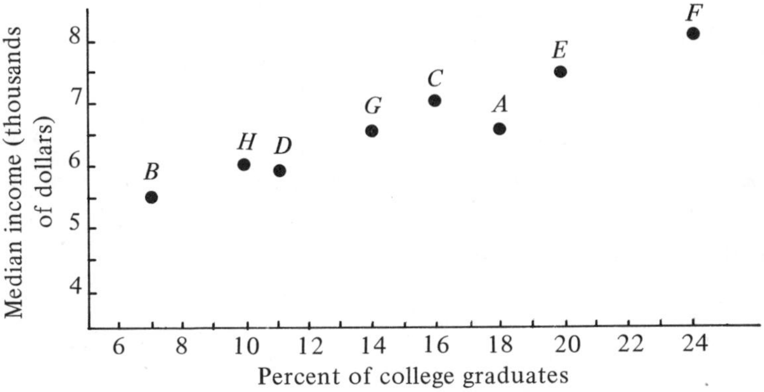

FIGURE 8–23 Scattergram showing relationship between education and income for a certain geographical region.

In the completed scattergram shown in Figure 8–23 the letter which identifies each of the cities is shown next to the dot for that city. (The dots on a scattergram are not usually identified; it was done this time only to aid in explaining the construction of the graph.)

You will notice two things about the scattergram:

1. The points are arranged approximately in a straight line. When this happens, we say that there is a **linear correlation** between the two variables.
2. The higher numbers in the income column correspond to the higher numbers in the percentage column. This causes the line to slope up to the right. This is called **positive correlation.**

TABLE 8–7

Temperature (in degrees Farenheit)	Coffee Sold (in hundreds of cups)	Donuts Sold (in dozens)
80	10	80
70	15	75
75	15	90
65	20	80
60	40	70
65	30	95
55	50	80
45	55	75
45	60	90
40	70	70

EXAMPLE 8.16 Table 8–7 shows temperatures at game time, the number of cups of coffee sold, and the number of donuts sold during one season at a football stadium.

A scattergram comparing sales of coffee with temperatures at game time is shown in Figure 8–24 and one comparing sales of donuts with temperatures at game time is shown in Figure 8–25.

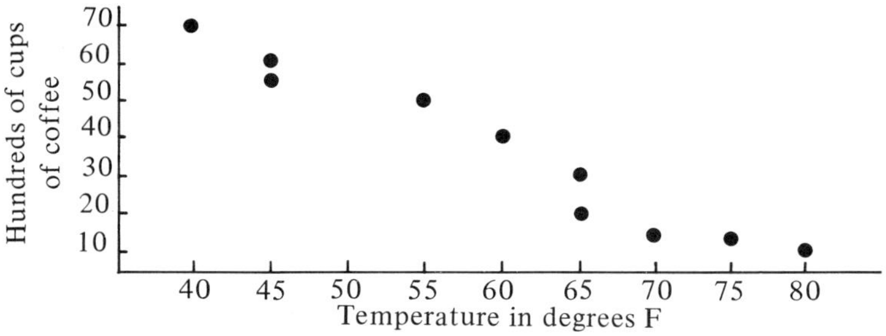

FIGURE 8–24 Comparison of sales of coffee with temperature at game time.

In Figure 8–24, the points tend to lie in a line that slopes down to the right. This is called a **negative correlation**. This happens because the higher temperatures correspond to the lower numbers of cups of coffee sold.

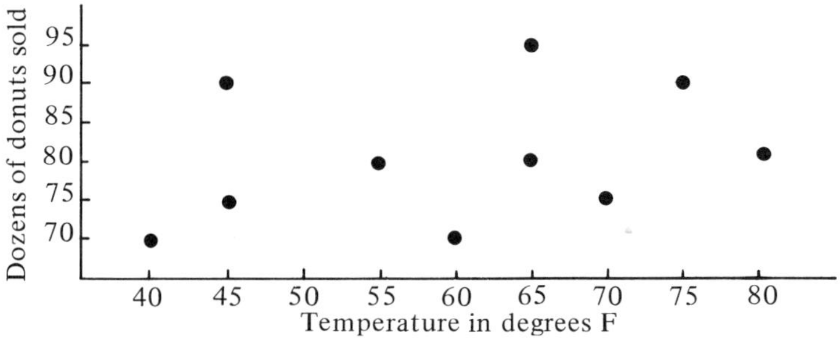

FIGURE 8–25 Comparison of sales of donuts with temperature at game time.

In Figure 8–25, there is no special tendency for the points to lie in a straight line. We say, in this case, that there is little linear correlation between temperatures at game time and the number of donuts sold.

8.6.2 Calculation of r

The preceding examples illustrate that between two sets of data there may be a positive linear correlation, a negative linear correlation, or there may be little or no linear correlation. Pearson defined a number r, called the **Pearson correlation coefficient**, to numerically describe how well the points on a scattergram "fit" a straight line. When the sample points lie exactly in a line sloping up to the right, $r = 1$ and we say that there is *perfect positive correlation.* When the sample points lie exactly in a line sloping down to the right, $r = -1$ and we say that there is *perfect negative correlation.* When there is no tendency for the sample points to lie in a straight line, $r = 0$ and we say there is *no linear correlation.* If r is near 1 or near -1, we say that the linear correlation is *high*; if r is near 0, we say that the linear correlation is *low.* We can expect that for Figure 8–23, r is nearly 1; for Figure 8–24, r is near -1; and for Figure 8–25, r is near 0.

Pearson's formula for computing r is:

$$r = \frac{n \sum xy - (\sum x)(\sum y)}{\sqrt{n \sum x^2 - (\sum x)^2}\sqrt{n \sum y^2 - (\sum y)^2}}$$

where

n = number of pairs of scores
x = label for variable on the horizontal axis
y = label for variable on the vertical axis
$\sum x$ = the sum of the x values
$\sum y$ = the sum of the y variables
$\sum xy$ = value found by multiplying the corresponding values of x and y and then adding all these products
$\sum x^2$ = value found by squaring all the x values and then adding all these products
$\sum y^2$ = value found by squaring all the y values and then adding all these products

The computation of r is illustrated in the following examples.

EXAMPLE 8.17 To compute r for the sets of data of Example 8.15, we first make the table of values shown as Table 8–8. The first two columns of this table are the ones given by Table 8–6. The other columns are used for the computation of $\sum xy$, $\sum x^2$, and $\sum y^2$. The table should be studied carefully

TABLE 8–8

Percent of College Grads	Median Incomes	xy	x^2	y^2
18	6.5	117.0	324	42.25
7	5.5	38.5	49	30.25
16	7.0	112.0	256	49.00
11	6.0	66.0	121	36.00
20	7.5	150.0	400	56.25
24	8.0	192.0	576	64.00
14	6.5	91.0	196	42.25
10	6.0	60.0	100	36.00
$\sum x = 120$	$\sum y = 53.0$	$\sum xy = 826.5$	$\sum x^2 = 2022$	$\sum y^2 = 356.00$

so that you may understand how each of the values to be substituted into the formula are found. When these values are substituted into

$$r = \frac{n \sum xy - (\sum x)(\sum y)}{\sqrt{n\sum x^2 - (\sum x)^2}\sqrt{n \sum y^2 - (\sum y)^2}}$$

we have

$$\begin{aligned} r &= \frac{8(826.5) - (120)(53)}{\sqrt{8(2022) - (120)^2}\sqrt{8(356) - (53)^2}} \\ &= \frac{6612 - 6360}{\sqrt{16{,}176 - 14{,}400}\sqrt{2848 - 2809}} \\ &= \frac{252}{\sqrt{1776}\sqrt{39}} \\ &= \frac{252}{(42.14)(6.25)} \\ &= \frac{252}{263.38} \\ &= 0.96 \text{ approx.} \end{aligned}$$

Note that $r = 0.96$ is close to 1, as we were led to expect from the scattergram of Figure 8–23.

EXAMPLE 8.18 Using the data of Example 8.16 comparing the temperatures at game time and cups of coffee sold, we have Table 8–9.

TABLE 8–9

Temperature x	Coffee Sold y	xy	x^2	y^2
80	10	800	6400	100
70	15	1050	4900	225
75	15	1125	5625	225
65	20	1300	4225	400
60	40	2400	3600	1600
65	30	1950	4225	900
55	50	2750	3025	2500
45	55	2475	2025	3025
45	60	2700	2025	3600
40	70	2800	1600	4900
$\sum x = 600$	$\sum y = 365$	$\sum xy = 19{,}350$	$\sum x^2 = 37{,}650$	$\sum y^2 = 17{,}475$

$$r = \frac{n \sum xy - (\sum x)(\sum y)}{\sqrt{n \sum x^2 - (\sum x)^2}\sqrt{n \sum y^2 - (\sum y)^2}}$$

Thus

$$\begin{aligned} r &= \frac{10(19{,}350) - (600)(365)}{\sqrt{10(37{,}650) - (600)^2}\sqrt{10(17{,}475) - (365)^2}} \\ &= \frac{19{,}350 - 219{,}000}{\sqrt{16{,}500}\sqrt{41{,}525}} \\ &= \frac{-25{,}500}{(128.45)(203.78)} \\ &= -0.97 \text{ approx.} \end{aligned}$$

Recall that the scattergram for this example led us to expect an r near -1, and -0.97 is nearly -1.

EXAMPLE 8.19 To compute r for the correlation between temperature and sales of donuts in Example 8.16, we note that $\sum x = 600$, $\sum y = 805$, $\sum xy = 48{,}600$, $\sum x^2 = 37{,}650$, and $\sum y^2 = 65{,}475$. Thus

$$\begin{aligned} r &= \frac{10(48{,}600) - (600)(805)}{\sqrt{10(37{,}650) - (600)^2}\sqrt{10(65{,}475) - (805)^2}} \\ &= 0.28 \text{ approx.} \end{aligned}$$

and this value, close to 0, shows that there is little correlation between sales of donuts and temperature at game time for this sample.

8.6.3 The Regression Line—Point Estimation

Now that we have learned how to compute r, how may it be used? Suppose that we examined an entire population and computed the correlation coefficient for two variables. If this coefficient equaled 1 or -1, we would say that there is perfect correlation between the variables in this population. When we examine a random sample taken from a population, a sample value of r near 1 or -1 is interpreted as reflecting correlation between the variables *in the population.* A sample value near 0 indicates that there is no correlation between the variables *in the population.* The statistician has tests which he may apply to help him to decide when a sample value of r is sufficiently far from 0 to reflect correlation in the population. We shall, unfortunately, not have time to consider any of these tests here.

If there is a significantly high linear correlation between two variables of a sample, we can try to represent the correspondence by a **regression line**—a line which best fits the dots on the scattergram. There is a mathematical formula for this line; however, we shall not use the formula in any of the work here. We may then use the regression line to estimate the value of y which corresponds to any given x from the population.

EXAMPLE 8.20 A random sample of students shows that there is a strong positive correlation between grade-point averages in high school English and general grade-point averages during the first year the students attend a certain college. The scattergram for the data is shown in Figure 8–26. To the scattergram has

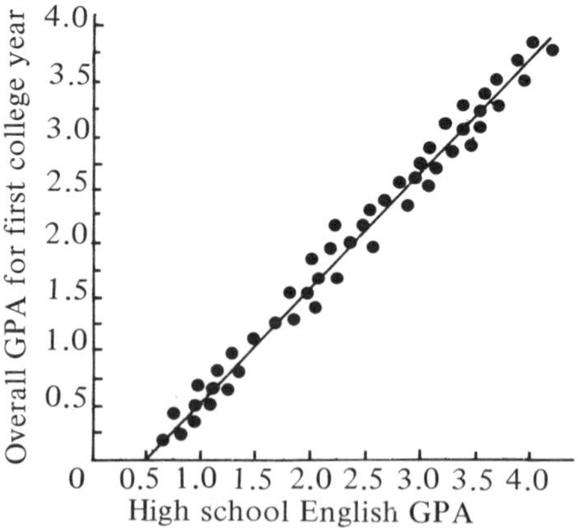

FIGURE 8–26 Ideal line added to scattergram.

been added the line which best represents a linear correspondence between the two variables. For any given value of $x > 0.5$, this line can be used to find a corresponding value of y. For example, if a high school graduate who applies for admission to the college has an English GPA of 3.0, the line can be used to estimate that his GPA for the first year of college will be about 2.5. This may be seen by finding the point on the prediction line which is directly above the 3.0 on the horizontal axis, then looking directly to the left of this point to find the number on the vertical axis which is at the same level as the point. (Follow the arrows on the dashed lines in Figure 8–27.)

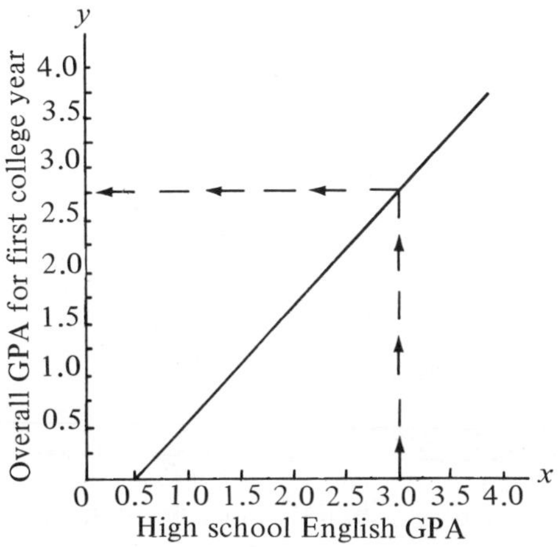

FIGURE 8–27 Use of the ideal line in predicting a y corresponding to x = 3.0.

In this section we have considered the possibility of a linear relationship between two sets of data. We have learned how to compute r, the Pearson correlation coefficient, which indicates how closely the points on a scattergram of the data come to lying in a straight line. We have used this "line of best fit" or regression line to estimate the value of one of the variables corresponding to a known value of the other variable.

A word of caution is in order here. The correlation coefficient is often misused by people who overlook the fact that *r measures only the strength of a linear relationship between two variables—it does not necessarily imply a cause-effect relationship.* For example, there is a high correlation between teachers' salaries and the average price of automobiles through the years. This does not mean that the price of automobiles has gone up because teachers' salaries have increased; instead it is due to the fact that both variables, teachers' salaries and the cost of automobiles, are dependent upon a third variable—the overall standard of living.

Exercise Set 8.6

1. State in each case whether you would expect to obtain a positive correlation, a negative correlation, or no correlation.
   (a) The number of hours a bowler practices and his average score
   (b) The thickness of the tread on a tire and the number of miles it has been driven
   (c) Amount of snow and the number of traffic accidents
   (d) A person's height and his income
   (e) The amount of alcohol consumed by drivers and the number of traffic accidents
   (f) Beef production and the cost of beef in the market

2. The length of time y, in hours, for a headache to disappear after the patient has been given different dosages x, in grains, of aspirin is:

Dosage x	1	2	3	4	5
Time y	1.5	1.3	1.4	1.1	1.0

   (a) Plot a scattergram.
   (b) Compute r. Use $\sqrt{50} = 7.07$ and $\sqrt{0.86} = 0.93$.

3. A random sample from a population of adult males yields the following information concerning years of schooling and annual income in thousands of dollars:

Schooling x	9	9	9	10	10	11	12	12	13	14	16	16
Salary y	3	4	10	7	5	8	9	6	15	12	15	13

   (a) Plot a scattergram.
   (b) Compute r. Use $\sqrt{776} = 27.86$ and $\sqrt{2267} = 47.61$.

4. A sample of 10 families in a community yields the following data:

Total income x in thousands of dollars	3	4	5	6	7	7	8	8	10	11
Percent of income spent on food	40	35	33	32	30	27	25	26	23	22

   The scattergram for these data is shown in Figure 8–28. To the scattergram has been added the regression line for the data. Use the regression line to estimate the percent of income spent on food by the average family whose annual income is:

(a) $9000

(b) $5500

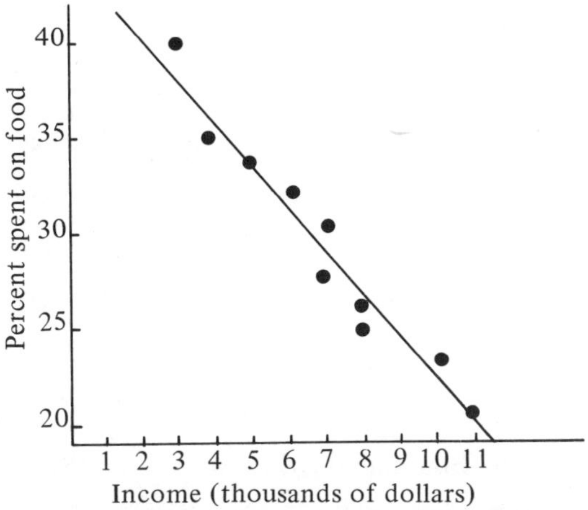

FIGURE 8–28 The regression line fits the scattergram with a correlation of $r = -0.97$.

5. The following data have been collected from a random sample of 10 fathers and their adult sons:

Height x of father in inches	72	65	62	64	69	61	71	65	69	68
Height y of son in inches	73	68	66	68	70	69	66	69	67	71

The scattergram and regression line for these data are shown in Figure 8–29. Use the regression line to estimate the average heights of the sons if the fathers are:

(a) 63 inches tall

(b) 67 inches tall

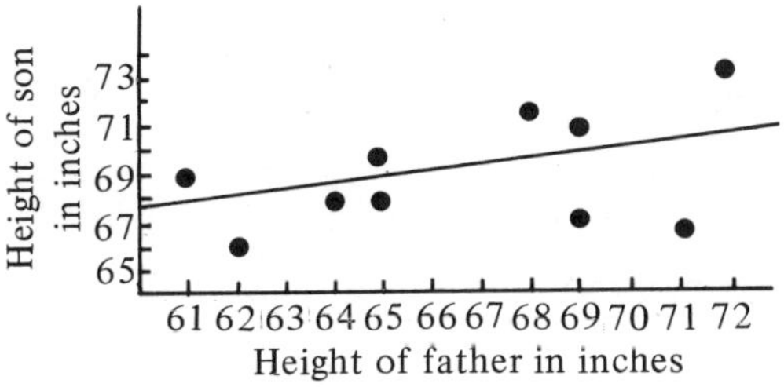

FIGURE 8–29 The regression line fits the scattergram with a correlation of 0.36.

6. Which of the two regression lines, the one in Figure 8–28 or the one in Figure 8–29, will be likely to give the more reliable predictions? Why?

SUMMARY

Terms you should know:

1. Data
2. Statistic
3. Score
4. Population
5. Sample
6. Random sample
7. Frequency table
8. Distribution
9. Histogram
10. Mode
11. Median
12. Mean
13. Dispersion
14. Standard deviation
15. Pascal's triangle
16. Binomial distribution
17. Normal distribution
18. Standard normal curve
19. z score
20. Motivated hypothesis
21. Null hypothesis
22. Hypothesis test
23. Significance level
24. Correlation coefficient
25. Scattergram
26. Linear correlation
27. Regression line

Formulas you should remember:

1. $\bar{x} = \dfrac{\sum x}{n}$ *(Mean)*

2. $s = \sqrt{\dfrac{\sum (x - \bar{x})^2}{n}}$ *(Standard deviation)*

3. $\mu = np$ *(Mean of binomial distribution)*

4. $\sigma = \sqrt{np(1 - p)}$ *(Standard deviation of binomial distribution)*

5. $z = \dfrac{x - \mu}{\sigma}$ *(z score)*

6. $r = \dfrac{n \sum xy - (\sum x)(\sum y)}{\sqrt{n \sum x^2 - (\sum x)^2}\,\sqrt{n \sum y^2 - (\sum y)^2}}$ *(Pearson correlation coefficient)*

REVIEW EXERCISES

1. Following are the speeds, in miles per hour, in which 20 cars were timed on a freeway. Make a frequency table and find the mean, median, and

mode of the distribution. Which of these best represents the speeds observed?

70	35	80	65	65
65	50	80	70	85
60	60	65	50	75
75	70	65	65	75

2. In 14 rounds of golf, a professional scored

   72, 69, 71, 76, 71, 70, 68, 73, 79, 70, 69, 71, 72, 71

   Find the mean, median, and mode of these scores. Which score is most representative of the professional's game?

3. An experiment is to roll a die six times. The experiment is repeated 36 times and the number of times that a 3 appears in each experiment is recorded. The results are

Number of 3's	Frequency
6	1
5	3
4	2
3	5
2	4
1	12
0	9
Total	36

   Draw a histogram for this data.

4. Consider the following grades on a mathematics test.

   75, 80, 81, 82, 83, 85, 85, 86, 86, 89, 90, 91, 90, 96

   Find the mean and the standard deviation for this distribution. How many scores are within one standard deviation of the mean?

5. In a study of transportation in a certain city it was found that the mean number of cars owned per family was 2 and the median number owned was 3. Which of the following is true?
   (a) Exactly half of the families own 2 or more cars
   (b) More than half the families own 2 or more cars
   (c) Less than half of the families own 2 or more cars

6. When a certain type of thumbtack is tossed in the air, it lands point up $\frac{1}{5}$ of the time. Three of these tacks are tossed. Find the probability that:

   (a) All land point up

   (b) All land point down

   (c) At least one lands point up

7. In a certain city, incompatibility is given as the legal reason for 70 percent of all divorces. Find the probability that three of five divorce cases are claimed to be due to incompatibility.

8. The burning time of an experimental rocket is normally distributed with $\mu = 4.26$ seconds and $\sigma = 0.04$ second. What is the probability that this kind of rocket will burn:

   (a) More than 4.20 seconds

   (b) Less than 4.20 seconds

   (c) From 4.20 to 4.32 seconds

9. The IQ's of the 5000 employees of a large corporation are normally distributed with a mean of 103 and a standard deviation of 15. If it is known that a certain job requires a minimum IQ of 90 and bores those with an IQ over 110, how many of the employees should be suitable for the job on the basis of IQ alone?

10. A multiple choice test consists of 50 questions, each with four possible answers. A student answers the test by flipping two fair coins and checks answer 1, 2, 3, or 4 on each question depending on whether he gets *HH*, *HT*, *TH*, or *TT*. Using the standard normal curve to approximate the binomial distribution for this problem, find the probability that he gets:

    (a) More than 15 correct answers

    (b) At least 20 correct answers

    (c) Fewer than 10 correct answers

    (d) More than 8 but fewer than 12 correct answers

11. A television network claims that its Monday night movie regularly has 30 percent of the total viewing audience. If the claim is correct, what is the probability that among 400 people who are watching TV on a particular Monday night more than 150 are watching the network movie?

12. A cancer research group wished to test the hypothesis that more than 40 percent of all college students smoke cigarettes. They sampled 500 students and found that 218 of them smoked cigarettes. Is the group's claim that more than 40 percent of the students smoke cigarettes justified at the 5 percent level?

13. To test the hypothesis that right-handed people put on their left shoe first, 20 right-handed persons were observed and it was found that 16 of them put on their left shoe first and that the other 4 consistently put on their right shoe first. At what significance level should one reject the hypothesis that the probability that a right-handed person puts on his left shoe first is $\frac{1}{2}$?

14. A meteorologist is considering the possibility of using the data on snowfall for January in predicting the snowfall for February. He finds that the snowfall for each of these months for the past 10 years has been:

Snowfall in Jan. in inches	4.0	5.0	4.5	6.0	6.6	5.0	10.0	4.2	8.0	2.0
Snowfall in Feb. in inches	3.0	3.5	3.0	4.0	4.0	3.1	5.5	3.0	5.0	2.1

(a) Plot a scattergram.

(b) Compute r.

(c) Do you think the meteorologist will be able to make reliable predictions from the data he has collected? Explain.

9 LINEAR PROGRAMMING

Linear programming is one of the newest branches of mathematics. It was first developed during World War II to help military strategists make the best use of existing weapons, manpower, and equipment. It was quickly realized that the same methods could apply to business and industry, for in these areas, as in war, success depends upon making the best use of available equipment, space, time, and manpower.

Some of the kinds of problems which may be solved using the methods of linear programming are:

1. How can a company assign personnel so they will be of maximum value?
2. How can a company minimize the cost of shipping its product?
3. How can one meet certain nutritional requirements while minimizing costs?
4. How can a varying sales demand be met with a minimum of inventory?

Unfortunately, in actual business or manufacturing situations the number of variables is usually so large that extensive clerical help or a computer is necessary to arrive at solutions. However, the simple examples discussed in this chapter should indicate some of the many possible problems that can be solved using linear programming.

*9.1 A LINEAR PROGRAMMING PROBLEM—CONSTRAINTS

As an example of a rather simple problem, consider the following:

EXAMPLE 9.1 A couple has \$22 to spend on recreation. They decide to spend the money on movies and concerts, by attending either a movie or a concert each evening until the money is spent. It costs them \$4 to attend a movie and \$2 to attend a concert. They decide to attend at least one more movie than concert. There are only 5 movies and only 6 concerts for them to attend. Assuming that they go at most once to any given movie or concert, how can they plan their attendance in order to have the maximum number of evenings out?

Solution If we let x represent the number of movies and y represent the number of concerts attended by the couple, then the cost of the x movies and y concerts is $4x + 2y$. Since the couple can spend no more than \$22, we write

$$4x + 2y \leq 22$$

in which the symbol $\leq$ is read "is less than or equal to."

The couple will attend *at least* one more movie than they do concerts. If they attend *only* one more movie than concerts, the difference between the number of concerts and the number of movies will be 1, that is,

$$x - y = 1$$

However, if the number of movies attended is greater than one more than the number of concerts attended, the difference between the number of movies and the number of concerts will be a number larger than 1, that is,

$$x - y > 1$$

($>$ is read "greater than.") We combine these two possibilities by writing

$$x - y \geq 1$$

There are only 5 movies and 6 concerts, and therefore

$$x \leq 5 \qquad \text{and} \qquad y \leq 6$$

We may assume that it is not possible to attend a negative number of either movies or concerts; thus we write

$$x \geq 0 \qquad \text{and} \qquad y \geq 0$$

These six relations

$$\begin{aligned} 4x + 2y &\leq 22 \\ x - y &\geq 1 \\ x &\leq 5 \\ x &\geq 0 \\ y &\leq 6 \\ y &\geq 0 \end{aligned}$$

are **constraints** placed on the variables by the problem.

The number n of evenings of entertainment possible is given by the equation

$$n = x + y$$

This equation is called the **objective function**. We wish to make n as large as possible and at the same time satisfy the six constraints.

Because it is not likely that the couple can buy a fraction of an admission to either a movie or a concert, we are safe in assuming that x and y must be integers. Thus the constraints $x \geq 0$ and $x \leq 5$ tell us that x must be one of the integers in the set $X = \{0, 1, 2, 3, 4, 5\}$, and from the constraints $y \geq 0$ and $y \leq 6$ we conclude that y must be one of the numbers in $Y = \{0, 1, 2, 3, 4, 5, 6\}$. Since the problem asks that we find a pair of numbers, one to take the place of x and the other to take the place of y in the objective function, the possible choices available to the couple can be generated by forming the cross product set $X \times Y$.

$$X \times Y = \left\{\begin{matrix} (0, 6) & (1, 6) & (2, 6) & (3, 6) & (4, 6) & (5, 6) \\ (0, 5) & (1, 5) & (2, 5) & (3, 5) & (4, 5) & (5, 5) \\ (0, 4) & (1, 4) & (2, 4) & (3, 4) & (4, 4) & (5, 4) \\ (0, 3) & (1, 3) & (2, 3) & (3, 3) & (4, 3) & (5, 3) \\ (0, 2) & (1, 2) & (2, 2) & (3, 2) & (4, 2) & (5, 2) \\ (0, 1) & (1, 1) & (2, 1) & (3, 1) & (4, 1) & (5, 1) \\ (0, 0) & (1, 0) & (2, 0) & (3, 0) & (4, 0) & (5, 0) \end{matrix}\right\}$$

The pair (x, y) of numbers we are looking for is one of the elements of $X \times Y$, but that pair must also satisfy the constraints

$$\begin{aligned} 4x + 2y &\leq 22 \\ x - y &\geq 1 \end{aligned}$$

and not all the pairs in $X \times Y$ will do this.

The constraint

$$x - y \geq 1$$

requires that x be at least one integral value larger than y. Thus we look at each pair of $X \times Y$ and delete every pair in which the first component x is either equal to or smaller than the second component y. When this is done, the set $X \times Y$ has been reduced to this array:

				(5, 4)
			(4, 3)	(5, 3)
		(3, 2)	(4, 2)	(5, 2)
	(2, 1)	(3, 1)	(4, 1)	(5, 1)
(1, 0)	(2, 0)	(3, 0)	(4, 0)	(5, 0)

The constraint

$$4x + 2y \leq 22$$

requires that the sum of four times the first component added to two times the second component is a number less than or equal to 22. We inspect each of the remaining pairs and delete all those which do not satisfy this condition. When this is done, the remaining pairs

			(4, 3)	
		(3, 2)	(4, 2)	
	(2, 1)	(3, 1)	(4, 1)	(5, 1)
(1, 0)	(2, 0)	(3, 0)	(4, 0)	(5, 0)

are the only ones left of $X \times Y$ which satisfy all the constraints on x and y.

We originally set out to find a pair (x, y) of numbers that would make the objective function

$$n = x + y$$

maximum and at the same time satisfy the constraints imposed upon x and y by the problem. That pair must be one of those in the above set. By adding the two numbers in each pair, the maximum is found to be with the pair (4, 3). Therefore we conclude that the couple can use their \$22 to finance 7 evenings out on the town by attending 4 movies and 3 concerts.

If x and y are not restricted to integral values, the method we have employed in Example 9.1 cannot be used, for in that case $X \times Y$ cannot be listed. Even when x and y have integral values, it may not be desirable to list $X \times Y$. Consider the situation in which the couple of our problem has \$100 to spend and live in a city with a large offering of movies and concerts. In that case $x \leq 25$ and $y \leq 50$, $X \times Y$ could contain as many as 1250 pairs, a most formidable number.

To deal with most linear programming problems, we need a more sophisticated line of attack than the "brute force" method we have used in this example. We will discuss one such method after we have reviewed some necessary algebraic concepts.

Whatever method is used, there are three preliminary steps which are the same. They are:

1. Identify the variables of the problem.
2. Identify the constraints on these variables.
3. Identify the objective function and decide if it is to be minimized or maximized.

EXAMPLE 9.2 A dietician must prepare a mixture of two foods A and B. Each unit of food A contains 10 grams of protein and 20 grams of carbohydrate and costs 15¢. Each unit of food B contains 20 grams of protein and 25 grams of carbohydrate and costs 20¢. What is the cheapest mixture that can be obtained if it must contain at least 500 grams of protein and 700 grams of carbohydrate? Write the constraints and the objective function for this problem. Is the objective function to be maximized or minimized?

Solution There are two variables in this problem, the number of units of food A and the number of units of food B. Represent the number of units of food A used by x and the number of units of food B used by y. Since food A contains 10 grams of protein per unit and food B contains 20 grams of protein per unit, the number of grams of protein in the mix is

$$10x + 20y$$

There must be at least 500 grams of protein in the mix, and so we write

$$10x + 20y \geq 500$$

Similarly, since there must be at least 700 grams of carbohydrate in the mix and each unit of food A supplies 20 of these and each unit of food B supplies 25, we write

$$20x + 25y \geq 700$$

We may assume that neither x nor y can be negative although one of them could possibly be 0; therefore we write

$$x \geq 0 \qquad y \geq 0$$

The constraints are then

$$\begin{aligned} 10x + 20y &\geq 500 \\ 20x + 25y &\geq 700 \\ x &\geq 0 \\ y &\geq 0 \end{aligned}$$

Since x units of food A are used and each unit costs 15¢, the cost of food A in the mix is $(15x)$¢. Similarly y units of food B will cost $(20y)$¢. Therefore the objective function is

$$\text{Cost} = 15x + 20y$$

and this function is to be minimized.

Exercise Set 9.1

Give the constraints and objective function for each of the following linear programming problems. In each case tell whether the objective function is to be maximized or minimized. Do not try to complete the problem at this time, however; you should save your work so that you may refer to it later.

1. A company produces vitamin A and vitamin B. The profit on vitamin A is \$150 a pound, and the profit on vitamin B is \$200 a pound. Vitamin A takes 5 hours per pound to manufacture and 3 hours per pound to package. Vitamin B takes 6 hours per pound to manufacture and 2 hours per pound to package. Fifty hours of manufacturing time is available, but only 30 hours of packaging time is available. How many pounds of vitamins A and B should be manufactured in order to maximize the profits?

2. A dealer wishes to stock at most 100 refrigerators and stoves. A refrigerator weighs 200 pounds and a stove 100 pounds, and he is limited to a total weight of 12,000 pounds. If he makes a profit of \$35 per refrigerator and \$20 per stove, how many of each should he stock for maximum profit?

3. A special diet requires a daily intake of at least 10 ounces of protein and 14 ounces of carbohydrates. The protein and carbohydrate are to come from two grain mixtures. Mixture *A* contains 3 ounces of protein and 6 ounces of carbohydrates per pound and costs 80¢ per pound. Mixture *B* contains 1 ounce of protein and 5 ounces of carbohydrate per pound and costs 60¢ per pound. How much of each mixture should be purchased in order to satisfy the dietary requirements at a minimum cost?

4. A dairy can deliver 250 gallons of homogenized or low fat milk. The dairy must deliver 150 gallons of homogenized milk and 80 gallons of low fat milk to satisfy certain regular customers. How many gallons of each should be delivered to maximize the profit, if the profit from a gallon of homogenized milk is 10¢ and the profit from a gallon of low fat milk is 20¢?

5. Suppose that the dealer in Problem 2 knows that he cannot sell more than 50 stoves. How many stoves and refrigerators should he stock for maximum profit?

*9.2 LINEAR RELATIONS

9.2.1 Open Sentences

The constraints on the variables of a linear programming problem are always stated in the form of open sentences. A sentence is an **open sentence** if it contains one or more variables and if specific numerical values must be given to the variables before the sentence can be judged as true or false. Examples of open sentences are:

$$3x + 1 = 7$$
$$x - 4 > 5$$
$$x = y$$

The first sentence is true if x is replaced by 2 but false if x is replaced by any other number. The second sentence is true if x is replaced by any number larger than 9, but it is false if x is replaced by 9 or by any number smaller than 9. The third sentence is true when x and y are given the same values, but it is false if x and y are assigned different values.

The first and third of the above sentences are called **equations**; the second sentence is an **inequality**. The constraints of a linear programming problem may be equations or inequalities; in the problems we shall consider they will always be inequalities.

Any value or values for the variables in an open sentence that make the sentence true when they replace the variables are called **solutions** of the sentence.

The solutions of a sentence are said to **satisfy** the sentence. The set of all solutions of a sentence is the **solution set** of the sentence.

The simplest open sentences in algebra are **linear equations** in one or two variables. A linear equation is one in which the variables appear to the first power only.

If a linear equation contains only one variable, then its single solution can be found by using the rules for forming equivalent equations:

Two equations are **equivalent** (have the same solution set) if one of them is formed by one or more of the following:

1. Adding the same number to both members of the other equation.
2. Multiplying both members of the other equation by any nonzero number.
3. Replacing one or more quantities by equal quantities.

EXAMPLE 9.3 Find the solution set for the equation

$$3x - 1 = 5$$

Solution

(1)	$3x - 1 = 5$	(*Given equation*)
(2)	$3x - 1 + 1 = 5 + 1$	(*Add 1 to each member of (1)*)
(3)	$3x = 6$	(*Replace each member of (2) by equivalent expressions*)
(4)	$\frac{1}{3}(3x) = \frac{1}{3}(6)$	(*Multiply each member of (3) by $\frac{1}{3}$*)
(5)	$x = 2$	(*Replace each member of (4) by equivalent expressions*)

When a linear equation contains two variables, there are infinitely many pairs of numbers in its solution set.

EXAMPLE 9.4 Find five solutions for

$$x + y = 5$$

Solution We wish to find two numbers such that their sum is 5. Five such pairs are listed in Table 9–1.

TABLE 9–1

x	-1	1	2	3	5
y	6	4	3	2	0

9.2.2 Graphs of Linear Equations

The simplest way to examine the solution set of an equation in two variables is to look at its geometrical model, which is a graph. To do this we make use of the concept of a **rectangular Cartesian coordinate system**.

We are already familiar with rulers, yardsticks, and metersticks. As you know, the numerals on such a measuring device indicate the positions of points or the measure of the distances of these points from the left end of the device. For instance, the numeral 3 on a ruler indicates a point which is 3 inches from the left end of the ruler. The mathematical idealization of a ruler is called a **number line**, a version of which is shown in Figure 9–1. Zero marks the beginning point or **origin**. The numerals 1, 2, 3,. . . are placed in order at marked points spaced at equal intervals proceeding to the right. The numerals $-1, -2, -3, \ldots$ are placed in order at marked points spaced at equal intervals to the left of the origin. The fact that the numbers increase toward the right is arbitrary but reflects a usual convention. The length of each interval is also arbitrary but remains the same in any one situation.

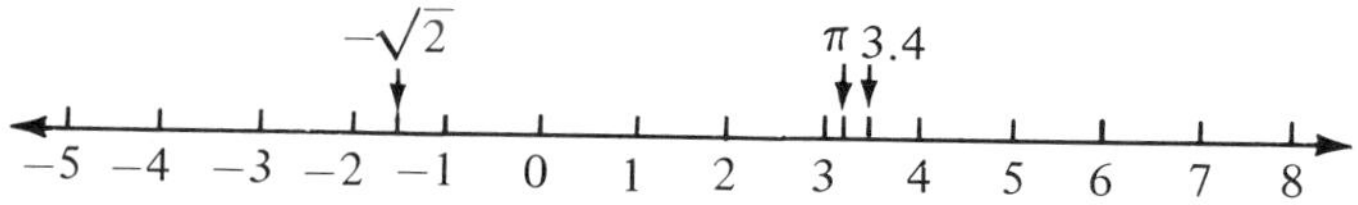

FIGURE 9–1 A number line.

We may locate points corresponding to rational nonintegral numbers by subdividing the intervals of the number line. To locate the point which is 3.4 units from the origin, for instance, we might divide the interval between the points labeled 3 and 4 by marking off nine equally spaced points in that interval. The endpoint of the fourth subinterval, counting from the 3, corresponds to 3.4. The location of points corresponding to the irrational numbers is more complicated, and we shall not be considering them here. The approximate location of an irrational number can be found, however, by using the first few decimals of the decimal approximation of the number. Thus π is located about 3.14 units from the origin between the points labeled 3 and 4, and $-\sqrt{2}$ is located about 1.4 units from the origin between the points labeled -1 and -2.

The most important fact about the representation of the real numbers on a number line is that there is a one-to-one correspondence between the real numbers and the points on a line. This correspondence is one which preserves the order relation; that is, if a, b, and c are any real numbers with $a < b < c$, then the points on the number line which correspond to a, b, and c are such that the point corresponding to b is between the points corresponding to a and c.

To construct a rectangular coordinate system, place two number lines at right angles with their origins coinciding. The number lines are called **coordinate axes**, and their intersection is the **origin** of the system. A rectangular

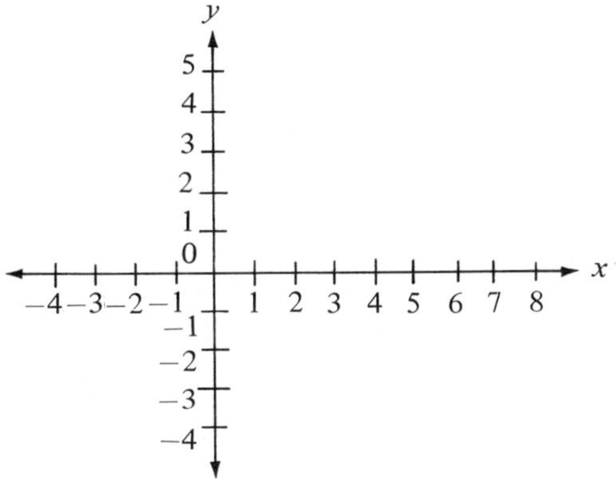

FIGURE 9–2 A rectangular coordinate system.

coordinate system is shown in Figure 9–2. If the variables in the open sentence under consideration are x and y, it is the usual practice to designate the horizontal axis as the **x axis** and the vertical axis as the **y axis**.

When a rectangular coordinate system has been established in a plane, it is possible to identify every point of the plane by means of an ordered pair of numbers called the **coordinates** of the point. The first number of a pair is the coordinate on the x axis at the point of intersection of that axis with a perpendicular from the given point. For this reason the first component of an ordered pair of numbers identifying a particular point is called the **x coordinate** of that point. The second number in the ordered pair of coordinates of a point is the number given by the intersection of the y axis with a line perpendicular to that axis through the given point. The second coordinate of a point is called its **y coordinate**.

EXAMPLE 9.4 Give the coordinates of the points A, B, C, and D in Figure 9–3.

Solution The coordinates are

$$\begin{aligned} A&: (2, 3) \\ B&: (-4, 1) \\ C&: (1, 0) \\ D&: (-5, -2) \end{aligned}$$

Since solutions to open sentences with two variables are in the form of pairs of numbers, it is natural to use the ordered-pair notation for such solutions. If the variables of the sentence are x and y, then a solution takes the form (x, y) where the actual numbers in the pair are the replacements for x and y which make the sentence true.

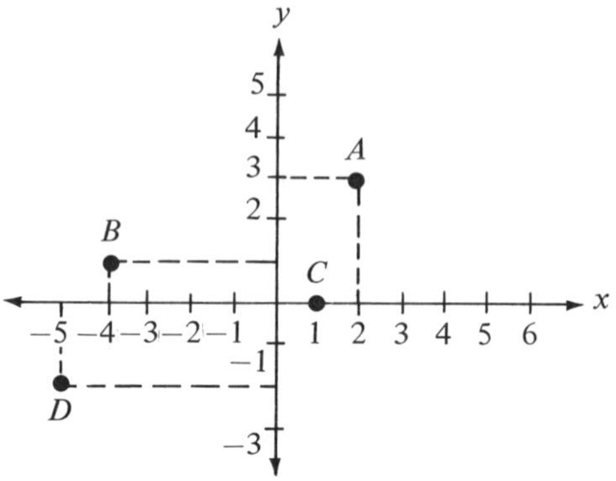

FIGURE 9–3

EXAMPLE 9.5 Find three of the solutions for the equation

$$3x + 4y = 12$$

Solution To find one solution, we pick any number as a replacement for one of the variables and after making that replacement, solve the resulting equation which now has only one variable. Thus, if we choose to replace x with 2, we have

$$\begin{aligned} 3(2) + 4y &= 12 \\ 6 + 4y &= 12 \\ 6 + 4y - 6 &= 12 - 6 \\ 4y &= 6 \\ \frac{4y}{4} &= \frac{6}{4} \end{aligned}$$

$$y = \tfrac{3}{2} \qquad \text{or} \qquad y = 1.5$$

Then $x = 2$ and $y = 1.5$ satisfies the original equation. We write this solution in ordered-pair form as

$$(2, 1.5)$$

Similarly, by selecting $x = 0$, we have

$$\begin{aligned} 3(0) + 4y &= 12 \\ 4y &= 12 \\ y &= 3 \end{aligned}$$

And (0, 3) is a solution also. Again, if we set $y = 0$, we have

$$\begin{aligned} 3x + 4(0) &= 12 \\ 3x &= 12 \\ x &= 4 \end{aligned}$$

Therefore three of the solution pairs of the equation

$$3x + 4y = 12$$

are

$$(2, 1.5) \qquad (0, 3) \qquad (4, 0)$$

The **graph** of an equation in the two variables x and y is the set of all points in the plane which can be put into a one-to-one correspondence with the ordered pairs of the solution set of the equation. The graph of a linear equation is always a straight line; this is the reason that the equation is called linear. A basic axiom of geometry is that two points determine a straight line; thus one need only find two solutions of a linear equation in order to draw its graph. It is a good idea, however, to find three solutions, two to establish the line and the third one as a check on the computations.

EXAMPLE 9.6 Graph the line which represents the equation

$$3x + 4y = 12$$

Solution In Example 9.5, we found three solutions of this equation. These are

$$(2, 1.5) \qquad (0, 3) \qquad (4, 0)$$

The points corresponding to these solutions are shown plotted on the coordinate axes in Figure 9–4. The line through these three points is the graph of the equation

$$3x + 4y = 12$$

EXAMPLE 9.7 Draw the graph of the linear equation

$$y = 2$$

Solution We may consider the equation $y = 2$ to be a linear equation in two variables x and y with the form

$$0x + y = 2$$

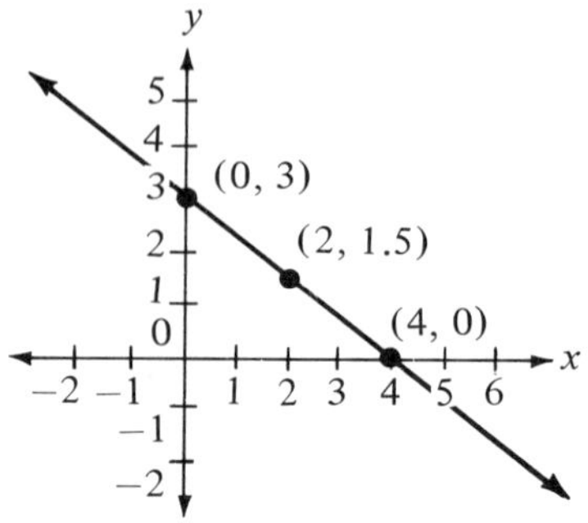

FIGURE 9–4 Graph of $3x + 4y = 12$.

The solution set of this equation is the set of ordered pairs

$$(a, 2)$$

in which a can be replaced by any real number. Three of these pairs are $(1, 2)$, $(0, 2)$, and $(-2, 2)$. Each of these graph as a point on a horizontal line two units above the x axis. That line is the required graph. It is shown in Figure 9–5.

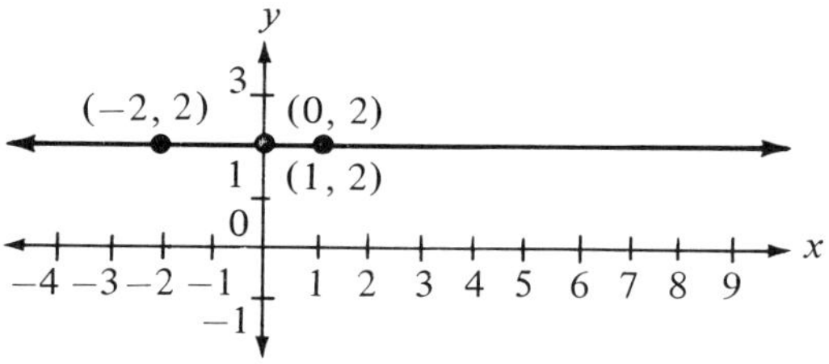

FIGURE 9–5 Graph of $y = 2$.

Exercise Set 9.2

1. Set up a rectangular coordinate system. On it graph the points whose coordinates are:

A:(3, 2)	C:(0, 4)	E:(−1, 0)	G:(2, −3)
B:(2, 3)	D:(3, 0)	F:(0, −4)	

2. Draw the graph of each of the following open sentences. Make a different coordinate system for each graph.

(a) $x = 3$	(f) $y = \frac{1}{4}x$
(b) $y = 4$	(g) $x + 2y = 6$
(c) $x + y = 4$	(h) $3x - y = 6$
(d) $x - y = 4$	(i) $5x + 4y = 20$
(e) $y = 2x$	(j) $3x - 5y = 15$

3. In each of the following problems two linear equations are given. For each pair set up a rectangular coordinate system, and graph both of the lines on it. Find the coordinates of the point in which the two lines intersect.

(a) $x = 3$, $y = 4$ (b) $x = y$, $x + y = 4$ (c) $2x + 3y = 5$, $3x - y = 2$

9.3 GRAPHS OF CONVEX REGIONS

9.3.1 Half-Planes

Because the constraints of a linear programming problem are often stated in the form of linear inequalities, we shall learn to graph that kind of open sentence.

The graph of the equation

$$2x + 3y = 6$$

is the line in Figure 9–6. That line separates the plane into two regions called **half-planes**. The line is said to be the **edge** of each of the half-planes.

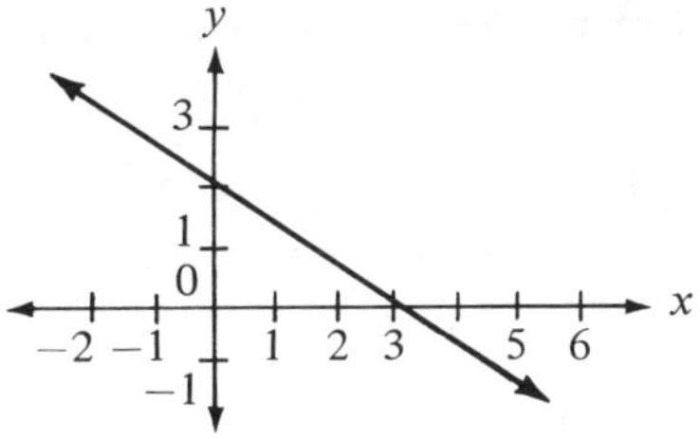

FIGURE 9–6 Graph of $2x + 3y = 6$.

It can be shown, although we shall not do so here, that one of the half-planes with edge

$$2x + 3y = 6$$

is the graph of the inequality

$$2x + 3y > 6$$

The other half-plane with edge $2x + 3y = 6$ is the graph of

$$2x + 3y < 6$$

It can be shown that the determination of the proper half-plane can be done in this way: Select any point in one of the half-planes. If its coordinates satisfy the inequality, then the half-plane which contains that point is the graph of the inequality; but if the coordinates of the point do not satisfy the inequality, then the other half-plane is the desired set of points. In the case of

$$2x + 3y > 6$$

the coordinates (2, 2) satisfy the inequality, and so the half-plane containing the point with those coordinates is the graph of

$$2x + 3y > 6$$

This graph is the shaded region in Figure 9–7. The dashed edge of the half-plane indicates that no point on that line has coordinates which satisfy the inequality.

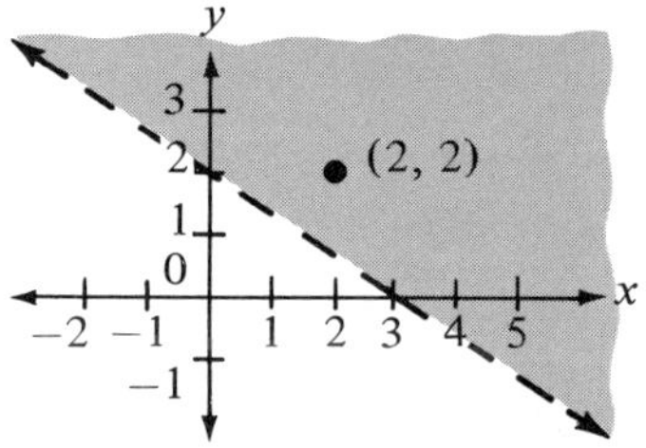

FIGURE 9–7 Graph of $2x + 3y > 6$. Since the point with coordinates $(2, \frac{2}{3})$ lies on the line $2x + 3y = 6$, and $3 > \frac{2}{3}$, the point (2, 3) is above the line.

The union of a half-plane with its edge is a set of points called a **closed half-plane**. The closed half-plane which is the union of the graphs of Figures 9–6 and 9–7 is the graph of

$$2x + 3y \geq 6$$

That graph is the one in Figure 9–8.

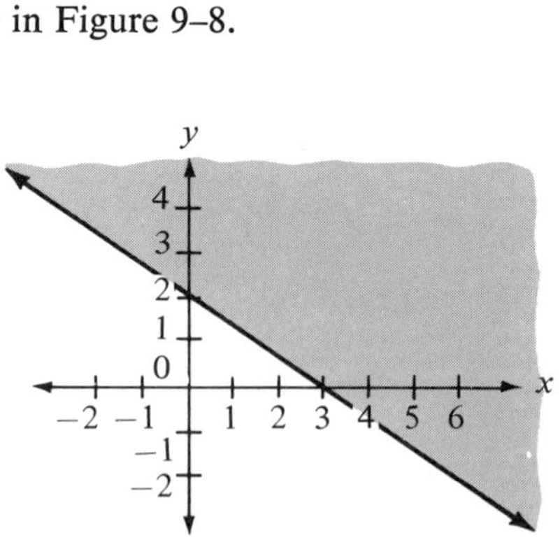

FIGURE 9–8 Graph of $2x + 3y \geq 6$.

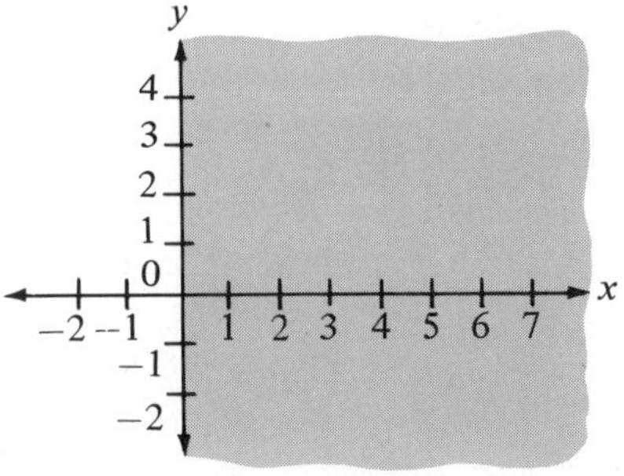

FIGURE 9–9 Graph of $x \geq 0$.

Usually there are several constraints on the variables of a linear programming problem. The intersection of the graphs of all the constraints contains every point whose coordinates simultaneously satisfy every one of the constraints.

EXAMPLE 9.8 Graph the region which is the intersection of the graphs of

$$\begin{aligned} x &\geq 0 \\ y &\geq 1 \\ x + y &\leq 4 \end{aligned}$$

Solution The graph of

$$x = 0$$

is the y axis, and the graph of

$$x > 0$$

is the half-plane on the right of the y axis since any point in that half-plane has an x coordinate greater than zero.

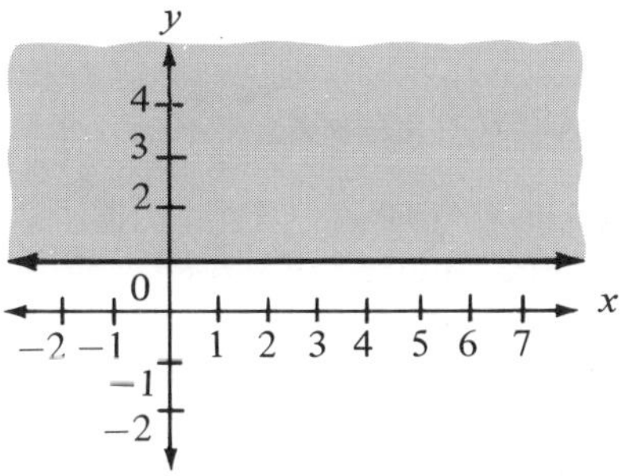

FIGURE 9–10 Graph of $y \geq 1$.

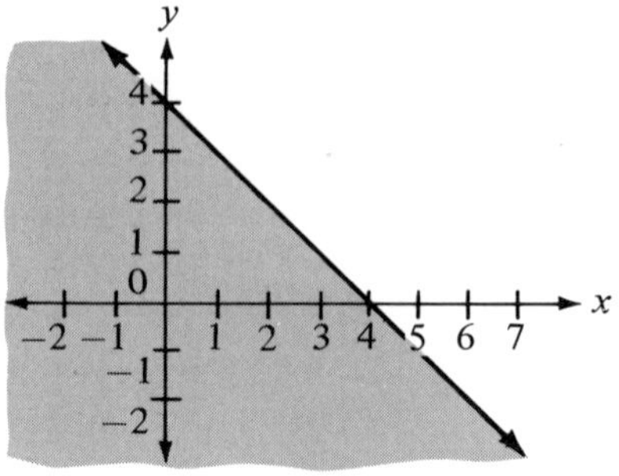

FIGURE 9–11 Graph of $x + y \leq 4$.

Therefore the graph of

$$x \geq 0$$

is the closed half-plane of Figure 9–9.

The graphs of $y \geq 1$ and $x + y \leq 4$ are the closed half-planes of Figures 9–10 and 9–11. The intersection of these three half-planes is shown in Figure 9–12.

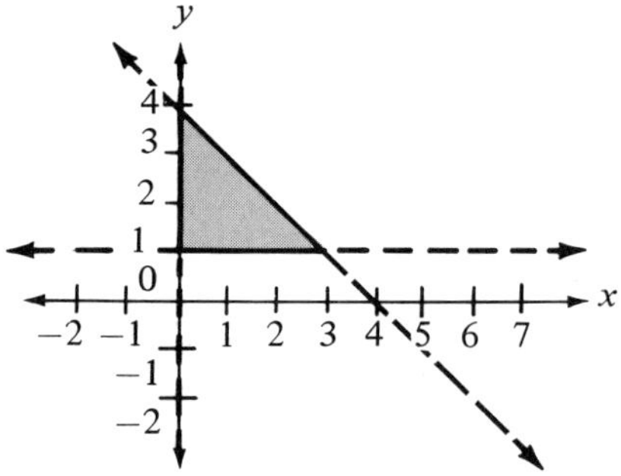

FIGURE 9–12 Convex polygonal region defined by $x \geq 0$, $y \geq 1$, $x + y \leq 4$.

9.3.2 Convex Polygonal Regions

The region graphed in Figure 9–12 is called a **convex polygonal region.** The intersection of a finite number of closed half-planes is always a convex polygonal region.

The line segments which are the boundaries of a convex polygonal region are called its **sides**, and the intersections of two sides is a **corner** of the polygonal region. A polygonal region may have a finite area as illustrated by the one in Figure 9–12, or it may have an infinite area as shown in Figure 9–13.

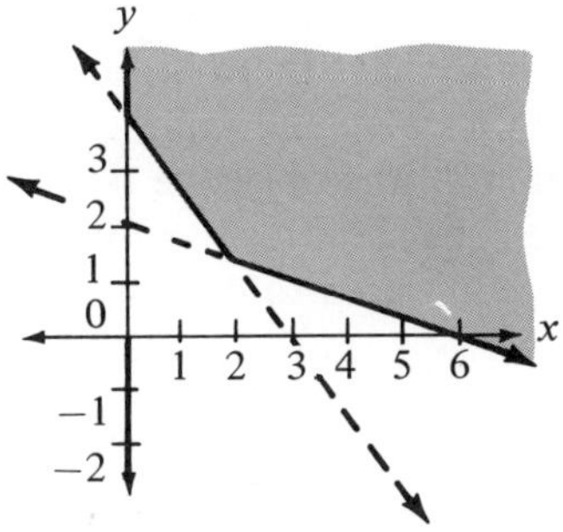

FIGURE 9–13 Polygonal region with infinite area. Region is defined by $x \geq 0$, $x + y \geq 3$, $x + 3y \geq 6$.

The intersection of the graphs of the constraints

$$\begin{aligned} 4x + 2y &\leq 22 \\ x - y &\geq 1 \\ x &\geq 0 \\ x &\leq 5 \\ y &\geq 0 \\ y &\leq 6 \end{aligned}$$

is the convex polygonal region of Figure 9–14. On that graph are identified the points whose coordinates are the pairs of integers we used earlier to solve the sample problem of Example 9.1. You are probably wondering why we bother to draw the graph of the constraints when it was really much easier to find the ordered pairs the way we did it before. Recall, however, that it was easy to find the pairs then because the problem required *only* integral values; if it had been possible for either x or y to take on nonintegral values, then it would have been impossible to list all possible ordered pairs. We can be sure though that all the real number replacements for both x and y are represented

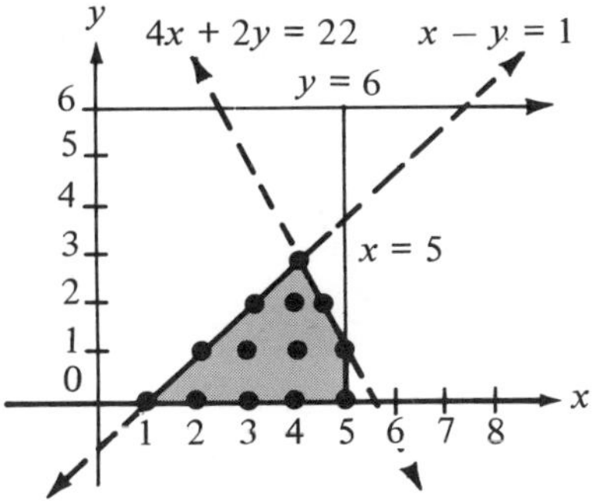

FIGURE 9–14 Region defined by $4x + 2y \leq 22$; $x - y \leq 1$; $x \geq 0$, $x \leq 5$; $y \geq 0$, $y \leq 6$.

as points on the sides or in the interior of the polygonal region in Figure 9–14. In the next section we shall see that there is an easy way to tell which of the infinite number of points in a polygonal region are likely to be the ones that will either maximize or minimize the objective function.

For referral purposes, the steps involved in graphing a closed half-plane are listed below. The inequality used to demonstrate the steps is

$$y - 2x \geq 1$$

However, the same steps should be used with any other inequality.

1. Find one solution of the equation

$$y - 2x = 1$$

by substituting an arbitrary value for one of the variables and then solving for the other variable. Thus if x is replaced by 1, we have

$$\begin{aligned} y - 2(1) &= 1 \\ y - 2 &= 1 \\ y &= 3 \end{aligned}$$

and one of the solutions of the equations is

$$(1, 3)$$

2. Repeat step 1 at least once obtaining a different solution each time. If y is replaced by 1, a second solution to the equation is

$$(0, 1)$$

3. Locate, on a rectangular coordinate system, the points whose coordinates are the solutions obtained in steps 1 and 2. Draw a straight line through these points. The line through the points (1, 3) and (0, 1) is shown in Figure 9–15.

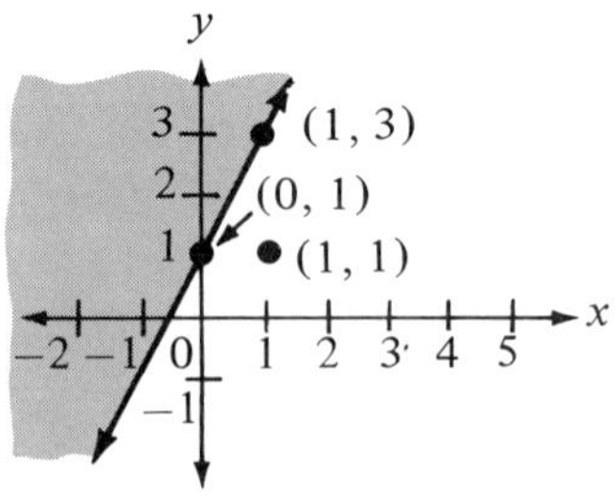

FIGURE 9–15 Graph of $y - 2x \geq 1$.

4. Select any point not on the line drawn in step 3. Substitute its coordinates for x and y in the inequality

$$y - 2x > 1$$

If the coordinates of the point make the inequality true, shade the half-plane containing the point. If the coordinates make the inequality false, shade the half-plane which does not contain the point. The point which was used to establish the half-plane shaded in Figure 9–15 was (1, 1). Since

$$1 - 2(1) > 1$$

or

$$-1 > 1$$

is not true, the half-plane opposite to the one which contains the point (1, 1) is the required set.

Exercise Set 9.3

In Problems 1 to 6, graph the solution set for each inequality.

1. $x \geq 1$
2. $y \leq 5$
3. $x + y \geq 4$
4. $x + y \leq 3$
5. $x \leq y$
6. $y \leq \frac{x}{2}$

In Problems 7 to 14, graph the polygonal region defined by the sets of inequalities. By inspecting the graphs, find the corner points of each region.

7. $x \geq 0$, $y \geq 0$
8. $x \geq 1$, $x \leq 4$
9. $x \geq 0$, $y \geq 0$, $x \leq 4$
10. $x \geq 2$, $y \leq 3$, $y \geq 1$
11. $x \leq 4$, $y \geq 1$, $x \leq y$
12. $y \leq 1$, $x \leq 4$, $y \leq x$
13. $x \geq 0$, $y \geq 0$, $x - 2y \leq 6$, $y - 2x \leq 10$
14. $x \geq 0$, $y \geq 0$, $x + 3y \geq 6$, $2x + y \geq 4$

9.4 THE GRAPHICAL METHOD

We now turn our attention back to the sample problem of Example 9.1. Recall that we were to find the maximum value the objective function

$$n = x + y$$

can attain subject to the restrictions on x and y imposed by the constraints

$$\begin{aligned} 4x + 2y &\leq 22 \\ x - y &\geq 1 \\ x &\geq 0 \\ x &\leq 5 \\ y &\geq 0 \\ y &\leq 6 \end{aligned}$$

Notice that the objective function is linear in x and y. Therefore we may use the following rule which we accept without proof:

If the objective function is a linear equation in two unknowns, then it attains its maximum and minimum values at corner points of the convex polygonal region described by its constraints.

The graph of the constraints of the problem of Example 9.1 is shown in Figure 9–16. The corner points of that polygon are (1, 0), (5, 0), (5, 1), and

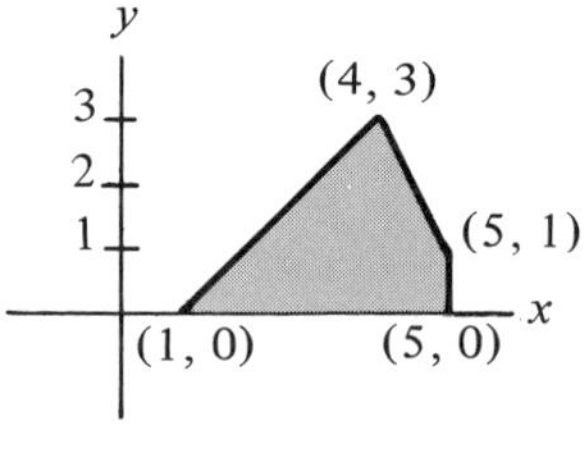

FIGURE 9–16 Graph of constraints of Example 9.1.

(4, 3). The rule stated above does not tell us which of these will make the objective function

$$n = x + y$$

maximum, but there are only four points and it is an easy matter to evaluate the object function $n = x + y$ at each of them. Thus

$$\begin{aligned} &\text{At } (1, 0): && n = 1 \\ &\text{At } (5, 0): && n = 5 \\ &\text{At } (5, 1): && n = 6 \\ &\text{At } (4, 3): && n = 7 \end{aligned}$$

Clearly, 7 is the greatest value n can have at any of these four points, and so 7 must be the answer to the problem of Example 9.1.

EXAMPLE 9.9 Solve the problem stated in Example 9.2.

Solution The constraints of that problem are

$$\begin{aligned} 10x + 20y &\geq 500 \\ 20x + 25y &\geq 700 \\ x &\geq 0 \\ y &\geq 0 \end{aligned}$$

The graph of the constraints is shown in Figure 9–17 on page 286. We are to minimize the objective function:

$$\text{Cost} = 15x + 20y$$

and so we evaluate the function at the corner points of the polygon. These are (0, 28), (50, 0), and (10, 20). The latter pair of coordinates was found by drawing parallels to the x and y axes through the corner and reading the coordinates from the intersections of these parallels with the axes.

$$\begin{aligned} &\text{At } (0, 28): && \text{Cost} = 15(0) + 20(28) \\ &&& \phantom{\text{Cost}} = 0 + 560 \\ &&& \phantom{\text{Cost}} = 560 \\ &\text{At } (10, 20): && \text{Cost} = 15(10) + 20(20) \\ &&& \phantom{\text{Cost}} = 150 + 400 \\ &&& \phantom{\text{Cost}} = 550 \\ &\text{At } (50, 0): && \text{Cost} = 15(50) + 20(0) \\ &&& \phantom{\text{Cost}} = 750 + 0 \\ &&& \phantom{\text{Cost}} = 750 \end{aligned}$$

The smallest of these numbers, 550¢ (or $5.50), is the minimum cost and it can be obtained by using 10 units of food A and 20 units of food B.

Clearly one's success in graphically solving a linear programming problem rests in one's ability to find the coordinates of the corners of a polygonal region. If these coordinates are integral and if the graphing is done accurately,

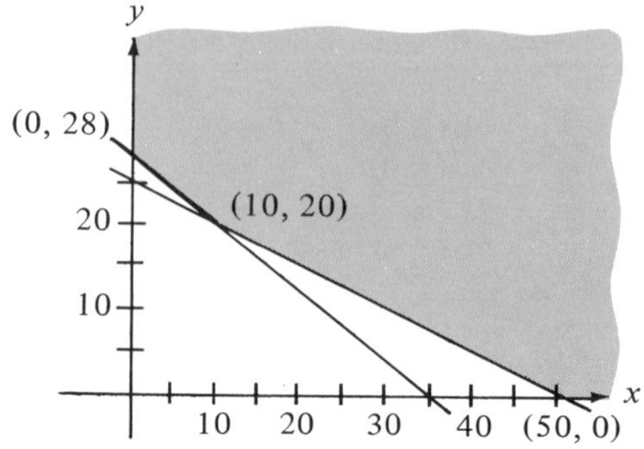

FIGURE 9–17 Region defined by $10x + 20y \geq 500$; $20x + 25y \geq 700$; $x \geq 0$, $y \geq 0$.

it is possible to read the coordinates directly from the graph. In practical problems, it is very unusual for the corners of a constraint polygon to have integral coordinates. In such cases we have two options; we may make a guess about the value of the coordinates by looking at the graph or we may find the exact values using algebraic methods to solve each pair of intersecting plane edges in the form of equations. For example, one of the corners of the region in Figure 9–17, the one at (10, 20), can be found algebraically by finding the solution to the system

$$\begin{aligned} 10x + 20y &= 500 \\ 20x + 25y &= 700 \end{aligned}$$

We learned one method for solving systems in Chapter 6; others can be found in any elementary algebra textbook. We shall not discuss the algebraic methods in this chapter, but instead we will use the graphical method and merely approximate any coordinates which are not integers.

EXAMPLE 9.10 The Sweet Tooth Candy Company makes two kinds of candy bars, Chewies and Gooies. Processing involves three stages: mixing, cooking, and packaging. Machinery for mixing is available for mixing 4 hours per day, cooking equipment is available 5.5 hours per day, and packaging equipment is available 2.5 hours per day. The average time per box, in hours, needed at each stage of the production is shown in the following table.

Average Processing Time in Hours

Candy	Mixing	Cooking	Packaging
Chewies	0.04	0.06	0.02
Gooies	0.05	0.05	0.04

If the profit per box of Chewies is \$0.60 and the profit per box of Gooies is \$0.40, how many boxes of each type of candy bar should the company make per day for a maximum profit?

Solution Let x represent the number of boxes of Chewies and let y represent the number of boxes of Gooies. Then

$$\begin{aligned} x &\geq 0 \\ y &\geq 0 \\ 0.04x + 0.05y &\leq 4 \\ 0.06x + 0.05y &\leq 5.5 \\ 0.02x + 0.04y &\leq 2.5 \end{aligned}$$

are the constraints.

The graph of the intersection of the solution sets of the constraints is the polygonal region in Figure 9–18. The corners of this polygonal region are the points identified on the graph as A, B, C, and D. Note that the coordinates of point B are only approximately the correct ones.

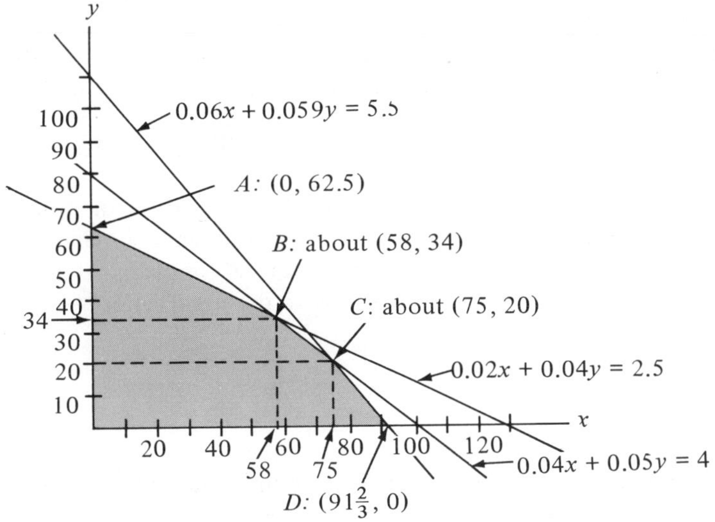

FIGURE 9–18 Polygonal region defined by $x \geq 0$, $y \geq 0$; $0.04x + 0.05y \leq 4$; $0.06x + 0.05y \leq 5.5$; $0.02x + 0.04y \leq 2.5$.

The objective function is

$$\text{Profit} = \$0.60x + \$0.40y$$

We evaluate this function at the four corners of the polygon.

$$\begin{aligned}
\text{At } A\text{:} \quad & \text{Profit} = 0.60(0) + 0.40(62.5) \\
& \phantom{\text{Profit}} = \$25 \\
\text{At } B\text{:} \quad & \text{Profit} = 0.60(58) + 0.40(34) \\
& \phantom{\text{Profit}} = 34.80 + 13.60 \\
& \phantom{\text{Profit}} = \$48.40 \\
\text{At } C\text{:} \quad & \text{Profit} = 0.60(75) + 0.40(20) \\
& \phantom{\text{Profit}} = 45.00 + 8.00 \\
& \phantom{\text{Profit}} = \$53.00 \\
\text{At } D\text{:} \quad & \text{Profit} = 0.60(91\tfrac{2}{3}) + 0.40(0) \\
& \phantom{\text{Profit}} = \$55.00
\end{aligned}$$

Thus we see that the maximum profit occurs at D, that is, when the Sweet Tooth Company produces 91 boxes of Chewies and no Gooies each day.

The steps in the graphical solution of a linear programming problem are summarized:

1. Graph the polygonal region defined by the constraints.
2. Find the coordinates of the corners of the polygon.
3. Evaluate the objective function at each corner of the polygon. If the function has a maximum or a minimum, these will occur at a corner point of the polygonal region.

When a linear programming problem involves three or more variables, the graphical method cannot be used. In those cases, the **simplex method** is often used. Unfortunately, this method is not as simple as its name implies, and we have neither the space nor the time to discuss it here.

Exercise Set 9.4

For the sets of Problems 1 to 6, (a) graph the polygonal region corresponding to the intersections of the half-planes, (b) from the graph in (a) give the coordinates of the corners of the region, (c) maximize the objective function given, and (d) minimize the objective function given.

1. $y \geq 1$
   $x \leq 4$
   $y \geq x$
   Objective function: $2x - y$

2. $y \geq 1$
   $x \geq 4$
   $y \leq x$
   Objective function: $3x + 9$

3. $x \geq 0$
   $y \geq 1$
   $x \leq 4$
   $x + 4y \leq 12$
   Objective function: $20x + 30y$

4. $x \geq 0$
   $y \geq 0$
   $x + y \geq 2$
   $3x + 5y \leq 15$
   Objective function: $x + 3y$

5. $x \geq 0$
   $x + 3y \leq 12$
   $2x + y \leq 9$
   $x + 2y \geq 4$
   Objective function: $0.25x + 0.30y$

6. $x \geq 0$
   $y \geq 0$
   $x + y \geq 4$
   $x + y \leq 12$
   $2x + 4y \leq 30$
   Objective function: $0.25x + y$

7. Complete Problem 1, Exercise Set 9.1.
8. Complete Problem 2, Exercise Set 9.1.
9. Complete Problem 3, Exercise Set 9.1.
10. Complete Problem 4, Exercise Set 9.1.
11. Complete Problem 5, Exercise Set 9.1.
12. Suppose that the Sweet Tooth Candy Company, in Example 9.10, finds that it must reduce the price of Chewies and therefore make a profit of only \$0.45 on each box of that kind of candy bar. How will this affect their production of candy bars?

SUMMARY

Terms you should know:

1. Linear programming
2. Constraint
3. Objective function
4. $x \leq y$ (either the number represented by x is less than the number represented by y or the two numbers are equal)
5. $x \geq y$ (either the number represented by x is greater than the number represented by y or the two numbers are equal)
6. Open sentence
7. Linear equation
8. Solution set of a linear equation

9. Rectangular coordinate system
10. Axes
11. Origin
12. x axis
13. y axis
14. Coordinates of a point
15. Graph of a linear equation
16. Half-plane
17. Closed half-plane
18. Convex polygonal region
19. Corner of a region

To solve a linear programming problem:

1. Identify the variables of the problem.
2. Identify the constraints on the variables.
3. Identify the objective function and determine if it is linear and if it is to be maximized or minimized.

If there are only two variables in the problem, the graphical method can be used:

4. Graph the polygonal region defined by the constraints.
5. Find the corners of the polygonal region.
6. Evaluate the objective function at each corner point of the polygon, and by comparing the values obtained, select either the maximum or the minimum.

REVIEW EXERCISES

1. Find the maximum value of $2x + y$ given that

$$\begin{aligned} x &\geq 2 \\ y &\geq 1 \\ x + y &\leq 8 \end{aligned}$$

2. Find the minimum value of $2x + 3y$ if

$$\begin{aligned} x &\geq 0 \\ y &\geq 0 \\ x + y &\geq 8 \\ 2x + 3y &\geq 19 \end{aligned}$$

3. Find the maximum and minimum values of $4x + 3y$ if

$$\begin{aligned} x + y &\leq 6 \\ x - y &\leq 2 \\ 3y - 5x &\geq 18 \end{aligned}$$

4. Find the maximum and minimum values of $2x + y + 3$ if

$$\begin{aligned} x + 3y &\geq 9 \\ 2x - y &\geq 4 \\ y &\geq 0 \\ x &\leq 12 \end{aligned}$$

5. A trucking company owns two types of trucks. Type A has 20 cubic yards of refrigerated space and 30 cubic yards of nonrefrigerated space. Type B has 20 cubic yards of refrigerated space and 10 cubic yards of nonrefrigerated space. A customer wants to haul some produce a certain distance and will require 160 cubic yards of refrigerated space and 120 cubic yards of nonrefrigerated space. The trucking company figures it will take 300 gallons of fuel for the type A truck to make the trip and 200 gallons of fuel for the type B truck. Find the number of trucks of each type that the company should allow for the job in order to minimize the fuel consumption.
6. A local television network is faced with the following problem. It has been found that program A with 20 minutes of music and 1 minute of advertising draws 30,000 viewers, while program B with 10 minutes of music and 1 minute of advertisement draws 10,000 viewers. Within one week the advertiser insists that at least 6 minutes be devoted to his advertisement, and the network can afford no more than 80 minutes of music. How many times per week should each program be given in order to obtain the maximum number of viewers?

10

GAMES FOR TWO PLAYERS

Game theory, like linear programming, is a comparatively new branch of mathematics. Both branches are, in fact, very closely related although on the surface they appear quite different. In linear programming, we consider a problem in which a single person tries to maximize or to minimize a function defined over a polygonal convex set of values. In game theory, we consider situations in which there are two or more persons, each of whose actions influence the outcome of a single event. Each person tries to choose his course of action so that, regardless of what his opponent does, the player can assure himself of a certain gain. The course of action he chooses is called his strategy.

Most recreational games such as checkers, chess, monopoly, poker, and other card games can be viewed as games of strategy. Game theory does not stop with parlor games, however. Many political, economic, and social activities have been brought within the scope of game theory.

Most games are too complex to allow complete analysis. There are a number of simple games, however, which are easy to work with and which do illustrate the theory. The ones we consider in this chapter involve only two players.

10.1 STRICTLY DETERMINED GAMES

10.1.1 Vocabulary

GAME A **game** will be a process which has these basic properties:

1. There are several moves; some may be random but others involve a decision by the player.
2. The random moves have well-defined probability distributions.
3. Each player has some knowledge of the outcome of prior moves.
4. There is a payoff at the end of each play of the game.

STRATEGY A player's **strategy** is a course of action, determined before the play starts, which anticipates every move of his opponent so well that the play could be carried out by proxy if necessary. We shall assume that each player has only a finite number of strategies at his disposal.

Any game specified by a finite set of rules may be represented by a matrix. Let us suppose in such a game that the strategies of one player R are numbered $1, 2, 3, \ldots, m$ and those of the second player C are numbered $1, 2, 3, \ldots, n$. Since each of the players must play according to one of his strategies, the game may proceed in any one of mn ways. Suppose that player R chooses strategy i and player C chooses strategy j. Player R wins an amount a_{ij}. We may then arrange the numbers a_{ij} into an $m \times n$ **game matrix**. When we do this, we think of a play of the game as consisting of a choice of a row by player R and a column by player C.

The construction of a game matrix is illustrated in the following example.

EXAMPLE 10.1 Each of the players R and C simultaneously extend one, two, or three fingers from a closed fist and C pays to R a sum equal to the total number of fingers shown.

This game can be represented by the 3×3 game matrix:

Player C

Number of fingers extended: 1 2 3

Player R

$$\begin{array}{c} 1 \\ 2 \\ 3 \end{array} \begin{bmatrix} 2 & 3 & 4 \\ 3 & 4 & 5 \\ 4 & 5 & 6 \end{bmatrix} = \mathbf{A}$$

The rows represent choices of player R; he can extend one finger, two fingers, or three fingers. The columns represent the possible choices of C. The number 2 in the first row and first column (position a_{11}) represents the payoff to R if R extends one finger and C also extends one finger. The number 3 in the second row and first column (position a_{21}) shows the payoff to R if he extends two fingers but C extends only one finger. The other entries were computed in a like manner.

In this book, payoffs to R will be shown by positive entries; payoffs from R to C will be indicated by negative numbers; and when R and C tie or break even, this will be indicated by an entry of 0 in the appropriate place in the game matrix.

10.1.2 Determining the Best Strategies

To determine how the players of the game of Example 10.1 should act, let us first consider R's viewpoint. He would like to win as much as possible. Therefore he will not play row 1 because the most he can win by playing that row is 4 units, and he can be sure of doing at least that well if he plays row 3. He will, accordingly, ignore the first row in his deliberations and consider only rows 2 and 3.

$$\begin{bmatrix} 2 & 3 & 4 \\ 3 & 4 & 5 \\ 4 & 5 & 6 \end{bmatrix}$$

Player R knows that his adversary, player C, will come to the same conclusion that R has, and so C will also consider only rows 2 and 3. Clearly C will not play the third column because that play would cost C at least 5 units, and he can be sure of losing less than that by playing column 1. Thus, R reasons, C will ignore column 3 and consider either column 1 or column 2.

$$\begin{bmatrix} 2 & 3 & 4 \\ 3 & 4 & 5 \\ 4 & 5 & 6 \end{bmatrix}$$

In the 2×2 matrix which is left, R sees that his best play is the bottom row. If he plays that one, he will win *at least* 4 units, whereas playing the other row (the middle row of the original matrix) the *most* he can expect to win is 4 units.

Player C's reasoning has led to the same conclusion as R's: R will play row 3, and C will play either column 1 or column 2. So C looks at those positions remaining:

$$\begin{bmatrix} 2 & 3 & 4 \\ 3 & 4 & 5 \\ 4 & 5 & 6 \end{bmatrix}$$

and decides that column 1 is his best play since it will assure that he loses only 4 units while playing column 2 would cost him 5 units.

We see that the best strategy for R is "extend three fingers" and C's best strategy is "extend one finger." If both players play these, R will win 4 units and C will lose 4 units; but if one player plays his best strategy and the other player does not, the first player will profit from the second player's indiscretion.

EXAMPLE 10.2 R and C play a game with this matrix:

$$\text{Player } R \quad \overset{\text{Player } C}{\begin{bmatrix} 7 & 3 & 1 \\ 5 & 4 & 6 \\ -1 & 2 & 7 \end{bmatrix}}$$

Suppose that R chooses row 1 as his strategy. The worst that can happen is that he will win only 1 unit. If he chooses row 2, the worst that can happen is he will win 4 units. If he chooses row 3, the worst that can happen is that he will lose 1 unit. Thus the *best of the worst* things that can happen will occur if he chooses strategy 2. Now let us consider the game from C's viewpoint. If he chooses column 1, he can lose as much as 7 units; if he chooses column 2, he can lose as much as 4 units; and if he chooses column 3, he can lose as much as 7 units. Therefore he should choose strategy 2 because, if he does, he will be assured that the *best of the worst* possible outcomes will be his, provided R plays rationally. Player R, of course, should play his second strategy, because if C plays rationally, this strategy will assure him, R, the maximum payoff.

Notice that game theory is basically pessimistic. Neither player sets out to win as much as he would like to win, but each chooses a course of action which will assure that he wins the most possible, if the other player behaves rationally, and will possibly win more if the other player does not behave rationally.

10.1.3 Strictly Determined Games

Each of the games discussed above is an example of a **strictly determined** game, so called because neither of the players would be inclined to alter his strategy even if his opponent's strategy were known.

STRICTLY DETERMINED GAME A matrix game is **strictly determined** if the matrix contains an entry v which is simultaneously the minimum of the row in which it occurs and the maximum of the column in which it occurs. The **value** of the game is v, and the location of v is called a **saddle point**. A game matrix may have more than one entry which is a saddle point, but if this is so all such entries will be numerically equal.

OPTIMAL STRATEGIES If a matrix game is strictly determined with value v, then the best strategies for the players are:

R: Play the row that contains v
C: Play the column that contains v

These *best* strategies are called the **optimal strategies**.

We shall see later that not all games are strictly determined. If a game is strictly determined, however, its solution is easy to determine, and so the first thing to try when solving any game is to look for a saddle point.

EXAMPLE 10.3 Find the optimal strategies and the value of this game.

					Row minimums	
	3	5	1	2	1	
	5	4	(3)	6	(3)	Maximum of the minimums
	4	8	3	0	0	
Column maximums	5	8	(3)	6		
			Minimum of the maximums			

Solution The maximum row minimum is 3, and minimum of the column maximums is also 3; therefore the game has a saddle point at the intersection of the second row and the third column. Optimal strategies are:

R: Play row 2
C: Play column 3

The value of the game is 3.

EXAMPLE 10.4 Two hamburger chains, Ritchie's Hamburgers and Chuck's Cheeseburgers, are trying to decide whether to build on the east side of the city or on the west side. They figure that they will split the sales if they both build on the same side of town, but if they build on opposite sides of town, the company that builds on a side of the town will get all the business from that side of town. The potential market is 4000 hamburgers a week on the east side of town and 6000 hamburgers a week on the west side of town. Where should the two companies build?

Solution Let the payoff to Ritchie's be:

(Number of hamburgers sold by Ritchie's)
− (number of hamburgers sold by Chuck's)

If both companies locate on the same side of town, they split the sales, and so the payoff to Ritchie's in that case is 0. If Ritchie's builds on the east side and Chuck's builds on the west side, then Ritchie's will sell 4000 hamburgers and Chuck's will sell 6000, and so the payoff for Ritchie's will be $4000 - 6000 = -2000$. If Ritchie's builds on the west side and Chuck's builds on the east side, then Ritchie's payoff will be $6000 - 4000 = 2000$. The game matrix is:

		Chuck's East	Chuck's West
Ritchie's	East	0	−2000
	West	2000	0

The largest row minimum is 0 and the smallest column maximum is also 0, and so this "game" is strictly determined with value 0. Both companies should build on the west side of town.

Exercise Set 10.1

Find the optimal strategies for each of the games. What is the value of each game?

1. $\begin{bmatrix} 3 & 1 \\ 4 & 0 \end{bmatrix}$

2. $\begin{bmatrix} 3 & 5 \\ 4 & 6 \end{bmatrix}$

3. $\begin{bmatrix} 4 & 3 & 6 \\ 8 & 1 & 0 \\ 2 & 3 & 0 \end{bmatrix}$

4. $\begin{bmatrix} 2 & 0 & 6 \\ 5 & 1 & 3 \\ 7 & -2 & 1 \end{bmatrix}$

5. $\begin{bmatrix} 4 & -8 & -3 & 0 \\ 7 & -6 & 1 & 2 \\ 0 & 0 & 4 & 6 \end{bmatrix}$

6. $\begin{bmatrix} 2 & -1 & -1 \\ -1 & 0 & -1 \\ -1 & 1 & -1 \\ 1 & -1 & -1 \end{bmatrix}$

7. Suppose that in Example 10.1, R and C decide that the payoff to R should be (number of fingers R shows) – (number of fingers C shows). Thus if R shows one finger and C shows two fingers, R's payoff is -2 units (he pays two units); but if R shows three fingers and C shows 1, then R receives 2 units. Make a game matrix and find the optimal strategies for both players. What is the value of the game to R?

8. Suppose that two players play the game of Example 10.1, only this time C pays R an amount equal to the sum of the fingers shown, while R pays to C an amount equal to the product of the fingers shown. Make the game matrix, and find the value of the game and the optimal strategies.

10.2 MIXED STRATEGIES

10.2.1 Pure Strategies—Probability Vectors

In the last section we considered strictly determined games. For such games, the optimal strategy for R is "play the row containing the saddle point of the game matrix," and the optimal strategy for C is "play the column containing the saddle point."

In Example 10.1 the game matrix was:

Number of fingers		Player C: 1	2	3
Player R	1	2	3	4
	2	3	4	5
	3	4 (Saddle point)	5	6

The optimal strategies are:

R: Extend three fingers
C: Extend one finger

If the game is played many times and R uses his optimal strategy on each play, it will not be long before C discovers R's strategy but, when he does discover it, C will not change his own strategy. Similarly, even after R discovers C's strategy he will not change his own. The strategies "R extend 3 fingers" and "C extend 1 finger" are examples of **pure strategies.**

PURE STRATEGY A **pure strategy** is a complete set of rules that describes every decision a player should make if a game is repeated several times.

We can represent R's pure strategy in Example 10.1 as a row vector $\mathbf{p} = [0, 0, 1]$ in which the first 0 is the probability that R plays the first row, the second 0 is the probability that he plays the second row, and the 1 in the third position is the probability that he plays the third row. The pure strategy for C can be represented by the column vector

$$\mathbf{q} = \begin{bmatrix} 1 \\ 0 \\ 0 \end{bmatrix}$$

in which the 1 gives the probability that C plays the first column and the 0's in the other two positions give the probability that he plays the other two columns. The vectors

$$\mathbf{p} = [0, 0, 1] \qquad \text{and} \qquad \mathbf{q} = \begin{bmatrix} 1 \\ 0 \\ 0 \end{bmatrix}$$

are called **probability vectors.**

PROBABILITY VECTOR A **probability vector** is a vector in which each component is a non-negative number and the sum of the components is 1. A probability vector may be either a row or a column vector.

EXAMPLE 10.5 Determine the pure strategies for the game:

$$\begin{bmatrix} 7 & 1 & 3 \\ 6 & 6 & \textcircled{4} \\ 1 & 7 & 2 \end{bmatrix} \quad \text{Saddle point}$$

Solution The game is strictly determined with value 4. The pure strategies are:

$$\text{For } R: \quad [0, 1, 0]$$

$$\text{For } C: \quad \begin{bmatrix} 0 \\ 0 \\ 1 \end{bmatrix}$$

EXAMPLE 10.6 R and C play a "coin matching" game. R writes one of the words "heads" or "tails" on a piece of paper in such a way that C cannot see what has been written. C tries to guess what R has written. If C guesses correctly, R pays him one unit, but if C guesses incorrectly, he pays R one unit. The game matrix is:

$$\begin{array}{cc} & C \\ & \begin{array}{cc} H & T \end{array} \\ R \begin{array}{c} H \\ T \end{array} & \begin{bmatrix} -1 & 1 \\ 1 & -1 \end{bmatrix} \end{array}$$

where -1 indicates a payment of 1 unit by R to C and a 1 indicates a payment of one unit by C to R. This game has no saddle point since the row minimums are both -1 and the column maximums are $+1$. Therefore the game is not strictly determined.

10.2.2 Mixed Strategies

It is easy to see that there is no optimal strategy for a single play of the game in Example 10.6. If R chooses to write "heads" and C discovers this, C can be sure of winning by saying "heads." If R chooses his second strategy, "tails," then C can win by saying "tails." Similarly, if R somehow finds out what C plans to call, then R can win by writing the opposite word.

It seems reasonable that if the game is repeated several times, R should play "heads" part of the time and "tails" the rest of the time. In fact, since the payoff is exactly the same for each of these strategies, it seems reasonable that he should play each strategy exactly half the time. We shall see later that this is the correct way to play the game. It also seems that R should switch from "heads" to "tails" by some chance method (perhaps by tossing a coin) so that there will be no pattern for C to discover. The same line of reasoning applies to C. He should call "heads" half of the time and "tails" half of the time and he should mix up the calls so that R has no way of determining what will be called next.

If our reasoning above is correct, the optimal strategies for R and C over several plays of the game should be:

For R: $\mathbf{p} = [\frac{1}{2}, \frac{1}{2}]$

For C: $\mathbf{q} = \begin{bmatrix} \frac{1}{2} \\ \frac{1}{2} \end{bmatrix}$

Strategies of this kind are called **mixed strategies.**

MIXED STRATEGIES For an $m \times n$ matrix game, a mixed strategy for player R is a command of the form "play row 1 with probability p_1, play row 2 with probability $p_2, \ldots,$ and play row m with probability p_m" where $p_1, p_2, \ldots, p_m$ are non-negative numbers with $p_1 + p_2 + \cdots + p_m = 1$. A mixed strategy for player C is a command of the form "play column 1 with probability q_1, play column 2 with probability $q_2, \ldots,$ and play column n with probability q_n" where $q_1, q_2, \ldots, q_n$ are non-negative numbers and $q_1 + q_2 + \cdots + q_n = 1$.

MIXED-STRATEGY VECTOR A mixed-strategy vector for player R is the probability row vector $\mathbf{p} = [p_1, p_2, \ldots, p_m]$, and a mixed-strategy vector for player C is the probability column vector

$$\mathbf{q} = \begin{bmatrix} q_1 \\ q_2 \\ \vdots \\ q_n \end{bmatrix}$$

For the game of Example 10.6, we decided on purely intuitive grounds that R's strategy should be

$$\mathbf{p} = [\tfrac{1}{2}, \tfrac{1}{2}]$$

and C's strategy should be

$$\mathbf{q} = \begin{bmatrix} \frac{1}{2} \\ \frac{1}{2} \end{bmatrix}$$

The next section will show that our intuition, in this case, is correct; these are actually the optimal strategies for the players. In order to verify this, it will be necessary that we use matrix multiplication. The following examples should help bring back the details of that operation.

EXAMPLE 10.7 Compute these products:

(a) $[1, 0, 2]\begin{bmatrix} 2 \\ 3 \\ 1 \end{bmatrix}$

(b) $[2, 3]\begin{bmatrix} 3 & 1 \\ 0 & 4 \end{bmatrix}$

(c) $\begin{bmatrix} 2 & 4 \\ 1 & 3 \end{bmatrix}\begin{bmatrix} 1 \\ 3 \end{bmatrix}$

(d) $[3, 2]\begin{bmatrix} 1 & 2 \\ 3 & 1 \end{bmatrix}\begin{bmatrix} 2 \\ 4 \end{bmatrix}$

Solution Recall that to multiply a row vector by a column vector, we multiply each element of the row vector by the corresponding element of the column vector and then add the products. The sum is the product of the vectors. Therefore:

$$(a)\ [1, 0, 2]\begin{bmatrix} 2 \\ 3 \\ 1 \end{bmatrix} = (1)(2) + (0)(3) + (2)(1)$$

$$= 2 + 0 + 2$$

$$= 4$$

To multiply a row vector by a matrix, we multiply the vector by each column of the matrix and the products form a row vector. Thus:

$$(b)\ [2, 3]\begin{bmatrix} 3 & 1 \\ 0 & 4 \end{bmatrix} = [2 \times 3 + 3 \times 0, 2 \times 1 + 3 \times 4]$$

$$= [6, 14]$$

$$(c)\ \begin{bmatrix} 2 & 4 \\ 1 & 3 \end{bmatrix}\begin{bmatrix} 1 \\ 2 \end{bmatrix} = \begin{bmatrix} 2 \times 1 + 4 \times 2 \\ 1 \times 1 + 3 \times 2 \end{bmatrix}$$

$$= \begin{bmatrix} 10 \\ 7 \end{bmatrix}$$

Recall that if **A**, **B**, and **C** are matrices, **A(BC)** = **(AB)C**. Thus:

$$\text{(d)}\ [3, 2]\left(\begin{bmatrix} 1 & 2 \\ 3 & 1 \end{bmatrix}\begin{bmatrix} 2 \\ 4 \end{bmatrix}\right) = \left([3, 2]\begin{bmatrix} 1 & 2 \\ 3 & 1 \end{bmatrix}\right)\begin{bmatrix} 2 \\ 4 \end{bmatrix}$$

$$= [9, 8]\begin{bmatrix} 2 \\ 4 \end{bmatrix}$$

$$= 50$$

10.2.3 Partial Ordering

PARTIAL-ORDERING PROPERTY FOR VECTORS A row (column) vector is said to be **less than or equal** to ($\leq$) another row (column) vector of the same dimension if each component of the first vector is less than or equal to the corresponding component of the second vector. Similarly, one row (column) vector is **greater than or equal to** ($\geq$) another row (column) vector having the same dimension if every element of the first vector is greater than or equal to the corresponding element of the second vector.

EXAMPLE 10.8

$$[1, 2, 3] \leq [1, 3, 3]$$
$$[0, 1, 4, 2] \leq [1, 5, 6, 3]$$

$$\begin{bmatrix} -1 \\ 0 \\ -1 \end{bmatrix} \leq \begin{bmatrix} -1 \\ 1 \\ 1 \end{bmatrix} \qquad \begin{bmatrix} 2 \\ 1 \end{bmatrix} \geq \begin{bmatrix} 1 \\ 0 \end{bmatrix} \qquad \begin{bmatrix} 1 \\ 0 \\ 1 \\ 1 \end{bmatrix} \geq \begin{bmatrix} 1 \\ 0 \\ 1 \\ 0 \end{bmatrix}$$

We shall use the order properties of vectors in the next section.

Exercise Set 10.2

1. For each of the following game matrices, (a) decide whether the game is strictly determined. (b) If a game is strictly determined, find the optimal strategies and the game value.

(a) $\begin{bmatrix} 1 & 4 \\ 2 & 3 \end{bmatrix}$ (c) $\begin{bmatrix} 5 & 0 \\ 0 & 2 \end{bmatrix}$ (e) $\begin{bmatrix} -1 & -1 \\ 0 & 2 \end{bmatrix}$

(b) $\begin{bmatrix} 0 & 1 \\ 2 & 3 \end{bmatrix}$ (d) $\begin{bmatrix} 3 & 1 \\ 4 & 0 \end{bmatrix}$ (f) $\begin{bmatrix} 0 & 1 \\ 1 & 0 \end{bmatrix}$

(g) $\begin{bmatrix} 6 & 2 & 1 \\ 4 & 3 & 5 \\ 0 & 1 & 6 \end{bmatrix}$ (h) $\begin{bmatrix} 2 & 2 & 1 & 4 \\ 0 & 2 & 5 & 3 \\ 4 & 2 & 3 & 2 \end{bmatrix}$

2. Find each of these products:

(a) $[0, 1, 0]\begin{bmatrix} 3 & 0 \\ 2 & 2 \\ 1 & 1 \end{bmatrix}$ (e) $[\frac{1}{2}, \frac{1}{2}]\begin{bmatrix} 1 & 0 \\ 0 & 1 \end{bmatrix}$

(b) $\begin{bmatrix} 3 & 0 \\ 2 & 2 \\ 1 & 1 \end{bmatrix}\begin{bmatrix} 0 \\ 1 \end{bmatrix}$ (f) $\begin{bmatrix} 1 & 0 \\ 0 & 1 \end{bmatrix}\begin{bmatrix} \frac{1}{2} \\ \frac{1}{2} \end{bmatrix}$

(c) $[2, 3]\begin{bmatrix} 1 & 4 \\ 3 & 1 \end{bmatrix}$ (g) $[\frac{1}{4}, \frac{3}{4}]\begin{bmatrix} 1 & 0 \\ 0 & 2 \end{bmatrix}\begin{bmatrix} \frac{4}{5} \\ \frac{1}{5} \end{bmatrix}$

(d) $\begin{bmatrix} 1 & 2 \\ 3 & 2 \end{bmatrix}\begin{bmatrix} 2 \\ 1 \end{bmatrix}$ (h) $[\frac{3}{4}, 0, \frac{1}{4}]\begin{bmatrix} 1 & 0 \\ 0 & 0 \\ -1 & 2 \end{bmatrix}\begin{bmatrix} \frac{1}{4} \\ \frac{3}{4} \end{bmatrix}$

3. Which of the symbols $=$, $\leq$, or $\geq$ should go in the blank to make each of the following true?

(a) $[2, 1, 1]$ ___ $[2, 0, 1]$

(b) $[3, 4]$ ___ $[1, 1]$

(c) $[1, 2]$ ___ $[2, 4]$

(d) $\begin{bmatrix} 1 \\ 1 \\ 2 \end{bmatrix}$ ___ $\begin{bmatrix} 1 \\ 1 \\ 2 \end{bmatrix}$

(e) $\begin{bmatrix} 1 \\ 1 \\ 0 \end{bmatrix}$ ___ $\begin{bmatrix} -1 \\ 1 \\ 0 \end{bmatrix}$

*10.3 NON-STRICTLY DETERMINED GAMES

In this section, we shall consider 2×2 matrix games which have no saddle points. These games have optimal mixed strategies with the following properties:

BASIC PROPERTIES OF OPTIMAL MIXED STRATEGIES If a game has the matrix

$$\mathbf{A} = \begin{bmatrix} a_{11} & a_{12} \\ a_{21} & a_{22} \end{bmatrix}$$

and if the value of the game is v, and the optimal strategies are

$$\mathbf{p} = [p_1, p_2] \quad \text{and} \quad \mathbf{q} = \begin{bmatrix} q_1 \\ q_2 \end{bmatrix}$$

then

$$\mathbf{pA} \geq [v, v]$$

and

$$\mathbf{Aq} \leq \begin{vmatrix} v \\ v \end{vmatrix}$$

If player R chooses strategy $\mathbf{p} = [p_1, p_2]$ and player C, independently, chooses strategy $\mathbf{q} = \begin{bmatrix} q_1 \\ q_2 \end{bmatrix}$, then R obtains the payoff a_{11} with probability p_1q_1. (Recall from Chapter 7 that the probability of the occurrence of two independent events is the product of the probabilities that either of the events occurs by itself.) Player R obtains payoff a_{12} with probability p_1q_2, he obtains payoff a_{21} with probability p_2q_1, and he obtains payoff a_{22} with probability p_2q_2. His expected value from the game (see Chapter 7) is

$$\text{Value} = a_{11}p_1q_1 + a_{12}p_1q_2 + a_{21}p_2q_1 + a_{22}p_2q_2$$

and as can easily be verified by performing the multiplications, this number is the same as the matrix product

$$\mathbf{pAq}$$

It can be shown that because $\mathbf{pA} \geq [v, v]$ and $\mathbf{Aq} \leq \begin{bmatrix} v \\ v \end{bmatrix}$ that

$$\mathbf{pAq} = v$$

Therefore the expected value of the game to R, as defined in Chapter 7, and the value of the game, as defined in this chapter, are one and the same.

The discussion above can be extended to matrix games of any dimension. That is, if **A** is the game matrix for an $m \times n$ game, and **p** and **q** are optimal strategies for the players, then the value of the game is

$$v = \mathbf{pAq}$$

In a 2×2 non-strictly determined game, with game matrix

$$\mathbf{A} = \begin{bmatrix} a_{11} & a_{12} \\ a_{21} & a_{22} \end{bmatrix}$$

the optimal strategies are given by the following formulas.

For R: $\quad \mathbf{p} = [p_1, p_2] \quad$ where $p_1 = \dfrac{a_{22} - a_{21}}{a_{11} + a_{22} - a_{12} - a_{21}}$

$$p_2 = \frac{a_{11} - a_{12}}{a_{11} + a_{22} - a_{12} - a_{21}}$$

For C: $\quad \mathbf{q} = \begin{bmatrix} q_1 \\ q_2 \end{bmatrix} \quad$ where $q_1 = \dfrac{a_{22} - a_{12}}{a_{11} + a_{22} - a_{12} - a_{21}}$

$$q_2 = \frac{a_{11} - a_{21}}{a_{11} + a_{22} - a_{12} - a_{21}}$$

The value of the game is

$$v = \frac{a_{11}a_{22} - a_{12}a_{21}}{a_{11} + a_{22} - a_{12} - a_{21}}$$

These formulas can be derived by making the strategy vectors

$$\mathbf{p} = [x_1, x_2]$$

and

$$\mathbf{q} = \begin{bmatrix} y_1 \\ y_2 \end{bmatrix}$$

conform to the conditions of the basic properties of optimal mixed strategies. The resulting system of six equations in the variables x_1, x_2, y_1, y_2, and v can be solved, but the computations are too involved to be considered here.

EXAMPLE 10.9 Find the optimal strategies for the "coin tossing" game of Example 10.6.

Solution The game matrix is:

$$\begin{bmatrix} -1 & 1 \\ 1 & -1 \end{bmatrix}$$

R's strategy is $\mathbf{p} = [p_1, p_2]$ in which

$$\begin{aligned} p_1 &= \frac{a_{22} - a_{21}}{a_{11} + a_{22} - a_{12} - a_{21}} \\ &= \frac{-1 - (1)}{-1 + (-1) - (1) - (1)} \\ &= \frac{-2}{-4} \\ &= \frac{1}{2} \end{aligned}$$

and

$$\begin{aligned} p_2 &= \frac{a_{11} - a_{12}}{a_{11} + a_{22} - a_{12} - a_{21}} \\ &= \frac{-1 - (1)}{-4} \\ &= \frac{-2}{-4} \\ &= \frac{1}{2} \end{aligned}$$

Similarly, C's strategy is $\mathbf{q} = \begin{bmatrix} q_1 \\ q_2 \end{bmatrix}$ in which

$$\begin{aligned} q_1 &= \frac{a_{22} - a_{12}}{a_{11} + a_{22} - a_{12} - a_{21}} \\ &= \frac{-1 - (1)}{-4} \\ &= \frac{-2}{-4} \\ &= \frac{1}{2} \end{aligned}$$

and

$$\begin{aligned} q_2 &= \frac{a_{11} - a_{21}}{a_{11} + a_{22} - a_{12} - a_{21}} \\ &= \frac{-1 - (1)}{-4} \\ &= \frac{-2}{-4} \\ &= \frac{1}{2} \end{aligned}$$

Thus $\mathbf{p} = [\frac{1}{2}, \frac{1}{2}]$ and $\mathbf{q} = \begin{bmatrix} \frac{1}{2} \\ \frac{1}{2} \end{bmatrix}$, and these are the values we determined intuitively when the problem first came up.

The value of the game is

$$\begin{aligned} v &= \frac{a_{11}a_{22} - a_{12}a_{21}}{a_{11} + a_{22} - a_{12} - a_{21}} \\ &= \frac{(-1)(-1) - (1)(1)}{-4} \\ &= \frac{1 - 1}{-4} \\ &= 0 \end{aligned}$$

EXAMPLE 10.10 Find the optimal strategies for the game

$$\begin{bmatrix} 1 & 0 \\ -1 & 2 \end{bmatrix}$$

Also, find the value of the game.

Solution We first check to see if the game is strictly determined.

			Row minimums
	1	0	0
	−1	2	−1
Column maximums	1	2	

Since there is no row minimum which is also a column maximum, the game is not strictly determined. Therefore R's strategy must be the mixed strategy $\mathbf{p} = [p_1, p_2]$, and C's strategy must be $\mathbf{q} = \begin{bmatrix} q_1 \\ q_2 \end{bmatrix}$ where

$$p_1 = \frac{2 - (-1)}{1 + 2 - (0) - (-1)} = \frac{2 + 1}{1 + 2 + 0 + 1} = \frac{3}{4}$$

$$p_2 = \frac{1 - 0}{4} = \frac{1}{4}$$

$$q_1 = \frac{2 - 0}{4} = \frac{2}{4} = \frac{1}{2}$$

$$q_2 = \frac{1 - (-1)}{4} = \frac{1 + 1}{4} = \frac{2}{4} = \frac{1}{2}$$

The strategies are

$$\mathbf{p} = [\tfrac{3}{4}, \tfrac{1}{4}] \qquad \text{and} \qquad \mathbf{q} = \begin{bmatrix} \frac{1}{2} \\ \frac{1}{2} \end{bmatrix}$$

The value of the game is

$$v = \frac{1(2) - (0)(-1)}{4}$$

$$= \frac{2}{4}$$

$$= \frac{1}{2}$$

The computations may be checked by using the fact that $\mathbf{pAq} = v$ when $\mathbf{p}$ and $\mathbf{q}$ are optimal strategies.

To play the game, player R should choose his first strategy $\frac{3}{4}$ of the time and his second strategy $\frac{1}{4}$ of the time. To decide how to "mix up" his strategies, player R might recall that the outcomes from tossing two coins are $\{(H, H), (H, T), (T, H), (T, T)\}$, and thus when two coins are tossed one can expect (T, T) $\frac{1}{4}$ of the time. He could therefore use this mixed strategy:

> "Toss two coins. If both land heads or one lands heads and the other tails, play row 1 of the game matrix. If both coins land tails, play row 2."

Player C could adopt this strategy:

> "Toss one coin. If it lands heads play column 1, but if it lands tails play column 2."

SOLUTION OF A GAME A **solution** of a matrix game consists of a listing of optimal strategies for each player and the value of the game.

Exercise Set 10.3

Solve the games in the following problems. First check to see if the game is strictly determined. If it is, find the pure strategies and the value of the game. If a game is not strictly determined, find the mixed strategies which are optimal and the game value by means of the formulas

$$p_1 = \frac{a_{22} - a_{21}}{a_{11} + a_{22} - a_{12} - a_{21}} \qquad p_2 = \frac{a_{11} - a_{12}}{a_{11} + a_{22} - a_{12} - a_{21}}$$

$$q_1 = \frac{a_{22} - a_{12}}{a_{11} + a_{22} - a_{12} - a_{21}} \qquad q_2 = \frac{a_{11} - a_{21}}{a_{11} + a_{22} - a_{12} - a_{21}}$$

$$v = \frac{a_{11}a_{22} - a_{12}a_{21}}{a_{11} + a_{22} - a_{12} - a_{21}}$$

1. $\begin{bmatrix} 3 & 5 \\ 8 & 1 \end{bmatrix}$

2. $\begin{bmatrix} 7 & -6 \\ 5 & 8 \end{bmatrix}$

3. $\begin{bmatrix} 1 & 2 \\ 3 & 4 \end{bmatrix}$

4. $\begin{bmatrix} 15 & 3 \\ -1 & 2 \end{bmatrix}$

5. $\begin{bmatrix} 3 & 0 \\ 2 & 4 \end{bmatrix}$

6. $\begin{bmatrix} 4 & 1 \\ 3 & 2 \end{bmatrix}$

7. $\begin{bmatrix} 1 & 10 \\ 9 & 12 \end{bmatrix}$

8. $\begin{bmatrix} 3 & 1 \\ 2 & 4 \end{bmatrix}$

9. $\begin{bmatrix} 2 & -1 \\ -1 & 2 \end{bmatrix}$

10. $\begin{bmatrix} 1 & 10 \\ -1 & 12 \end{bmatrix}$

11. A game is played as follows: Each player holds up either one or two fingers. If the sum of the number of fingers shown is even, R gets an amount equal to the sum; and if the sum is odd, C gets the amount. Make the game matrix; find optimal strategies for each player and find the value of the game.

12. A secret service agent must carry a message from his headquarters to a friend's office. He can take either of two routes, one of them deserted, the other leading through the center of a city. His opponent wishes to prevent him from delivering the message and must guess which route to guard. If the opponent catches the secret agent on the deserted road, he will kill him; if he catches him on the city street, he will merely be able to detain him for some time. Give the agent 5 points if he delivers the message, -1 point if he is detained and -100 points if he is killed. Make the game matrix, and find optimal strategies for the "players."

13. In Example 10.10, optimal mixed strategies for R were $[\frac{3}{4}, \frac{1}{4}]$ and those for C were $\begin{bmatrix} \frac{1}{2} \\ \frac{1}{2} \end{bmatrix}$. R's pure strategy might be "toss two coins and play row 1 if the result is two heads or a head and a tail, but play row 2 if the coins both land tails up." A pure strategy for C might be "toss one coin and play column 1 if it lands heads, but play column 2 if it lands tails." Find similar rules for a player whose mixed strategy is

    (a) $[\frac{3}{8}, \frac{5}{8}]$ HINT: Use three coins.
    (b) $[\frac{1}{6}, \frac{5}{6}]$ HINT: Use a die.
    (c) $[\frac{1}{3}, \frac{2}{3}]$
    (d) $[\frac{2}{9}, \frac{7}{9}]$
    (e) $[\frac{1}{4}, \frac{1}{2}, \frac{1}{4}]$

*10.4 $m \times 2$ AND $2 \times n$ GAMES

10.4.1 Row and Column Dominations

The methods of the previous sections can be extended to games in which one of the players has only two strategies from which to choose.

EXAMPLE 10.11 Player R extends two fingers from a closed fist, and at the same time player C extends three fingers. C pays to R an amount equal to the difference of the number of fingers extended. The game matrix is:

$$\begin{array}{lccc} & & \text{Player } C & \\ \text{Number of fingers extended} & 1 & 2 & 3 \\ \text{Player } R \quad \begin{matrix} 1 \\ 2 \end{matrix} & \multicolumn{3}{c}{\begin{bmatrix} 0 & 1 & 2 \\ 1 & 0 & 1 \end{bmatrix}} \end{array}$$

This is an example of a 2×3 game. If C is on his toes, he will not play his third strategy because that one may cause him to lose 2 units while neither of the other strategies can cost him more than 1 unit. Therefore he will ignore column 3 and consider the 2×2 matrix game:

$$\begin{bmatrix} 0 & 1 \\ 1 & 0 \end{bmatrix}$$

Optimal strategies for this game are, using the formulas of Section 10.3,

$$[\tfrac{1}{2}, \tfrac{1}{2}] \qquad \text{and} \qquad \begin{bmatrix} \frac{1}{2} \\ \frac{1}{2} \end{bmatrix}$$

and the value of the game is $\frac{1}{2}$.

To extend these strategies to the original game, we note that the probability that C uses his third strategy is 0. Therefore his strategy for the original game is

$$\mathbf{q} = \begin{bmatrix} \frac{1}{2} \\ \frac{1}{2} \\ 0 \end{bmatrix}$$

and R's strategy is

$$\mathbf{p} = [\tfrac{1}{2}, \tfrac{1}{2}]$$

The process used above to change the 2×3 matrix game to a 2×2 game is known as **column domination.**

COLUMN DOMINATION If in a game matrix, every element of one of the columns is less than or equal to the corresponding element of another column, then the first column **dominates** the second column. Any column of a matrix which is dominated by another column may be deleted from the matrix without having any effect on the value of the game, or on the optimal strategies.

Sometimes we may simplify a matrix by means of **row domination.**

ROW DOMINATION If in a game matrix every element of one row is greater than or equal to the corresponding element of another row, then the first row **dominates** the second one. Any row of a matrix which is dominated by another row may be deleted from the matrix without having any effect on the value of the game or on the optimal strategies.

It is useful, in deleting dominated rows and columns, to recall that the components of the game matrix are payoffs by C to R. Therefore C will never choose the column with the larger entries, while R will always choose the row with the greater entries.

10.4.2 Derived Games

DERIVED GAMES A **derived** game is one formed by eliminating rows or columns from the original game matrix.

EXAMPLE 10.12 Find optimal strategies for the matrix game:

$$\begin{bmatrix} 2 & 1 \\ 1 & 0 \\ -1 & 2 \end{bmatrix}.$$

Solution Row 1 dominates row 2 since $2 \geq 1$ and $1 \geq 0$. Therefore row 2 can be deleted leaving

$$\begin{bmatrix} 2 & 1 \\ -1 & 2 \end{bmatrix}$$

Optimal solutions for this 2×2 game are

$$[\tfrac{3}{4}, \tfrac{1}{4}] \qquad \text{and} \qquad \begin{bmatrix} \frac{1}{4} \\ \frac{3}{4} \end{bmatrix}$$

and the value of the game is $\frac{5}{4}$.

The optimal strategies for the original game are

$$\mathbf{p} = [\tfrac{3}{4}, 0, \tfrac{1}{4}] \qquad \text{and} \qquad \mathbf{q} = \begin{bmatrix} \frac{1}{4} \\ \frac{3}{4} \end{bmatrix}$$

and the value of the game is $\frac{5}{4}$.

Because the second row of the original matrix was deleted to form the 2×2 matrix, when we expand the strategy vector $[\frac{3}{4}, \frac{1}{4}]$, we must insert a 0 between the two components (position p_{12}).

EXAMPLE 10.13 Solve this game:

$$\begin{bmatrix} 1 & 3 & 3 & 2 \\ 2 & 2 & 1 & 2 \\ 2 & 0 & 1 & 1 \end{bmatrix}$$

Solution This game has no saddle point; thus it is not strictly determined. However, row 2 dominates row 3, and thus row 3 may be deleted.

$$\begin{bmatrix} 1 & 3 & 3 & 2 \\ 2 & 2 & 1 & 2 \\ 2 & 0 & 1 & 1 \end{bmatrix}$$

In the 2 × 4 matrix which remains, column 1 dominates both of columns 2 and 4 and these may be deleted.

$$\begin{bmatrix} 1 & 3 & 3 & 2 \\ 2 & 2 & 1 & 2 \\ 2 & 0 & 1 & 1 \end{bmatrix}$$

The solution of the 2 × 2 game which remains after the above deletions is, using formulas of Section 10.3,

$$\mathbf{p} = [\tfrac{1}{3}, \tfrac{2}{3}] \qquad \mathbf{q} = \begin{bmatrix} \tfrac{2}{3} \\ \tfrac{1}{3} \end{bmatrix} \qquad \text{and} \qquad v = \tfrac{5}{3}$$

Thus the solution to the original game is

$$\mathbf{p} = [\tfrac{1}{3}, \tfrac{2}{3}, 0] \qquad \mathbf{q} = \begin{bmatrix} \tfrac{2}{3} \\ 0 \\ \tfrac{1}{3} \\ 0 \end{bmatrix} \qquad v = \tfrac{5}{3}$$

10.4.3 Games Without Dominating Rows or Columns

There are methods by which $m \times n$ games, with $m > 2$ or $n > 2$, and no row or column dominances may be solved. We shall discuss two such methods, one here and the other one in the next section.

Consider the game matrix:

$$\mathbf{A} = \begin{bmatrix} 1 & 3 & 0 \\ 3 & 2 & 3 \end{bmatrix}$$

We may derive three 2 × 2 games from this one by deleting columns. These games are:

$$\mathbf{B} = \begin{bmatrix} 3 & 0 \\ 2 & 3 \end{bmatrix} \qquad \mathbf{C} = \begin{bmatrix} 1 & 0 \\ 3 & 3 \end{bmatrix} \qquad \mathbf{D} = \begin{bmatrix} 1 & 3 \\ 3 & 2 \end{bmatrix}$$

It can be shown that the optimal strategies for one of these derived games, when expanded, will be optimal for the original game. To find which strategies are optimal, we find the optimal strategies for each of the derived games. These are

For game **B**: $[\frac{1}{4}, \frac{3}{4}]$ and $\begin{bmatrix} \frac{3}{4} \\ \frac{1}{4} \end{bmatrix}$

For game **C**: $[0, 1]$ and $\begin{bmatrix} 0 \\ 1 \end{bmatrix}$

For game **D**: $[\frac{1}{3}, \frac{2}{3}]$ and $\begin{bmatrix} \frac{1}{3} \\ \frac{2}{3} \end{bmatrix}$

We expand these strategies to fit matrix **A** by inserting 0's in the column vectors. Game **B** was derived from game **A** by deleting column 1 of **A**. Therefore we place a 0 above the entries in the column strategy for game **B** and obtain

$$\begin{bmatrix} 0 \\ \frac{3}{4} \\ \frac{1}{4} \end{bmatrix}$$

Column 2 was deleted from **A** to make **C**; thus we put a 0 between the two entries in the column strategy for **C** and obtain

$$\begin{bmatrix} 0 \\ 0 \\ 1 \end{bmatrix}$$

Finally, because **D** was formed by the deletion of column 3 of **A**, we expand the column strategy of game **D** by inserting a 0 after its two entries:

$$\begin{bmatrix} \frac{1}{3} \\ \frac{2}{3} \\ 0 \end{bmatrix}$$

When the strategies for games **B**, **C**, and **D** have been so altered, we have:

From game **B**: $[\frac{1}{4}, \frac{3}{4}]$ and $\begin{bmatrix} 0 \\ \frac{3}{4} \\ \frac{1}{4} \end{bmatrix}$

From game **C**: $[0, 1]$ and $\begin{bmatrix} 0 \\ 0 \\ 1 \end{bmatrix}$

From game **D**: $[\frac{1}{3}, \frac{2}{3}]$ and $\begin{bmatrix} \frac{1}{3} \\ \frac{2}{3} \\ 0 \end{bmatrix}$

We know that if **p** and **q** are optimal strategies for game **A**, then

$$\mathbf{pA} \geq [v, v, v] \qquad \text{and} \qquad \mathbf{Aq} \leq \begin{bmatrix} v \\ v \end{bmatrix}$$

where v is the value of the game.

R wishes to use a strategy which will ensure that he wins a certain amount. If we compute the possible values to *R*, using the three strategies above:

$$[\tfrac{1}{4}, \tfrac{3}{4}]\begin{bmatrix} 1 & 3 & 0 \\ 3 & 2 & 3 \end{bmatrix} = [2\tfrac{1}{2}, 2\tfrac{1}{4}, 2\tfrac{1}{4}]$$

$$[0, 1]\begin{bmatrix} 1 & 3 & 0 \\ 3 & 2 & 3 \end{bmatrix} = [3, 2, 3]$$

$$[\tfrac{1}{3}, \tfrac{2}{3}]\begin{bmatrix} 1 & 3 & 0 \\ 3 & 2 & 3 \end{bmatrix} = [2\tfrac{1}{3}, 2\tfrac{1}{3}, 2]$$

We see that in the first case *R* can be sure of a payoff of at least $2\frac{1}{4}$, but he can be sure of a payoff of only 2 in the second and third cases. Therefore *R*'s optimal strategy is the first strategy,

$$\mathbf{p} = [\tfrac{1}{4}, \tfrac{3}{4}]$$

C wishes to use a strategy which will ensure that he loses as little as possible. The products

$$\begin{bmatrix} 1 & 3 & 0 \\ 3 & 2 & 3 \end{bmatrix}\begin{bmatrix} 0 \\ \frac{3}{4} \\ \frac{1}{4} \end{bmatrix} = \begin{bmatrix} 2\frac{1}{4} \\ 2\frac{1}{4} \end{bmatrix}$$

$$\begin{bmatrix} 1 & 3 & 0 \\ 3 & 2 & 3 \end{bmatrix}\begin{bmatrix} 0 \\ 0 \\ 1 \end{bmatrix} = \begin{bmatrix} 0 \\ 3 \end{bmatrix}$$

$$\begin{bmatrix} 1 & 3 & 0 \\ 3 & 2 & 3 \end{bmatrix}\begin{bmatrix} \frac{1}{3} \\ \frac{2}{3} \\ 0 \end{bmatrix} = \begin{bmatrix} 2\frac{1}{3} \\ 2\frac{1}{3} \end{bmatrix}$$

show that with the first strategy C can be sure of losing only $2\frac{1}{4}$; with the second strategy he may lose as much as 3; and with the third strategy he may lose as much as $2\frac{1}{3}$. Therefore he will choose the first strategy.

$$\mathbf{q} = \begin{bmatrix} 0 \\ \frac{3}{4} \\ \frac{1}{4} \end{bmatrix}$$

The value of the game, to R, is $\mathbf{pAq} = 2\frac{1}{4}$.

The method illustrated above may be summarized as follows. When a matrix with dimensions greater than 2×2 has no row or column dominations:

1. Form all possible 2×2 matrices which can be derived by deleting rows or columns or both rows and columns.
2. Compute optimal strategies for each of the 2×2 games.
3. Expand the strategies to "fit" the original matrix by inserting 0's for the rows and columns which were deleted in step 1.
4. The optimal strategy for R is the row vector of step 3 which makes R's minimal winnings as large as possible regardless of which strategy C may choose. The optimal strategy for C is the column vector of step 3 which makes the maximum that he can expect to lose be the smallest possible no matter what strategy R chooses.

The solution of any game should follow this pattern:

1. Check for saddle points.
2. If there are no saddle points, eliminate all dominated rows and columns. Check the resulting matrix for saddle points.
3. If the game which remains, after eliminating dominated rows and columns, is not strictly determined and:
   (a) It is 2×2, solve it by means of the formulas of Section 10.3.
   (b) It is $m \times 2$ or $2 \times n$, the optimal strategies may be found among the optimal strategies of the 2×2 games derived by eliminating rows or columns from that matrix.

EXAMPLE 10.14 Solve the game:

$$\begin{bmatrix} 3 & 5 & 1 \\ 4 & 6 & 1 \\ 2 & 1 & 2 \end{bmatrix}$$

Solution Row 2 dominates row 1, and so we may delete row 1. We then have the 2 × 3 matrix game:

$$\begin{bmatrix} \cancel{3} & \cancel{5} & \cancel{1} \\ 4 & 6 & 1 \\ 2 & 1 & 2 \end{bmatrix}$$

Column 3 of the new matrix dominates column 1. When column 1 has been deleted, the game matrix is 2 × 2:

$$\begin{bmatrix} \cancel{3} & \cancel{5} & \cancel{1} \\ \cancel{4} & 6 & 1 \\ \cancel{2} & 1 & 2 \end{bmatrix}$$

and its optimal strategies and value are

$$[\tfrac{1}{6}, \tfrac{5}{6}] \qquad \begin{bmatrix} \frac{1}{6} \\ \frac{5}{6} \end{bmatrix} \qquad \tfrac{11}{6}$$

respectively.

Therefore the optimal strategies for the original game are

$$\mathbf{p} = [0, \tfrac{1}{6}, \tfrac{5}{6}] \qquad \text{and} \qquad \mathbf{q} = \begin{bmatrix} 0 \\ \frac{1}{6} \\ \frac{5}{6} \end{bmatrix}$$

and the value of the game is $v = \frac{11}{6}$.

Exercise Set 10.4

Find optimal strategies for each player and the value of each game. Check first to see if the game is strictly determined; next check row and column dominances.

1. $\begin{bmatrix} 0 & 2 \\ 1 & 3 \\ -1 & 0 \\ 2 & 0 \end{bmatrix}$

2. $\begin{bmatrix} 1 & 5 & 7 \\ 6 & 2 & 8 \end{bmatrix}$

3. $\begin{bmatrix} 4 & 6 & 3 \\ 1 & 2 & 5 \end{bmatrix}$

4. $\begin{bmatrix} 24 & 3 \\ 10 & 2 \end{bmatrix}$

5. $\begin{bmatrix} 10 & 5 & 4 & 6 \\ 12 & 3 & 3 & 4 \end{bmatrix}$

6. $\begin{bmatrix} 2 & 2 & 1 & 4 \\ 0 & 2 & 5 & 3 \\ 4 & 2 & 3 & 2 \end{bmatrix}$

7. $\begin{bmatrix} 0 & 13 \\ 7 & 0 \\ -10 & 10 \\ 10 & 12 \end{bmatrix}$

8. $\begin{bmatrix} 5 & 10 & 6 & 15 \\ 5 & 7 & 8 & 25 \\ 0 & 5 & 6 & 5 \end{bmatrix}$

9. $\begin{bmatrix} 1 & 3 & 0 & 2 & 1 \\ 2 & 1 & 2 & 4 & 3 \end{bmatrix}$

10. $\begin{bmatrix} 3 & 1 \\ 5 & 2 \\ -1 & 0 \\ -1 & 3 \\ 1 & 0 \\ -1 & 2 \end{bmatrix}$

11. $\begin{bmatrix} 6 & 2 & 2 & 1 \\ 3 & 4 & 5 & 5 \end{bmatrix}$

12. $\begin{bmatrix} 1 & 0 \\ 2 & 1 \\ 1 & 2 \\ 1 & -1 \end{bmatrix}$

13. Player R is given three cards, a black 7, a black 5, and a red 3. Player C is given two cards, a black 5 and a red 7. Each player shows a card simultaneously. If the cards match in color, R wins the difference between the numbers on the cards. If the cards do not match in color, then C wins the difference between the numbers on the cards. Solve this game.

14. In the card game of Problem 13, suppose that R has black 8, black 5, and red 4, while C has black 7 and red 5. Solve this game.

15. Suppose that R conceals in his hand one, two, or three coins and C guesses "odd" or "even." If C's guess is correct, he wins the amount which R holds; otherwise he must pay R this amount. What strategies should the players use? What is the game value?

*10.5 RELATION TO LINEAR PROGRAMMING

In the last section we considered the game

$$\begin{bmatrix} 1 & 3 & 0 \\ 3 & 2 & 3 \end{bmatrix}$$

which is not strictly determined. There we showed that the game can be solved by finding which of the strategies from the derived games

$$\begin{bmatrix} 3 & 0 \\ 2 & 3 \end{bmatrix} \quad \begin{bmatrix} 1 & 0 \\ 3 & 3 \end{bmatrix} \quad \begin{bmatrix} 1 & 3 \\ 3 & 2 \end{bmatrix}$$

give the largest game value when extended to the original game.

Another way to solve this problem is to use what we know of linear programming. Recall that when R plays his optimal strategy, whatever it may be, his payoff will be at least as large as v, the game value. R's optimal strategy will be a row probability vector with two components. Suppose the first component of R's strategy vector is x where $0 \leq x \leq 1$. Then because the sum of the components of a probability vector must be 1, the second component of R's strategy vector must be $1 - x$. Then if C plays column 1, R's payoff will be

$$[x, 1 - x]\begin{bmatrix}1\\3\end{bmatrix} = 3 - 2x$$

and this must be at least as large as v. That is,

$$3 - 2x \geq v$$

Similarly, if C plays column 2,

$$[x, 1 - x]\begin{bmatrix}3\\2\end{bmatrix} = x + 2 \geq v$$

and if C plays column 3,

$$[x, 1 - x]\begin{bmatrix}0\\3\end{bmatrix} = 3 - 3x \geq v$$

If we graph the inequalities

$$\begin{aligned} x &\geq 0 \\ x &\leq 1 \\ 3 - 2x &\geq v \\ x + 2 &\geq v \\ 3 - 3x &\geq v \end{aligned}$$

on a rectangular coordinate system having as its vertical axis the v axis, we obtain the polygonal region of Figure 10–1.

We wish v to be maximum, and this is true at the intersection of the lines whose equations are

$$\begin{aligned} x + 2 &= v \\ 3 - 3x &= v \end{aligned}$$

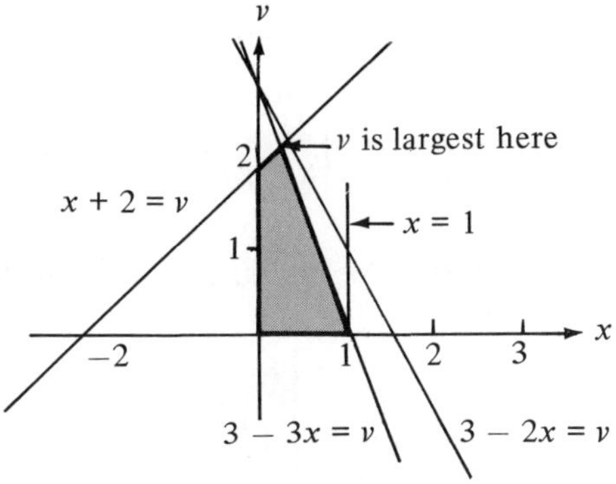

FIGURE 10–1

To find the coordinates of this point, equate the left members of the two equations and solve for x:

$$\begin{aligned} x + 2 &= 3 - 3x \\ x + 2 - 2 &= 3 - 3x - 2 \\ x &= 1 - 3x \\ x + 3x &= 1 - 3x + 3x \\ 4x &= 1 \\ x &= \tfrac{1}{4} \end{aligned}$$

Then from $x + 2 = v$ we get

$$\tfrac{1}{4} + 2 = v$$

or

$$v = 2\tfrac{1}{4}$$

If $x = \frac{1}{4}$, $1 - x = \frac{3}{4}$, and therefore R's strategy is

$$\mathbf{p} = [\tfrac{1}{4}, \tfrac{3}{4}]$$

To find C's strategy, we note that R's strategy was determined by the intersection of the lines which were computed from the second and third columns of the game matrix:

$$\begin{bmatrix} 1 & 3 & 0 \\ 3 & 2 & 3 \end{bmatrix}$$

Consider just these two columns:

$$\begin{bmatrix} 3 & 0 \\ 2 & 3 \end{bmatrix}$$

Then by the method of Section 10.3, C's strategy for this 2×2 game is

$$\begin{bmatrix} \frac{3}{4} \\ \frac{1}{4} \end{bmatrix}$$

and when this is extended to the original game, we have C's strategy for that game:

$$\mathbf{q} = \begin{bmatrix} 0 \\ \frac{3}{4} \\ \frac{1}{4} \end{bmatrix}$$

It can easily be checked by computing

$$[\tfrac{1}{4}, \tfrac{3}{4}]\begin{bmatrix} 1 & 3 & 0 \\ 3 & 2 & 3 \end{bmatrix}\begin{bmatrix} 0 \\ \frac{3}{4} \\ \frac{1}{4} \end{bmatrix}$$

that the game value is $2\frac{1}{4}$.

Note that the matrix

$$\begin{bmatrix} 1 & 0 \\ 3 & 3 \end{bmatrix}$$

determined by the intersection of the lines whose equations are

$$\begin{aligned} x + 2 &= v \\ 3 - 2x &= v \end{aligned}$$

is precisely the matrix $\mathbf{C}$ which we used to solve this problem back in Section 10.4.

EXAMPLE 10.15 Solve the game:

$$\begin{bmatrix} 2 & -1 \\ 1 & 1 \\ 0 & 2 \end{bmatrix}$$

Solution This game is not strictly determined and it has no dominant rows. The graphical method of solving a linear programming problem restricts us to the use of only two variables. Therefore in this problem we shall first find

C's optimal strategy since it must be a two-component column vector. We take as C's strategy the vector

$$\begin{bmatrix} x \\ 1 - x \end{bmatrix}$$

where $x \geq 0$ and $x \leq 1$. Then if R chooses row 1, R's expected value is

$$[2, -1]\begin{bmatrix} x \\ 1 - x \end{bmatrix} = 3x - 1$$

and we assume that C wishes to make this at most v, the value of the game. Thus

$$3x - 1 \leq v$$

Similarly, if R chooses row 2,

$$[1, 1]\begin{bmatrix} x \\ 1 - x \end{bmatrix} = 1 \leq v$$

and if R plays row 3,

$$[0, 2]\begin{bmatrix} x \\ 1 - x \end{bmatrix} = 2 - 2x \leq v$$

The intersection of the graphs of the above inequalities is the polygonal region of Figure 10–2.

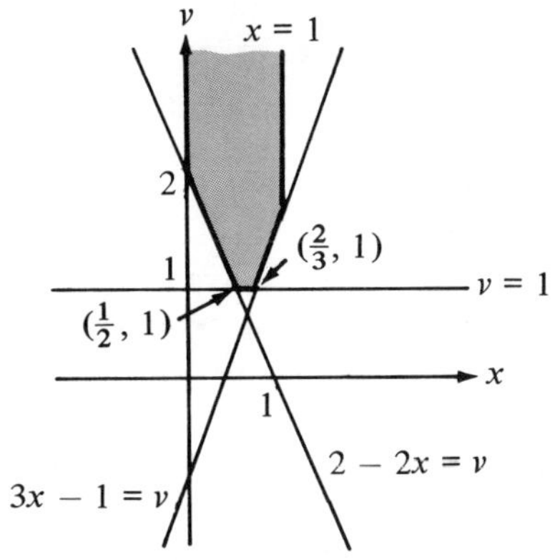

FIGURE 10–2

Player C wishes to minimize R's payoff, and so he will choose a strategy which will guarantee that v is as small as possible. From the graph we see that there are two corners of the polygonal region which make $v = 1$, the smallest it can be on the given region. C may choose either of these as his strategy. One corner is at $(\frac{1}{2}, 1)$ and the other is at $(\frac{2}{3}, 1)$. Thus C may choose either

$$\begin{bmatrix} \frac{1}{2} \\ \frac{1}{2} \end{bmatrix} \quad \text{or} \quad \begin{bmatrix} \frac{2}{3} \\ \frac{1}{3} \end{bmatrix}$$

as his strategy for the game. [Actually because $(\frac{1}{2}, 1)$ and $(\frac{2}{3}, 1)$ are on a horizontal line, C may choose strategy $\begin{bmatrix} x \\ 1 - x \end{bmatrix}$ such that x is any number between $\frac{1}{2}$ and $\frac{2}{3}$ inclusive.] R's strategy, in either case, is $[0, 1, 0]$ and it is found by solving either of the games

$$\begin{bmatrix} 2 & -1 \\ 1 & 1 \end{bmatrix} \quad \text{or} \quad \begin{bmatrix} 1 & 1 \\ 0 & 2 \end{bmatrix}$$

The first of these consists of the rows of the original game which were used to compute the inequalities

$$\begin{aligned} 3x - 1 &\leq v \\ 1 &\leq v \end{aligned}$$

and the second consists of the rows which were used to compute the inequalities

$$\begin{aligned} 1 &\leq v \\ 2 - 2x &\leq v \end{aligned}$$

The linear programming method can be extended to games whose matrices have any dimension; however for games other than $m \times 2$ and $2 \times n$, the graphical method of solution cannot be used.

In this chapter, we have considered only games having two participants. We have also considered only "cutthroat" games in which one player profits at the expense of the other. There are many situations, particularly in economics, in which it may be possible for both players to obtain larger payoffs if they work together. Game theory has been extended to such games, but at the present time it has not provided satisfactory answers to all the questions which arise.

Game theory has also been applied to games with more than two players. In these games, almost always, some of the players join together to the detriment of other players. One of the most important tasks of game theory is to try to determine why and when such coalitions will occur.

Exercise Set 10.5

1. Solve the following games using the method of this section.

(a) $\begin{bmatrix} 6 & 5 & 7 & 1 \\ 2 & 4 & 3 & 5 \end{bmatrix}$ (b) $\begin{bmatrix} 0 & 2 \\ 1 & 3 \\ -1 & 0 \\ 2 & 0 \end{bmatrix}$

(c) $\begin{bmatrix} 1 & 2 & 3 \\ 3 & 2 & 1 \end{bmatrix}$ (There is more than one optimal strategy for C).

(d) $\begin{bmatrix} 1 & 3 & 1 & 5 \\ 2 & 0 & 1 & -1 \end{bmatrix}$

2. Solve the following games. Check first for saddle points and then try to reduce the game by row and column domination.

(a) $\begin{bmatrix} 4 & 1 & 2 \\ 3 & 5 & 1 \\ 2 & 1 & 1 \end{bmatrix}$ (c) $\begin{bmatrix} 3 & 5 & 1 & 2 \\ 4 & 8 & 3 & 0 \\ 5 & 4 & 3 & 6 \end{bmatrix}$

(b) $\begin{bmatrix} 3 & 6 & 4 \\ 1 & 0 & 8 \end{bmatrix}$ (d) $\begin{bmatrix} 0 & 2 \\ 1 & 3 \\ -1 & 1 \end{bmatrix}$

(e) R and C decide to play the following game. R is to hide a half-dollar in one hand and a quarter in the other. If C correctly guesses the coin held, he receives the coin. What are the strategies and what is the game value?

(f) General N. T. Witt, commanding two divisions, must capture a pass defended by one division. To do this, he must have a numerical superiority of one division when he attacks the pass. It may be, however, that while the General attacks, his opponent simultaneously attacks the General's camp. If so, a superiority of one division is also sufficient to win. The General gains 1 unit if he captures the opponents camp without losing his own; he loses 1 unit if the opponent captures his camp. Otherwise the game is a tie. What is the General's optimal strategy?

(g) Two players play this game. Each player shows from one to three fingers. R always pays C an amount equal to the number of fingers that C shows; if C shows exactly one more or two fewer fingers than R, then C pays R $\frac{1}{2}$ unit. Find the strategies and the game value.

(h) In the game of Problem 2(g) suppose that R pays C an amount equal to the number of fingers C shows and if C shows exactly one more or two fewer fingers than R, then C pays R two units. Solve this game.

SUMMARY

Terms you should know:

1. Game
2. Game matrix
3. Payoff
4. Strategy
5. Optimal strategy
6. Saddle point
7. Strictly determined
8. Probability vector
9. Mixed strategy
10. Dominance

To solve two-player games:

1. Set up a game matrix.
2. Test for saddle points.
3. Check for row or column dominance.
4. If after removing dominated rows and columns the remaining matrix has no saddle point and:

   (a) It is 2×2, find the optimal strategies

$$\mathbf{p} = [p_1, p_2] \qquad \text{and} \qquad \mathbf{q} = \begin{bmatrix} q_1 \\ q_2 \end{bmatrix}$$

by means of the formulas

$$p_1 = \frac{a_{22} - a_{21}}{a_{11} + a_{22} - a_{12} - a_{21}}$$

$$p_2 = \frac{a_{11} - a_{12}}{a_{11} + a_{22} - a_{12} - a_{21}}$$

$$q_1 = \frac{a_{22} - a_{12}}{a_{11} + a_{22} - a_{12} - a_{21}}$$

$$q_2 = \frac{a_{11} - a_{21}}{a_{11} + a_{22} - a_{12} - a_{21}}$$

and the game value is

$$v = \frac{a_{11}a_{22} - a_{12}a_{21}}{a_{11} + a_{22} - a_{12} - a_{21}}$$

(b) It is $m \times 2$ or $2 \times n$, optimal strategies may be found by the method of derived games (Section 10.4) or by the linear programming method (Section 10.5).

REVIEW EXERCISES

Find optimal strategies and the game value for each game.

1. $\begin{bmatrix} 15 & 3 \\ -1 & 2 \end{bmatrix}$

2. $\begin{bmatrix} -7 & 4 & 0 \\ 6 & 5 & 7 \\ 15 & 2 & -3 \end{bmatrix}$

3. $\begin{bmatrix} 0 & 5 & 6 & -3 \\ 1 & -1 & 2 & 3 \\ 1 & 2 & 3 & 4 \\ -1 & 0 & 7 & 5 \end{bmatrix}$

4. $\begin{bmatrix} -1 & -2 & 0 \\ 8 & -5 & 7 \end{bmatrix}$

5. $\begin{bmatrix} 1 & 0 & 2 \\ 0 & 3 & 2 \end{bmatrix}$

6. $\begin{bmatrix} 1 & 0 & 1 & 1 & 2 \\ 0 & -1 & -2 & -3 & -10 \end{bmatrix}$

7. $\begin{bmatrix} 3 & 5 \\ 4 & 2 \\ 2 & 1 \end{bmatrix}$

8. $\begin{bmatrix} 4 & 1 & 0 \\ 1 & 4 & 3 \\ 0 & 3 & 2 \end{bmatrix}$

9. A game is played as follows. Player R is given the aces of spades and diamonds and the 2 of hearts. Player C is given the aces of hearts and clubs and the 2 of spades. Each then shows a card simultaneously. If both cards are the same suit, then the ace beats the 2. If the cards are of different suits and different denominations, then the black card beats the red. If they are of the same denomination, then the red card beats the black. If they are both black aces or both red aces, there is a tie. The player winning will win as many units as there are spots on his opponent's card. Solve this game.

10. Al has invitations to two parties on the same evening, one at Joe's pad and one at Fred's place. The only problem is that his ex-girlfriend also has invitations to the same party. If both of them attend the same party,

there could be a messy scene. Al would like to go to Joe's party because he has a yen for Joe's sister and she will be at the party. Give Al 5 points if he avoids his ex and 0 if both of them attend the same party. Also give Al 2 extra points for going to Joe's party. Find optimal strategies for both Al and his former girlfriend.

11. Two companies, one large and one small, manufacturing the same product, wish to build a new store in one of four towns located on a given highway. The populations and locations of the towns are shown in Figure 10–3. If the large company is nearer a town, it will capture 80 percent of the business of that town; if both stores are equally distant from a town, then the large company will capture 60 percent of the business of that town; and if the small store is nearer a town, then the large company will capture 40 percent of the business of that town. Assume every inhabitant of a town is a consumer of the product. Make a game matrix whose entries are sales by the large company, and determine in which locations the companies should build.

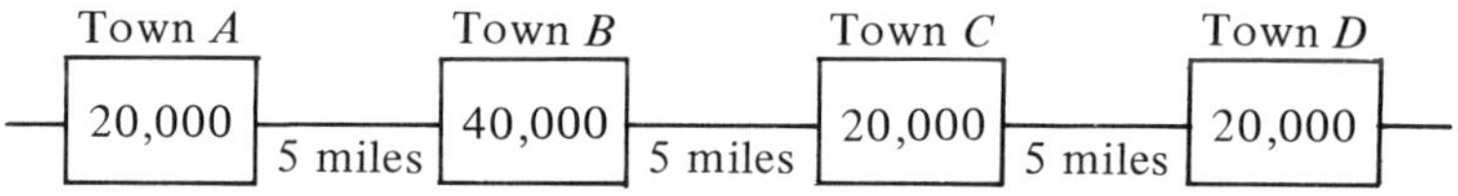

FIGURE 10–3

12. Two players play this card game. Player R has a black 9, a black 5, a red 7, and a red 3. Player C has a black 8, a red 4, and a red 3. Both players show a card simultaneously. If the colors match, C receives an amount equal to the difference of the numbers on the cards. If the colors are not the same, R collects the difference in points. Solve this game.

APPENDICES

Appendix A
A REVIEW OF SOME ALGEBRAIC CONCEPTS

Certain sections of this book require a knowledge of some of the ideas of algebra. This appendix reviews these ideas.

VOCABULARY—OPERATIONS

VARIABLE We use symbols to represent numbers. When a symbol can represent only one number it is called a **constant**, but when it is free to take on several different values it is called a **variable**. Ten, 3, π, and IX are constants but x, y, $\square$, and θ are symbols often used as variables.

ALGEBRAIC OPERATIONS The algebraic operations are:

1. *Addition*, denoted by +, that is,

$$3 + 4 \qquad x + y \qquad 4 + x + w \qquad \text{etc.}$$

2. *Subtraction*, denoted by $-$, that is,

$$5 - 3 \qquad x - y \qquad \text{etc.}$$

3. *Multiplication* In algebra it is common practice not to use $\times$ as the symbol for multiplication since there is a possibility that it will be confused with the variable x. Instead we use four different methods:

   (a) To indicate the product of constants such as 3, 5, 15, $\frac{1}{2}$, etc., we use parentheses around the symbols for the numbers. Thus (3)(5) means "multiply 3 by 5," and (2)(7)(6) means "the product of the numbers 2, 7, and 6."

   (b) To indicate the product of a constant by a variable or the product of two or more variables, we write the symbols in a line next to each other. Thus $3x$ indicates the product of the constant 3 by the variable x, and xy means "multiply the number represented by x by the number represented by y."

   (c) An alternative way to indicate the product of a single constant by a variable or the product of two or more variables is to write the symbols for the numbers in a line and place a raised dot between them. Thus we may write $4 \cdot x$ or $m \cdot n \cdot p$ instead of $4x$ or mnp.

   (d) If the numbers to be multiplied all have the same value, we use an **exponent**. Thus the 3 in 2^3 is an exponent. It tells us that we are to compute the product (2)(2)(2). Similarly 5^4 means (5)(5)(5)(5) and x^7 means $x \cdot x \cdot x \cdot x \cdot x \cdot x \cdot x$. When an exponent is used, the product is called a **power**. Thus 3^2 is the *second power of 3* and 2^{10} is the *tenth power of 2.*

4. *Division* Two methods are used to denote division. The first of these is the ordinary division sign $\div$. The second method, which is more commonly used in algebra, is to write the two numbers in fractional form. Thus $\frac{3}{4}$ means "divide 3 by 4," $x/7$ means "divide x by 7," and $2/y$ means "divide 2 by y."

5. *Roots* If x is a positive number, and if:

$$\begin{array}{lll} x^2 = y & \text{then} & \sqrt{y} = x \\ x^3 = y & \text{then} & \sqrt[3]{y} = x \\ x^4 = y & \text{then} & \sqrt[4]{y} = x \end{array}$$

and so on.

$\sqrt{y}$ is called the **square root** of y, $\sqrt[3]{y}$ is called the **cube root** of y, $\sqrt[4]{y}$ is called the **fourth root** of y, etc.

ALGEBRAIC EXPRESSIONS An **algebraic expression** is any combination of variables and constants in which the variables and constants are combined by means of one or more of the algebraic operations.

$$3 - 2 \qquad 2x + y \qquad x^2 - xy \qquad \frac{3x + 1}{x - 2}$$

are examples of algebraic expressions.

EVALUATING AN EXPRESSION To **evaluate an expression** one replaces each of the variables by a constant and then performs all the indicated operations.

EXAMPLE

$$3x + y^2 \div 2 - 9$$

Evaluate if $x = 5$ and $y = 4$.

Solution Replace the x and y in $3x + y^2 \div 2 - 9$ by 5 and 4, respectively.

$$3(5) + 4^2 \div 2 - 9$$

Raise 4 to the power indicated:

$$3(5) + 16 \div 2 - 9$$

Go through the resulting expression from left to right performing all the multiplications and divisions *as you come to them.*

$$15 + 8 - 9$$

Finally go through the expression left after the multiplications and divisions are completed, and do all additions and subtractions *in the order in which you come to them.* The sum of 15 and 8 is 23 and $23 - 9$ is 14. Therefore when $x = 5$ and $y = 4$, the value of $3x + y^2 \div 2 - 9$ is 14.

The order in which the operations were performed in the example above is one which has been adopted by all mathematicians so that everyone will obtain the same answer to a given problem.

ORDER OF OPERATIONS

1. Raise all numbers to their indicated powers. (If roots are involved, one computes these in this step also.)

2. Working from left to right, perform all multiplications and divisions in the order in which they occur.
3. Working from left to right, perform all additions and subtractions in the order in which they occur.

PARENTHESES **Parentheses** are used as indicated above to show multiplication. They are also used to show exceptions to the order in which operations shall be performed. For example,

$$(8 + 2) \div 5$$

means "first add the 8 and the 2 and *then* divide the sum by 5." When used in this way, parentheses always demand "do the operation inside the parentheses first." Other symbols which are used in the same way as parentheses are the brackets, [ ], and the braces, { }.

Exercises for Practice

1. Make the following computations:

   (a) $6 - 3(3 - 1)$ (b) $2^2 + 3^3 - 1^5$ (c) $12 + 4 \div 8$

   Answers (a) 0 (b) 30 (c) $12\frac{1}{2}$

2. Evaluate each expression if $x = 2$, $y = 3$, and $z = 5$:

   (a) $2xy^2 + 4$ (b) $\dfrac{3x - y}{z}$ (c) $x^3 + y^2 - (x + y)$ (d) $(4x - y)^2$

   Answers (a) 40 (b) $\frac{3}{5}$ (c) 12 (d) 25

THE NUMBER LINE

Figure App–1 shows a number line. Each point of the line to the right of the point labeled with a 0 is associated with one of the positive numbers, and each point of the line to the left of the 0 is associated with one of the negative numbers.

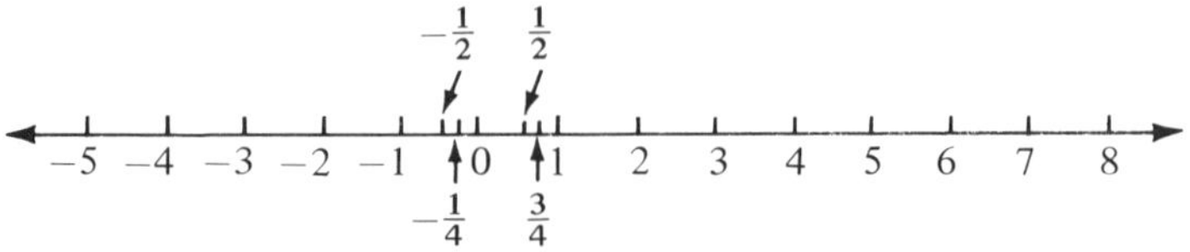

FIGURE App–1 A number line.

Recall that -3 is read "negative 3," and it denotes the number which is associated with a point on the "negative side" of a number line and 3 units from the point labeled 0. Similarly $+5$ is the number associated with a point on the "positive side" of a number line and 5 units from the 0 point. (In practice we often drop the + sign when writing a positive number.)

ABSOLUTE VALUE Geometrically, the **absolute value** of a number is the distance of that number from 0 on the number line. If x is a symbol for a number, then $|x|$ denotes the absolute value of that number. For example,

$|+3| = 3$ (*because* $+3$ *is located* 3 *units from* 0 *on the number line*)
$|-3| = 3$ (*because* -3 *is located* 3 *units from* 0 *on the number line*)
$|0| = 0$ (*because* 0 *is no distance from* 0)

OPERATIONS WITH SIGNED NUMBERS

ADDITION We may use the number line to define addition of signed numbers. To perform the addition indicated by $(-3) + (+5)$, start at -3 on the number line and count 5 units in the positive direction (to the right in Figure App–1). This takes us to the point labeled $+2$. Therefore

$$(-3) + (+5) = +2$$

To add $(-2) + (-4)$ using the number line, we start at -2 and count 4 units in the negative direction, ending at -6. Therefore

$$(-2) + (-4) = -6$$

EXAMPLE Use the number line to verify each of these sums:

$(+1) + (-4) = -3$ $\quad$ $(-4) + (+6) = +2$
$(+2) + (+2) = 4$ $\quad$ $(-3) + (+3) = 0$
$(-1) + (-2) = -3$ $\quad$ $5 + [(-2) + (+4)] = +7$

Using the number line as an aid in addition becomes cumbersome if the absolute values of the numbers are great. Therefore we have the following two-part rule.

RULE FOR ADDITION OF SIGNED NUMBERS

1. If two numbers have the same sign, both positive or both negative, their sum can be found by adding their absolute values and giving this sum the sign of the two numbers. As an example,

$$(-3) + (-18) = -21 \qquad \textit{because } |-3| = 3, |-18| = 18$$

and

$$3 + 18 = 21$$

2. If the two numbers have opposite signs, one negative and one positive, find the difference of their absolute values (smaller absolute value subtracted from larger absolute value) and give this difference the sign of the number with the larger absolute value. Thus

$$(-3) + (+2) = -1$$

since

$$|-3| - |+2| = 1$$

and because -3 is farther away from 0 than is $+2$. Similarly,

$$(-1) + (+3) = +2$$

since

$$|+3| - |-1| = 2$$

and $+3$ is farther from 0 than is -1.

SUBTRACTION Subtraction can be performed on the number line in the following way.

If a and b are signed numbers, to compute $a - b$, start at a on the number line and count b units in the opposite direction as the sign of b. For example, to compute $+3 - (-2)$, start at $+3$ and count 2 units in the positive direction (since the sign of -2 is a negative sign) ending on $+5$. Therefore

$$+3 - (-2) = +5$$

An algebraic rule for subtraction is:

RULE FOR SUBTRACTING SIGNED NUMBERS To compute $a - b$, change the sign of b and then add the resulting number to a, *using the rules for adding signed numbers.*

EXAMPLE

$$(+2) - (-3) = (+2) + (+3) = +5$$
$$(-3) - (+2) = (-3) + (-2) = -5$$
$$(-3) - (-2) = (-3) + (+2) = -1$$

As a consequence of the above rule we may interpret problems such as $2 - 3$ to mean $2 + (-3)$.

There is no simple geometric interpretation for either of the operations multiplication or division. The easiest thing to do is to memorize the following rules.

RULE FOR MULTIPLICATION To multiply two numbers having the same sign, multiply their absolute values and give this product a positive sign. To multiply two numbers with opposite signs, multiply their absolute values and give this product a negative sign.

EXAMPLES

$$(+3)(+5) = +15$$
$$(-3)(-5) = +15$$
$$(-3)(+5) = -15$$
$$(+3)(-5) = -15$$

RULE FOR DIVISION To divide a signed number by another signed number when both have the same sign, divide their absolute values and give the quotient a positive sign. If the numbers have opposite signs, divide their absolute values and give the quotient a negative sign.

EXAMPLE

$$(-4) \div (-2) = +2 \qquad \frac{-4}{+2} = -2$$

$$(+4) \div (+2) = +2 \qquad \frac{+4}{-2} = -2$$

Exercises for Practice

1. Perform these computations:
   (a) $(-3) + (-5)$ (c) $4 + (-2)$
   (b) $(+2) + (+3)$ (d) $-6 + (-8)$

(e) $[(-2) + (-3)] + (-4)$

(f) $[4 + (-7)] + (+5)$

(g) $(-1) - (-2)$

(h) $(-5) - (+3)$

(i) $4 - 7$

(j) $56 - (-8)$

(k) $(-2)(-5)$

(l) $(-1)(-18)$

(m) $(+6)(-30)$

(n) $(-12) \div (+6)$

(o) $(-40) \div (-10)$

(p) $15 \div 3$

(q) $\dfrac{-45}{-5}$

Answers

(a) -8 (b) $+5$ (c) $+2$ (d) -14

(e) -9 (f) $+2$ (g) $+1$ (h) -8

(i) -3 (j) $+64$ (k) $+10$ (l) $+18$

(m) -180 (n) -2 (o) $+4$ (p) $+5$

(q) $+9$

2. Evaluate the following expressions if $a = 2$, $b = -3$, and $c = -1$.

(a) $3b + 4$

(b) $a + [b + c]$

(c) $a(bc)$

(d) $a^2 - b$

(e) $a + bc$

(f) $(a - b) - c$

Answers (a) -5 (b) -2 (c) 6 (d) 7 (e) 5 (f) 6

SIMPLIFYING EXPRESSIONS

TERMS In an expression, numerals such as $2x$, $3y$, and xyz which are separated from one another by signs of addition or subtraction are called **terms** of the expression. There are three terms in the expression $3xy + 2z + 5$. They are $3xy$, $2z$, and 5. The numbers which are multiplied to make a term are called the **factors** of the term. The factors of $3xy$ are 3, x, and y. If one of the factors of a term is a constant such as 2, 3, or -5, it is called the **numerical coefficient** of the term. In the term $3xy$, the 3 is the numerical coefficient of the term. Similarly -2 is the numerical coefficient of the term $-2y$. If no constant is written as a factor of a term, the numerical coefficient is understood to be $+1$. Thus each of the terms, x, xy, and x^2 has $+1$ as its numerical coefficient. On the other hand, the numerical coefficient of terms such as $-x$, $-y^3$, and $-yz$ is understood to be -1.

ADDITION OF TERMS Terms that have exactly the same variable factors are called **like terms**. The variable factors in two like terms must have the same exponents. Thus $3x^2$ and $4x^2$ are like terms, but $3x^2$ and $4x$ are not like terms. To add like terms, add their numerical coefficients and write this sum as the numerical coefficient of a term that has exactly the same variable factors as the original terms.

EXAMPLE

$$2x + 3x = 5x$$
$$3xy + (-xy) = 2xy$$

If two terms have different variable factors, then their sum is indicated by placing a + sign between them, but they *cannot be combined into a single term.*

MULTIPLICATION OF TERMS To multiply terms, multiply their numerical coefficients and write the product as the numerical coefficient of a term that has as its variable factors the product of the variable factors of the original terms.

EXAMPLE

$$(3x)(5y) = 15xy$$
$$(-2x)(3xy) = -6x^2y$$
$$(2)(4x)(yz) = 8xyz$$
$$-4(abc)(ac) = -4a^2bc^2$$

We call the process of adding like terms and multiplying terms **simplification.**

EXAMPLE Simplify $3x - 4y + (2)(3x) - x$.

Solution Using the order of operations discussed earlier, we have

$$\begin{aligned} 3x - 4y + (2)(3x) - x &= 3x - 4y + 6x - x \\ &= 8x - 4y \end{aligned}$$

EQUATIONS

An **equation** is an expression containing one or more variables which becomes true or false when the variables are replaced by constants. The following are examples of equations:

$$3x + 4 = 12$$
$$x^2 + x - 1 = 0$$
$$x + y - 2 = 0$$

We shall be concerned here only with equations in one variable in which the exponent of the variable is 1. Examples of this type of equation are:

$$x + 4 = 9$$
$$2x = 8$$

The two expressions of an equation that are separated by the equal sign (=) are called **members** of the equation. The member on the left side of the equal sign is the **left member**, and the one on the right side of the equal sign is the **right member**.

The solution of an equation of the type we are discussing is the number that, when the variable of the equation is replaced by it, makes the equation true. In the equation $x - 4 = 7$, for instance, the solution is 11 because when the x is replaced by 11 the equation is true.

EQUIVALENT EQUATIONS Two equations are **equivalent** if both have the same solution.

$x + 2 = 8$ and $x = 6$ are equivalent equations because each is true when the x is replaced by 6.

An equation can be transformed to an equivalent equation by:

1. Adding the same number to both of its members
2. Multiplying both of its members by the same nonzero number
3. Replacing expressions by other expressions having the same value

To find the solution of an equation, we transform it step by step into a sequence of equivalent equations ending with one which has the variable with coefficient +1 as one member and a constant as its other member.

EXAMPLE Solve the equation $2x - 4 = 6$.

Solution

Step 1.	$2x - 4 = 6$	(*Write the original equation*)
Step 2.	$2x - 4 + 4 = 6 + 4$	(*Add 4 to both members so that the left member will contain only an x term*)
Step 3.	$2x = 10$	(*Simplify both members of the last equation*)
Step 4.	$(\frac{1}{2})(2x) = (\frac{1}{2})(10)$	(*Multiply both members of the equation in step 3 by $\frac{1}{2}$ so that after simplification there will be only an x term in the left member*)
Step 5.	$x = 5$	(*Simplify both members of the equation resulting from doing step 4*)

Step 6. $2(5) - 4 = 10 - 4$ *(Replace the x in the original equation with a 5 and simplify to check on any possible errors in computations in steps 2 to 6)*
$= 6$

The order of the steps used in solving an equation will vary from equation to equation and also to suit the preference of the individual working the problem. In general the best approach is to transform the given equation to one of the form

$$a\square = b$$

in which the $\square$ is the variable of the equation and a and b are constants. Then one multiplies each member of this equation by $1/a$ to obtain the solution.

EXAMPLE One method of finding the solution to the equation $3x - 5x = x - 4$ is outlined below.

$3x - 5x = x - 4$

$-2x = x - 4$ *(The terms in the left member were combined by addition)*

$-3x = -4$ *($-x$ was added to both members so that there would be no x term on the right)*

$x = \frac{4}{3}$ *(Both members were multiplied by $-\frac{1}{3}$)*

Exercises for Practice

1. For each of the following tell what you would do to both members to transform the given equation to one in the form $x =$ (some constant).

(a) $3x = 4$
(b) $-2x = -6$
(c) $\frac{1}{3}x = 2$
(d) $-\frac{1}{2}x = 5$
(e) $x - 7 = 4$
(f) $x - 3 = -5$
(g) $x + 6 = 7$
(h) $10 + x = 2$

Answers

(a) Multiply by $\frac{1}{3}$
(b) Multiply by $-\frac{1}{2}$
(c) Multiply by 3
(d) Multiply by -2
(e) Add 7
(f) Add 3
(g) Add -6
(h) Add -10

2. Solve each of the following equations.

(a) $3x = 12$	(g) $2x - 4 = 8$
(b) $2x = 8$	(h) $3x - 5 = 4$
(c) $x - 4 = -2$	(i) $4x + 1 = 7$
(d) $x - 6 = 4$	(j) $5 = 2x - 6$
(e) $x + 5 = 7$	(k) $2x - 1 = x + 4$
(f) $x + 10 = -1$	(l) $x - 3x = 6$

Answers

(a) 4	(e) 2	(i) $\frac{3}{2}$ or $1\frac{1}{2}$
(b) 4	(f) -11	(j) $\frac{11}{2}$
(c) 2	(g) 6	(k) 5
(d) 10	(h) 3	(l) -3

Appendix B

TABLE I Square Roots

n	$\sqrt{n}$	n	$\sqrt{n}$
1	1.00	26	5.10
2	1.41	27	5.20
3	1.73	28	5.29
4	2.00	29	5.39
5	2.24	30	5.48
6	2.45	31	5.57
8	2.65	32	5.66
8	2.83	33	5.75
9	3.00	34	5.83
10	3.16	35	5.92
11	3.32	36	6.00
12	3.46	37	6.08
13	3.61	38	6.16
14	3.74	39	6.25
15	3.87	40	6.32
16	4.00	41	6.40
17	4.12	42	6.48
18	4.24	43	6.56
19	4.36	44	6.63
20	4.47	45	6.71
21	4.58	46	6.78
22	4.69	47	6.86
23	4.80	48	6.93
24	4.90	49	7.00
25	5.00	50	7.07

TABLE II Areas of a Standard Normal Distribution. An entry in the table is the proportion of the area under the entire curve which is between z = 0 and a positive value of z.

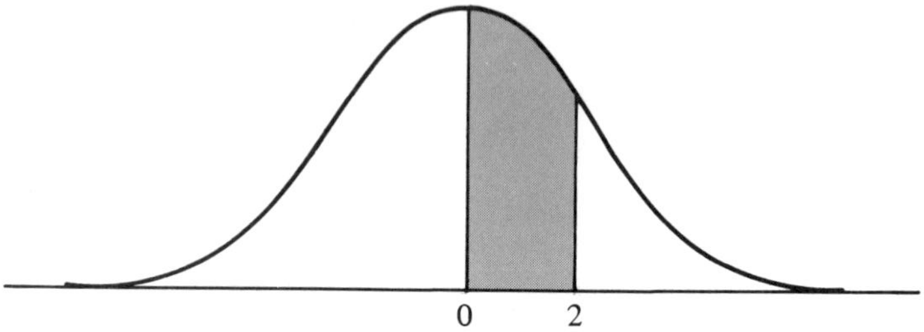

z	Area	z	Area
0.0	0.000	1.6	0.445
0.1	0.040	1.7	0.455
0.2	0.079	1.8	0.464
0.3	0.118	1.9	0.471
0.4	0.155	2.0	0.477
0.5	0.191	2.1	0.482
0.6	0.226	2.2	0.486
0.7	0.258	2.3	0.489
0.8	0.288	2.4	0.491
0.9	0.316	2.5	0.494
1.0	0.341	2.6	0.495
1.1	0.364	2.7	0.496
1.2	0.385	2.8	0.497
1.3	0.403	2.9	0.498
1.4	0.419	3.0	0.499
1.5	0.433		

ANSWERS TO SELECTED PROBLEMS

Exercise Set 2.1

1. Does not define a set.
3. Does not define a set.
5. Defines a set.
7. {4}
9. {0, 2, 4, 6, . . .}
11. $\varnothing$
13. The last four letters of the English alphabet
15. Positive numbers which are divisible by 3
17. $\in$
19. $\in$
21. Members of the Congress of the United States

Exercise Set 2.2

1. Addition, subtraction, multiplication, and division are binary operations. Two unary operations are square root and absolute value.
3. (a) 16 (c) 2
5. {6, 8}
7. {1, 4, 5, 6, 7, 8, 10}
9. {2, 3, 4, 6, 8, 9}
11. {2, 3, 4, 9}
13. U
15. $A \cap B$ is {5, 10}.
17. $A \cap (B \cap C)$ is $\varnothing$.
19. $(A \cap B) \cup (B \cap C)$ is {5, 6, 8, 10}.
21. $\tilde{B}$ is {1, 2, 3, 4, 7, 9}.
23. $\widetilde{A \cap B}$ is {1, 2, 3, 4, 6, 7, 8, 9}.
25. $\widetilde{\tilde{A} \cap \tilde{B}}$ is {1, 5, 6, 7, 8, 10}.

Exercise Set 2.3

1.

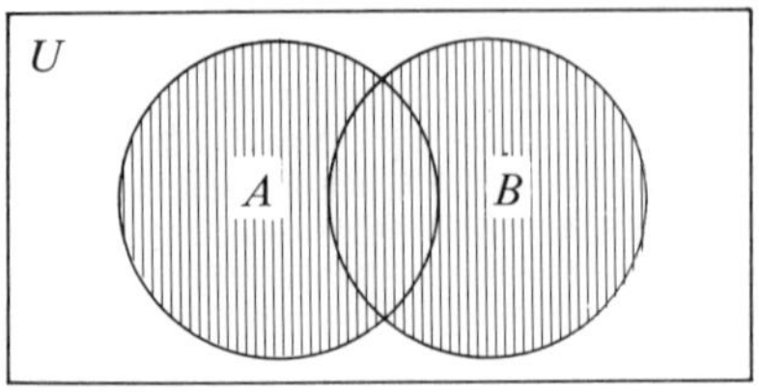

FIGURE A–1

3.

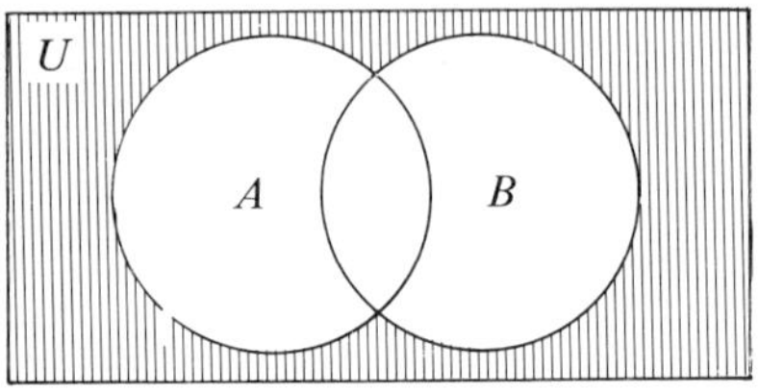

FIGURE A–2

5.

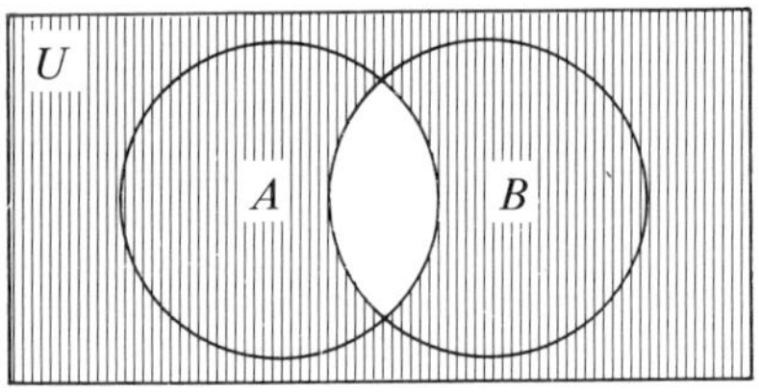

FIGURE A–3

7.

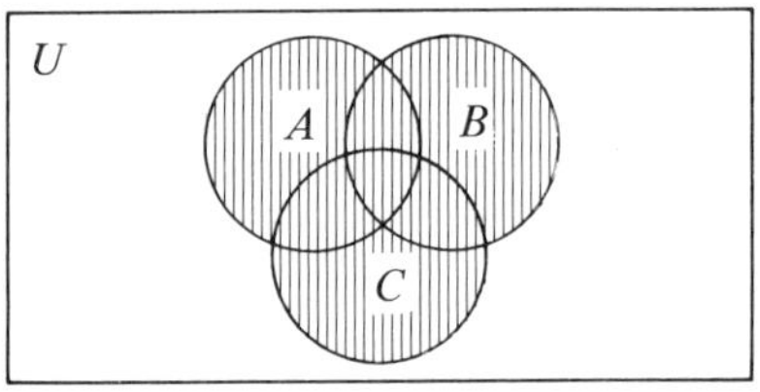

FIGURE A–4

9.

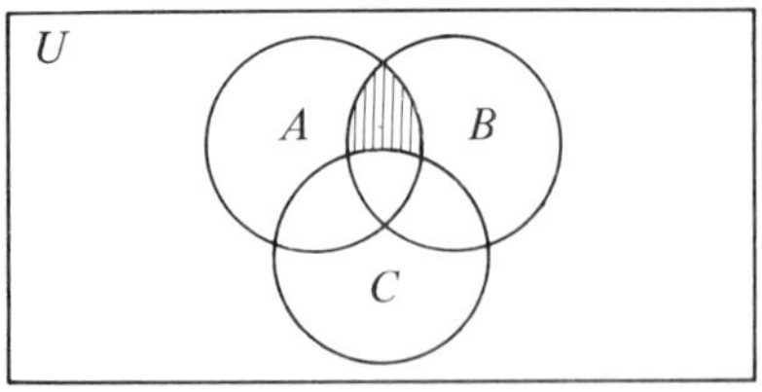

FIGURE A–5

11.

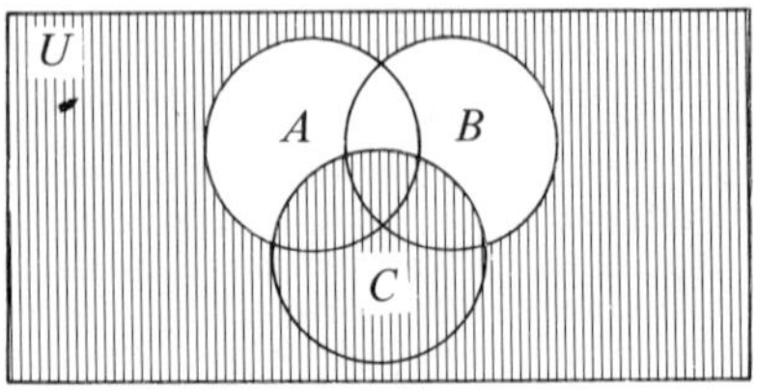

FIGURE A–6

13.

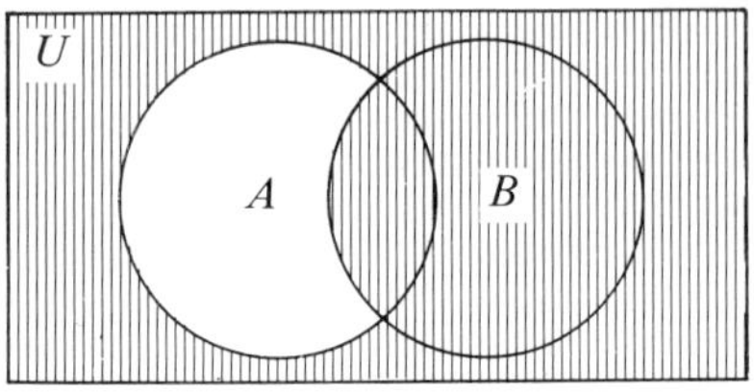

FIGURE A–7

15.

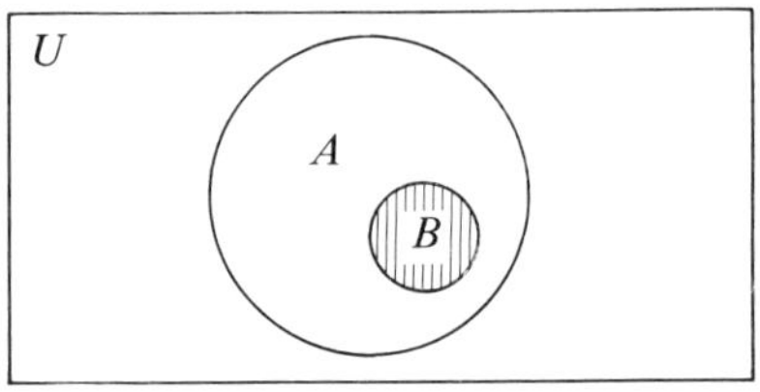

FIGURE A–8

17.

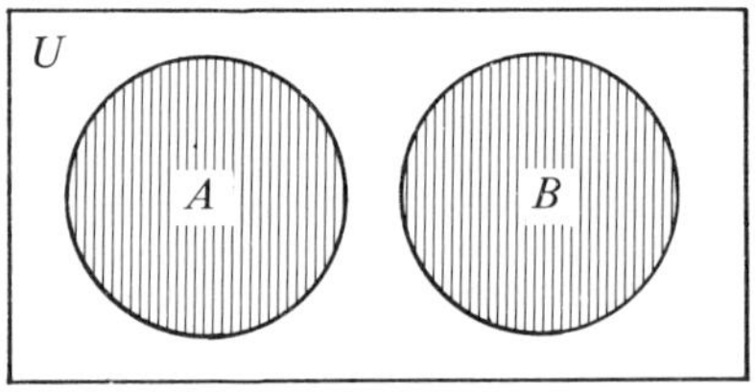

FIGURE A–9

19.

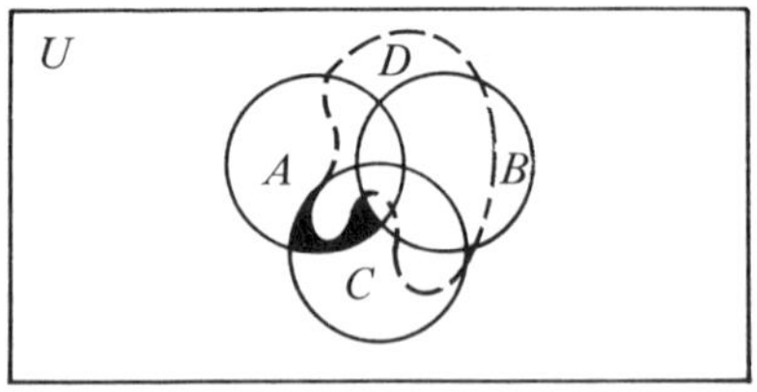

FIGURE A–10

Exercise Set 2.4

1. Figure A–11. $A \cup B$ is the area shaded either vertically or horizontally or both vertically and horizontally. $B \cup A$ is the area shaded either horizontally or vertically or both horizontally and verticallv. Both of these shadings are identical, and therefore $A \cup B = B \cup A$.

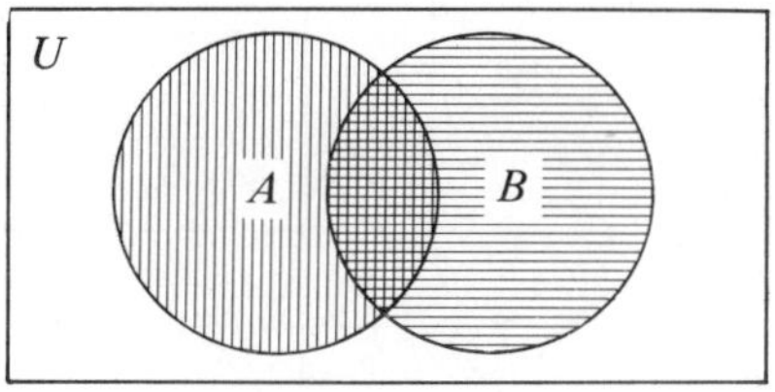

FIGURE A–11

3. Figure A–12. A is shaded horizontally, and $\tilde{A}$ is shaded vertically. $A \cup \tilde{A}$ consists of all regions which have shading on them. Since all of U is shaded, this means that $A \cup \tilde{A} = U$.

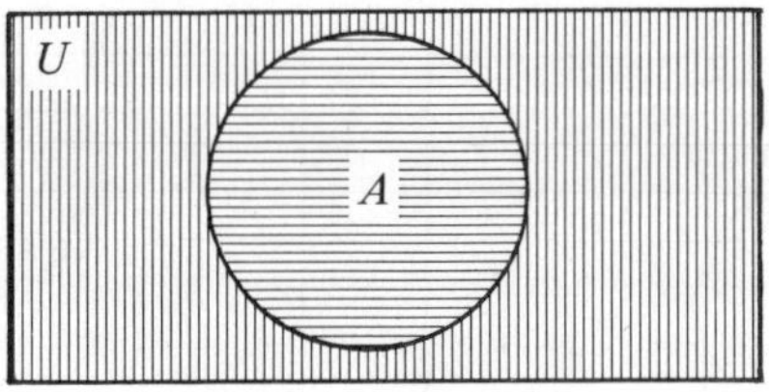

FIGURE A–12

5. Figure A–13. $\tilde{A} \cap B$ is shaded horizontally, and A is shaded vertically. $A \cup (\tilde{A} \cap B)$ consists of all regions with any kind of shading. That region is $A \cup B$. Therefore $A \cup (\tilde{A} \cap B) = A \cup B$.

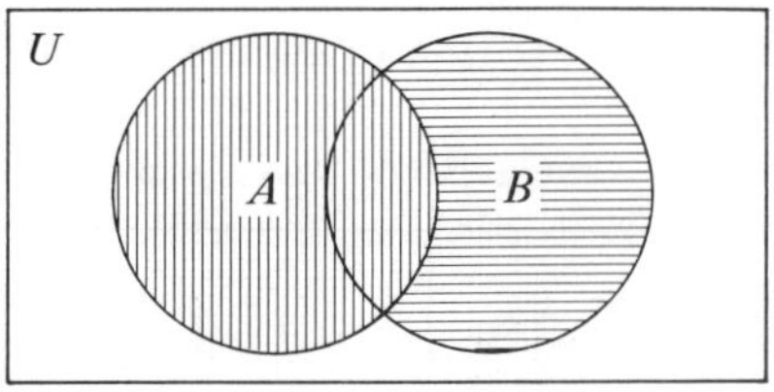

FIGURE A–13

7. Figure A–14. In Figure A–14(a), $A \cap B$ has vertical shading, C has horizontal shading, and the region with both shadings is $(A \cap B) \cap C$. In Figure A–14(b), A has horizontal shading, $B \cap C$ has vertical shading, and

$A \cap (B \cap C)$ has both shadings. Since the shaded region representing $(A \cap B) \cap C$ is exactly the same as the one representing $A \cap (B \cap C)$, we know that $(A \cap B) \cap C = A \cap (B \cap C)$.

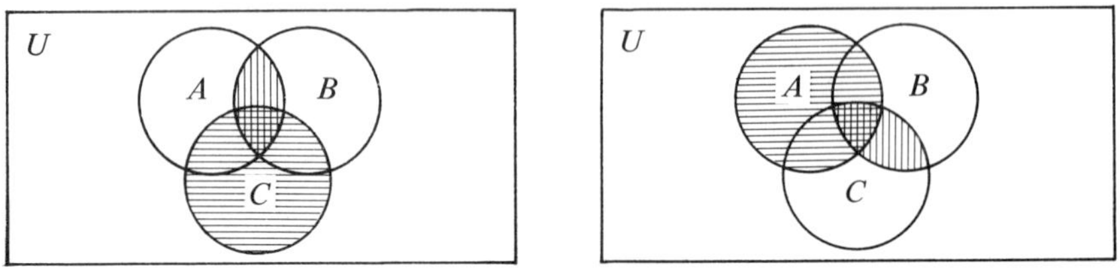

(a) (b)

FIGURE A–14

9. Figure A–15. $A \cup (B \cap C)$ is the entire shaded region of Figure A–15(a), and $(A \cup B) \cap (A \cup C)$ is the region of Figure A–15(b) which has both horizontal and vertical shadings. $A \cup (B \cap C) = (A \cup B) \cap (A \cup C)$ since each of these sets is represented by the same region.

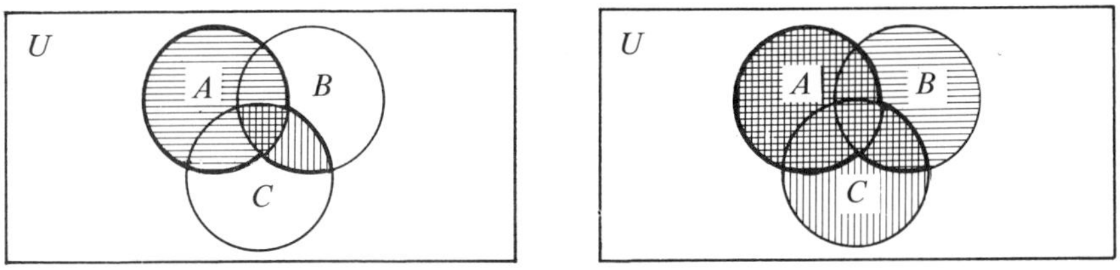

(a) (b)

FIGURE A–15

11. Figure A–16. $\widetilde{A \cap B}$ is the region shaded vertically in Figure A–16(a.) $\tilde{A} \cup \tilde{B}$ is the entire shaded region of Figure A–16(b). $\widetilde{A \cap B} = \tilde{A} \cup \tilde{B}$.

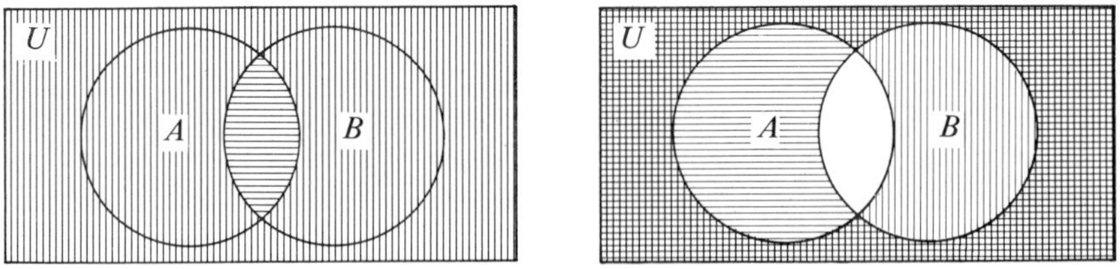

(a) (b)

FIGURE A–16

Exercise Set 2.5

1. $\begin{array}{cc} a, & b \\ \updownarrow & \updownarrow \\ x, & y \end{array}$ or $\begin{array}{c} a, b \\ \times \\ x, y \end{array}$

3. One set is {1, 2, 3, 4}.

5. 4

7. 21

9. 2

11. 1

13. 0

15. Figure A–17
   (a) 6 (b) 0 (c) 3

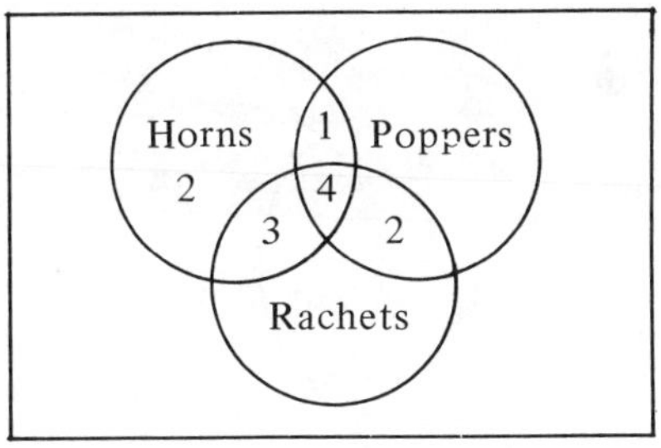

FIGURE A–17

17. One possible correspondence is

$$\begin{array}{cccccc} 1, & 2, & 3, & 4, & \ldots, & n, \ldots \\ \updownarrow & \updownarrow & \updownarrow & \updownarrow & & \updownarrow \\ 0.5, & 1, & 1.5, & 2, & \ldots, & \tfrac{1}{2}n, \ldots \end{array}$$

where n takes on the values 1, 2, 3, 4, ...

19. One possible correspondence is

$$\begin{array}{cccccc} 1, & 3, & 5, & 7, & \ldots, & 2n-1, \ldots \\ \updownarrow & \updownarrow & \updownarrow & \updownarrow & & \updownarrow \\ 2, & 4, & 6, & 8, & \ldots, & 2n \quad, \ldots \end{array}$$

where n takes on the values 1, 2, 3, 4, . . .

Exercise Set 2.6

1. $\{(a, b), (a, c), (a, d), (b, b), (b, c), (b, d), (c, b), (c, c), (c, d)\}$
3. $\{(a, x), (b, x), (c, x)\}$
5. $\{(a, a), (a, b), (a, c), (b, a), (b, b), (b, c), (c, a), (c, b), (c, c)\}$
7. $$\begin{aligned} N(A) + N(B) &= N(A \cup B) + N(A \cap B) \\ &= N(\{a, b, c, d\}) + N(\{b, c\}) \\ &= 4 + 2 \\ &= 6 \end{aligned}$$
9. $N(A)$ multiplied by $N(B) = N(A \times B) = 9$.
11. Yes. Since $A \cup B = B \cup A$ and $A \cap B = B \cap A$, $N(A \cup B) + N(A \cap B)$ must equal $N(B \cup A) + N(B \cap A)$.
13. Figure A–18. $N(A) = 10$, $N(B) = 3$, and $N(A \mid B) = N(A \cap \tilde{B}) = 7$. $N(A) - N(B) = 7 = N(A \mid B)$.

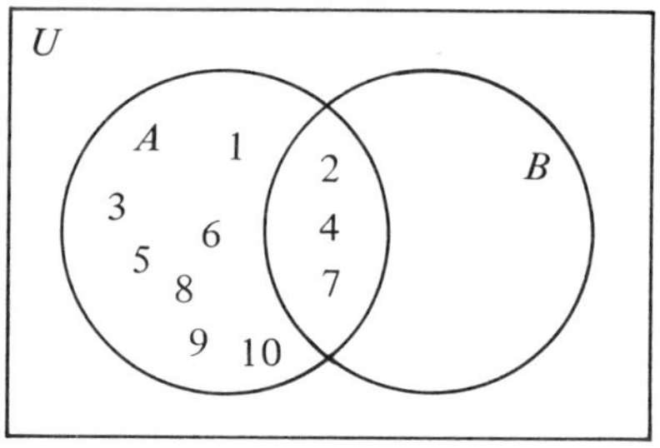

FIGURE A–18

Exercise Set 2.7

1. Finite
3. Finite
5. Infinite
7. $$\begin{array}{ccccccccc} \ldots, & -3, & -2, & -1, & 0, & 1, & 2, & 3, & \ldots \\ & \updownarrow & \updownarrow & \updownarrow & \updownarrow & \updownarrow & \updownarrow & \updownarrow & \\ \ldots, & 6, & 4, & 2, & 1, & 3, & 5, & 7, & \ldots \end{array}$$
9. Let $A = \{1, 2, 3, \ldots\}$ and $B = \{a\}$. Then $A \cup B = \{a, 1, 2, 3, \ldots\}$. Since A and B are disjoint sets, it seems that $A \cup B$ must contain one more element than A. If this is so, $N(A \cup B) = N(A) + 1$; that is, $N(A \cup B) =$

$\aleph_0 + 1$. However, $A \cup B$ can be put into a one-to-one correspondence with A in the following way:

$$\begin{array}{l} a, 1, 2, 3, \ldots \\ \updownarrow \ \updownarrow \ \updownarrow \ \updownarrow \ \cdots \\ 1, 2, 3, 4, \ldots \end{array}$$

Therefore we must conclude that $\aleph_0 + 1 = \aleph_0$.

11. Consider $A \times A$ in which $A = \{1, 2, 3, \ldots\}$. If the first number in each ordered pair of $A \times A$ is made the numerator of a rational number and the second number in the ordered pair is made the denominator, the ordered pairs in $A \times A$ have a one-to-one correspondence with the positive rational numbers. We know $N(\text{positive rationals}) = \aleph_0$. Therefore we conclude $\aleph_0 \times \aleph_0 = \aleph_0$.

REVIEW EXERCISES FOR CHAPTER 2

1. (b) only
2. (a) and (c) through (i) are true.
3. (a) $\subseteq$ (b) $\subseteq$
4. (a) $\subset$. (b) $\not\subset$
5. (a) a, b, c, d
   (b) $\varnothing$, $\{a\}$, $\{b\}$, $\{c\}$, $\{a, b\}$, $\{a, c\}$, $\{b, c\}$, $\{a, b, c\}$
   (c) All but $\{a, b, c\}$
6. (a) The set of green sea horses.
   (b) The set of creatures which are green or horses or live in the sea.
   (c) The set of horses which do not live in the sea.
   (d) The set of creatures which do not live in the sea or are not horses.
7.

(a) 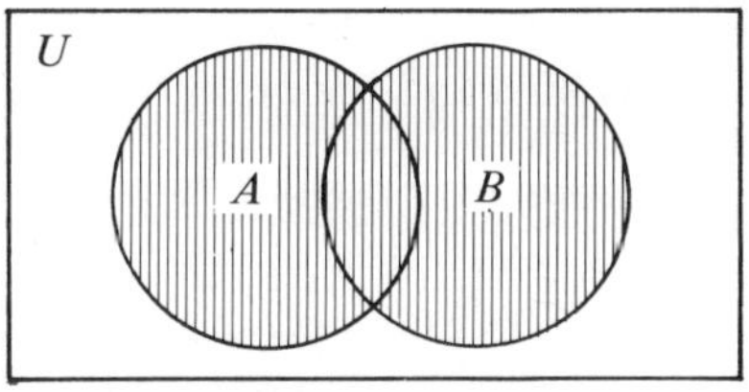

FIGURE A–19

(b) 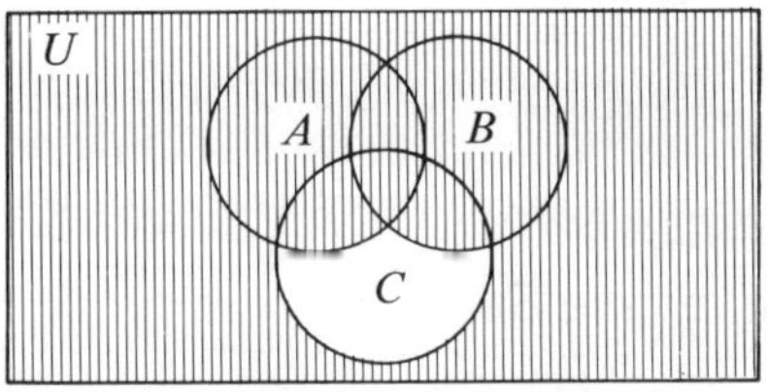

FIGURE A–20

(c)

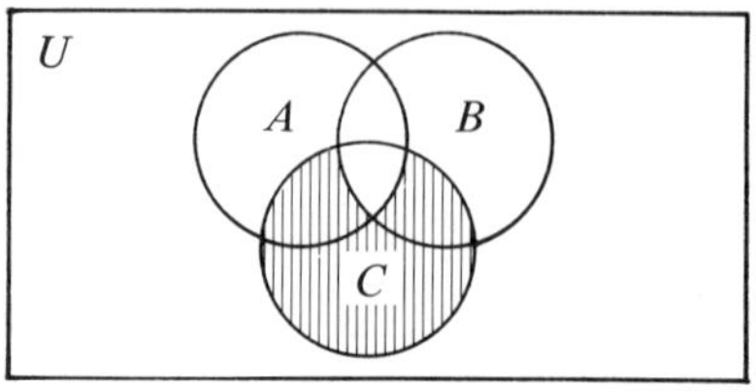

FIGURE A–21

(d)

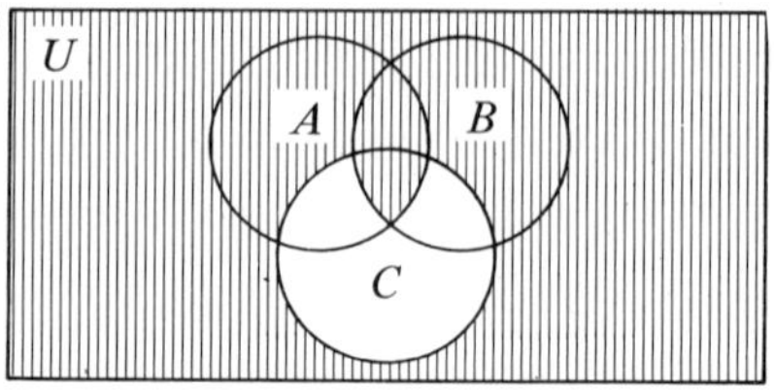

FIGURE A–22

8.

(a)

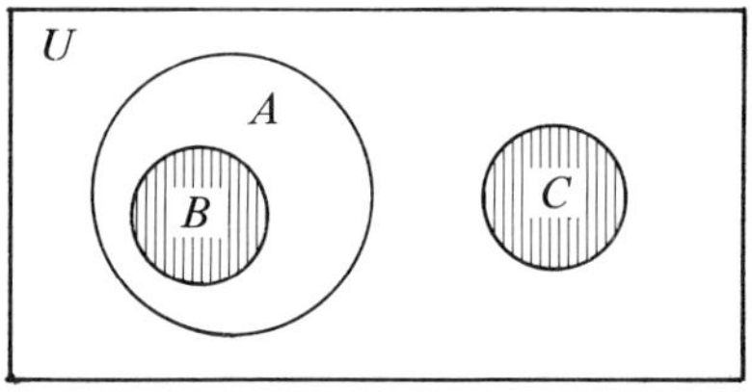

FIGURE A–23

(b)

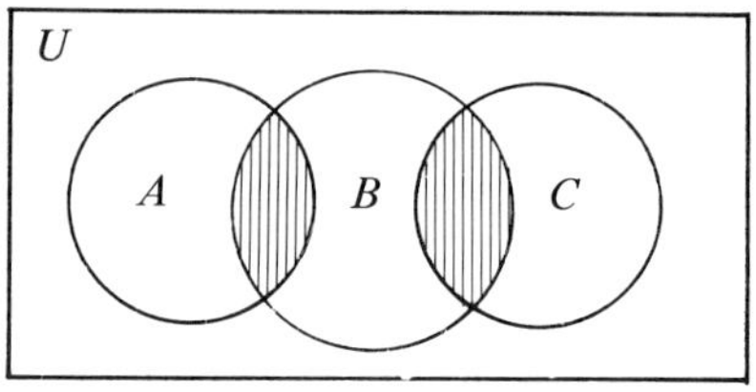

FIGURE A–24

9. (a) is a one-to-one correspondence, (b) and (c) are not.

10. (a) 5 (b) 3 (c) m

11. (a) (x, b) (b) (a, y)

Exercise Set 3.1

1. (a) Simple proposition (e) Not a proposition
   (c) Not a proposition (g) Compound proposition

2. (a) Birds do not fly.
   (c) Birds fly or the sky is blue.
   (e) Birds fly and the sky is blue and the grass is green.
   (g) Birds don't fly or the sky isn't blue.
   (i) If birds fly, then the sky is blue or the grass is green.
   (k) Birds do not fly if and only if the sky is blue and the grass is green.

Exercise Set 3.2

1. (a) F (c) T (e) F (g) T (i) T (k) T (m) F

2.

		(a)	(c)	(e)	(g)	(i)	(k)
p	q	$\sim p \wedge q$	$\sim p \wedge \sim q$	$p \rightarrow \sim q$	$\sim(p \rightarrow q)$	$p \rightarrow (q \rightarrow p)$	$(p \wedge q) \leftrightarrow p$
T	T	F	F	F	F	T	T
T	F	F	F	T	T	T	F
F	T	T	F	F	F	T	T
F	F	F	T	T	F	T	T

3.

		(a)	(b)
p	q	$p \veebar q$	$(p \veebar q) \rightarrow (p \vee q)$
T	T	F	T
T	F	T	T
F	T	T	T
F	F	F	T

Exercise Set 3.3

1. (a)

p	q	$p \wedge q$	$p \vee q$	$(p \wedge q) \rightarrow (p \vee q)$
T	T	T	T	T
T	F	F	T	T
F	T	F	T	T
F	F	F	F	T
①	②	③	④	⑤

(c)

p	q	$p \rightarrow q$	$\sim q$	$(p \rightarrow q) \wedge \sim q$
T	T	T	F	F
T	F	F	T	F
F	T	T	F	F
F	F	T	T	T
①	②	③	④	⑤

(e)

p	q	$\sim p$	$\sim q$	$p \rightarrow q$	$\sim p \vee \sim q$	$(p \rightarrow q) \leftrightarrow (\sim p \vee \sim q)$
T	T	F	F	T	F	F
T	F	F	T	F	T	F
F	T	T	F	T	T	T
F	F	T	T	T	T	T
①	②	③	④	⑤	⑥	⑦

(g)

p	q	r	$q \vee r$	$\sim(q \vee r)$	$p \wedge \sim(q \vee r)$
T	T	T	T	F	F
T	T	F	T	F	F
T	F	T	T	F	F
T	F	F	F	T	T
F	T	T	T	F	F
F	T	F	T	F	F
F	F	T	T	F	F
F	F	F	F	T	F
①	②	③	④	⑤	⑥

(i)

p	q	r	$\sim r$	$p \wedge q$	$(p \wedge q) \wedge \sim r$	$(p \wedge q) \rightarrow r$	$[(p \wedge q) \rightarrow r] \leftrightarrow [(p \wedge q) \wedge \sim r$
T	T	T	F	T	F	T	F
T	T	F	T	T	T	F	F
T	F	T	F	F	F	T	F
T	F	F	T	F	F	T	F
F	T	T	F	F	F	T	F
F	T	F	T	F	F	T	F
F	F	T	F	F	F	T	F
F	F	F	T	F	F	T	F
①	②	③	④	⑤	⑥	⑦	⑧

2. (a) You drink your milk or you do not drink your milk and your bones are not strong.
   (c) You drink milk and your bones are strong or you don't drink milk and your bones are not strong.
   (e) You do not drink milk and your bones are strong.

(g) You do not drink milk, you do take calcium tablets, and your bones aren't strong.

3. Sixteen rows.

Exercise Set 3.4

1. (a) Logically true
   (c) Logically false
   (e) Logically false

3. (a)

p	$\sim p$	$\sim(\sim p)$	$p \leftrightarrow \sim(\sim p)$
T	F	T	T
F	T	F	T

Since $p \leftrightarrow \sim(\sim p)$ is logically true, p and $\sim(\sim p)$ are equivalent.

(c) $\sim(q \vee \sim p)$ is equivalent to $\sim q \wedge \sim(\sim p)$ (*DeMorgan's Law*)
$\sim q \wedge \sim(\sim p)$ is equivalent to $\sim q \wedge p$ (*Part (a) of this exercise*)
$\sim q \wedge p$ is equivalent to $p \wedge \sim q$ (*Commutative Property*)
Therefore $\sim(q \vee \sim p)$ is equivalent to $p \wedge \sim q$.

(e)

p	q	$\sim p$	$\sim p \wedge q$	$p \vee (\sim p \wedge q)$	$p \vee q$	$[p \vee (\sim p \wedge q)] \leftrightarrow (p \vee q)$
T	T	F	F	T	T	T
T	F	F	F	T	T	T
F	T	T	T	T	T	T
F	F	T	F	F	F	T

Since $[p \vee (\sim p \wedge q)] \leftrightarrow (p \vee q)$ is logically true, $p \vee (\sim p \wedge q)$ is equivalent to $p \vee q$.

4. (a) The sun is not shining or it is raining.
   (c) That geometric figure is not a square or its area is not 10 square inches.

5. (b), (c), and (e)

Exercise Set 3.5

1. (a) $(p \wedge q) \rightarrow r$
   (c) $q \rightarrow r$
   (e) $q \rightarrow r$

2. (a) *Converse* If I do stay indoors, it will be because it rains.
Inverse If it does not rain, I shall not stay indoors.
Contrapositive If I do not stay indoors, it will not rain.

(c) *Converse* If a tree is a conifer, it is a pine.
Inverse If a tree is not a pine, it is not a conifer.
Contrapositive If a tree is not a conifer, it is not a pine.

3. (a) F (c) F (e) T (g) T

4. (a) Not equivalent (e) Not equivalent
(c) Equivalent (g) Equivalent

5. Parts (a), (b), and (d) have true converses.

Exercise Set 3.6

1. Valid. Detachment.

3. Invalid. $\sim p \rightarrow q$ and p are both true when q is false and p is true.

5. Invalid. Let p be false, q be true, and r be false; then $p \rightarrow q$ and $p \rightarrow r$ are true but $q \rightarrow r$ is false.

7. Invalid. If r is true and s is false, $r \vee s$ is true.

9. Using only the two premises given, this argument is invalid. Suppose "Yesterday was Tuesday" is true but "Today is Wednesday" is false. Then "If today is Wednesday, yesterday was Tuesday" is a true proposition.

11. Invalid. "Yesterday was Tuesday only if today is Wednesday" is equivalent to "If yesterday was Tuesday, today is Wednesday," and this form is true when "Yesterday was Tuesday" is false and "Today is Wednesday" is true.

Exercise Set 3.7

1. (a) Valid (e) Valid
(c) Valid (g) Invalid. Let p be false, q be false, and r be true.
(i) Invalid. "Wishes are horses" F and "Beggars would ride" T.
(k) Valid
(m) Invalid. "This is December" and "Next month will be January" both false and all the other simple propositions true makes all the premises true and the conclusion false.

2. (a) r
(c) r

(e) No conclusion possible

(g) These premises are contradictory; if they are all true then the proposition "Seven-Year-Itch will tie for first" is both true and false.

REVIEW EXERCISES FOR CHAPTER 3

1. (a) Mathematics is easy and it is fun.
   (b) Mathematics is easy or it is fun.
   (c) Mathematics is not easy or it is not fun.
   (d) Mathematics is not easy and it is not fun. (It is not true that mathematics is easy or fun.)
   (e) Mathematics is easy, fun, and useful.
   (f) If mathematics is not easy, then it is useful.
   (g) If mathematics is easy and fun, then it is not useful.
   (h) Mathematics is fun if and only if it is easy and useful.

2. (a) F (c) F (e) T (g) F
   (b) T (d) F (f) F (h) T

3. (a) Logically false
   (b) Logically true
   (c) Neither logically true nor logically false
   (d) Neither logically true nor logically false.
   (e) Logically true
   (f) Logically true

4. (a) *Converse* If I get wet, then it will be because it rains.
   Inverse If it does not rain, I will not get wet.
   Contrapositive If I do not get wet, it will be because it does not rain.
   (b) *Converse* If I stay indoors, the wind will blow or it will rain.
   Inverse If the wind does not blow and it does not rain, I shall not stay indoors.
   Contrapositive If I do not stay indoors, the wind will not blow and it will not rain.

5. (a) $p \to q$ (b) $q \to p$ (c) $q \to p$ (d) $p \leftrightarrow q$

6. (a) Valid (d) Valid (g) Valid
   (b) Valid (e) Valid (h) Invalid
   (c) Valid (f) Valid (i) Valid

Exercise Set 4.1

1. A

3. A

5. The system is closed relative to $\times$ because the symbols in the body of the table are all elements of S.

7. $(1 \times 4) \times 3 = 4 \times 3 = 5$ and $1 \times (4 \times 3) = 1 \times 6 = 6$. Since $(1 \times 4) \times 3 \neq 1 \times (4 \times 3)$, $\times$ is not an associative operation.

9. Not closed. Is commutative. Not associative because $(3 + 1) + 4 = 8$, but $3 + (1 + 4) = 3 + 5$ and $3 + 5$ is not defined by the table.

11. Closed, commutative, associative

13. (a) Propositions
    (b) $\wedge$ and $\vee$
    (c) $\sim$
    (d) Equivalence

14. (a)

$\sim$	0	1	2	3	4	5
0	0	1	2	3	4	5
1	1	0	1	2	3	4
2	2	1	0	1	2	3
3	3	2	1	0	1	2
4	4	3	2	1	0	1
5	5	4	3	2	1	0

Exercise Set 4.2

1. (a) D is the identity. (b) Each element is its own inverse.

3. (a) # is the identity. (b) Each element is its own inverse.

5. (a) The system is closed relative to #.
   (b) Yes
   (c) $(3 \# 3) \# 4 = 0 \# 4 = 4$ but $3 \# (3 \# 4) = 3 \# 1 = 2$
   (d) 0
   (e) Each element is its own inverse.

7. Identity is 1. Inverse of $\frac{3}{4}$ is $\frac{4}{3}$.

Exercise Set 4.3

1. Yes. The system is closed relative to +; it is commutative and associative. The + identity element is 0, and every element has an inverse relative to +.

3.

+	0	1	2	3	4
0	0	1	2	3	4
1	1	2	3	4	0
2	2	3	4	0	1
3	3	4	0	1	2
4	4	0	1	2	3

×	0	1	2	3	4
0	0	0	0	0	0
1	0	1	2	3	4
2	0	2	4	1	3
3	0	3	1	4	2
4	0	4	3	2	1

The system with operation + is a group. The system with operation × is not a group because 0 has no inverse relative to ×. If 0 is left out, the remaining elements form a group with operation ×.

5.

&	0	1	2	*F*	*G*	*H*
0	0	1	2	*F*	*G*	*H*
1	1	2	0	*H*	*F*	*G*
2	2	0	1	*G*	*H*	*F*
F	*F*	*G*	*H*	0	1	2
G	*G*	*H*	*F*	2	0	1
H	*H*	*F*	*G*	1	2	0

(a) 0 (c) No

7. Yes

Exercise Set 4.4

1. A_1; each of the operations is commutative.
   A_2; the identity for + is a; the identity for $\cdot$ is b.
   A_3; $a' = b$, $b' = a$, $c' = d$, and $d' = c$.
   A_4; each operation distributes over the other.

3. $$\begin{aligned}(a\cdot b\cdot c) + (a'\cdot b'\cdot c') &= (a\cdot b\cdot c) + (b\cdot a\cdot d)\\ &= (a\cdot b)\cdot c + (b\cdot a)\cdot d\\ &= a\cdot c + a\cdot d\\ &= a + a\\ &= a\end{aligned}$$

5. A_4; A_3; A_2

Exercise Set 4.5

1. $(x + y' + z)(yz' + x + y)$
3. $(xy + xz + x'y')(z')(x' + y + z')$
5. $x'(x + y) + (x + y)(x' + y)$
7.

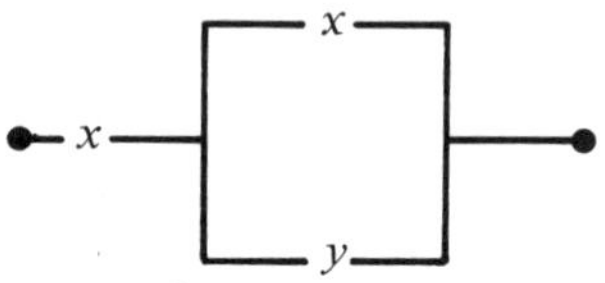

FIGURE A–25

9.

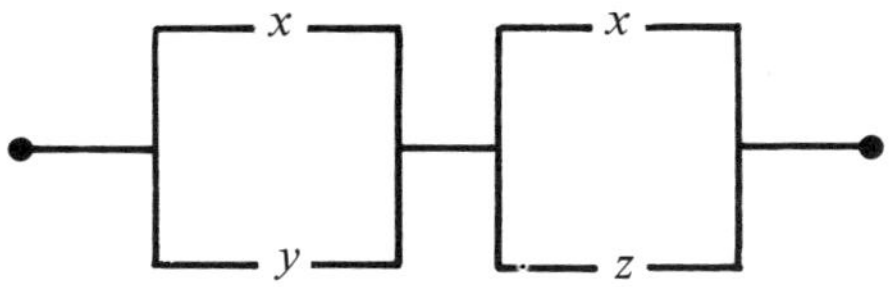

FIGURE A–26

11.

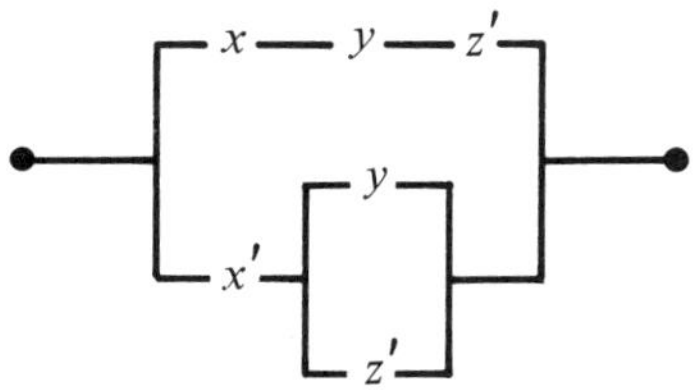

FIGURE A–27

13.

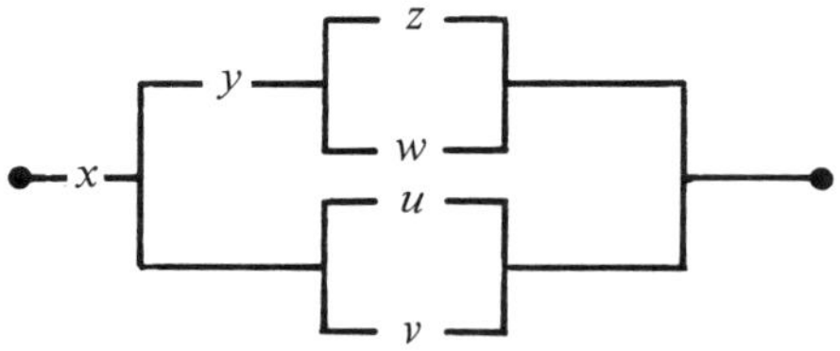

FIGURE A–28

15. x	y	y'	x'	xy'	$x' + y$	$x(x' + y)$	$xy' + x(x' + y)$
1	1	0	0	0	1	1	1
1	0	1	0	1	0	0	1
0	1	0	1	0	1	0	0
0	0	1	1	0	1	0	0

17.

x	w	z	$x + x(wz + x)$
1	1	1	1
1	1	0	1
1	0	1	1
1	0	0	1
0	1	1	0
0	1	0	0
0	0	1	0
0	0	0	0

19.

x	y	$x(x + y)$
1	1	1
1	0	1
0	1	0
0	0	0

21.

x	y	z	$xyz' + x'(y + z')$
1	1	1	0
1	1	0	1
1	0	1	0
1	0	0	0
0	1	1	1
0	1	0	1
0	0	1	0
0	0	0	1

23. $x + xz + x'yz = (x + xz) + x'yz$ (by T_5)
$= x + x'yz$ (since by T_4, $x + xz = x$)

Exercise Set 4.6

1. $xyz + xy'z' + x'yz'$

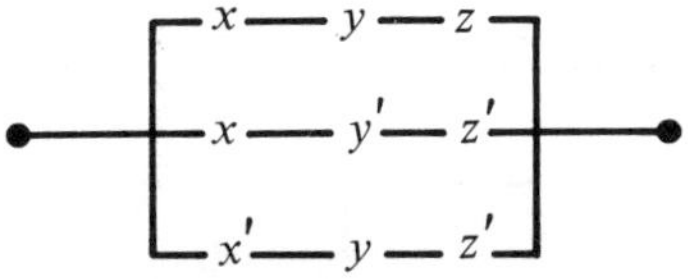

FIGURE A–29

3. $(x' + y' + z)(x + y' + z')(x + y' + z)$

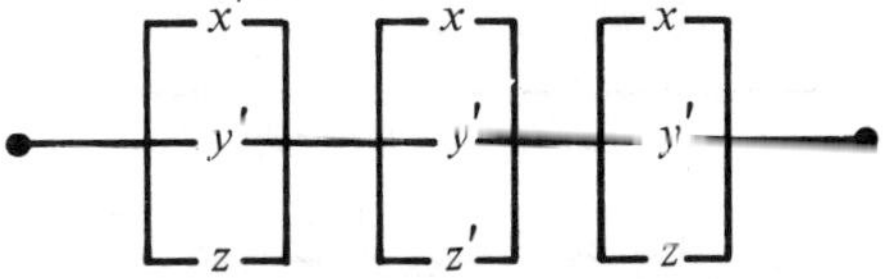

FIGURE A–30

5. $xyz + xy'z' + x'yz' + x'y'z$

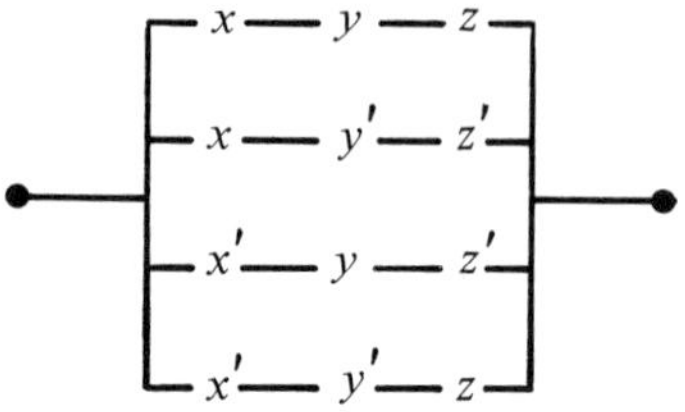

FIGURE A–31

7. $xy' + x'y$

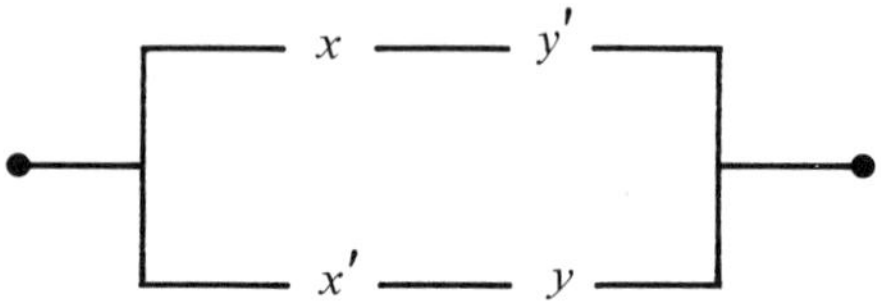

FIGURE A–32

Exercise Set 4.7

1. x
3. No switches at all
5. $x'y$

REVIEW EXERCISES FOR CHAPTER 4

1. (a) $5 \# 4 = 3$ and $4 \# 5 = 3$. No, there are many other pairs which must be tested before one can say that the system is commutative.
   (b) $(3 \# 2) \# 5 = 5$ and $3 \# (2 \# 5) = 3$. The system is not associative.
   (c) 1
   (d) 1 is its own inverse, 4 is its own inverse.
2.

+	I	II	III	IV
I	I	II	III	IV
II	II	I	IV	III
III	III	IV	II	I
IV	IV	III	I	II

This system has all the group properties.

3.

(a)

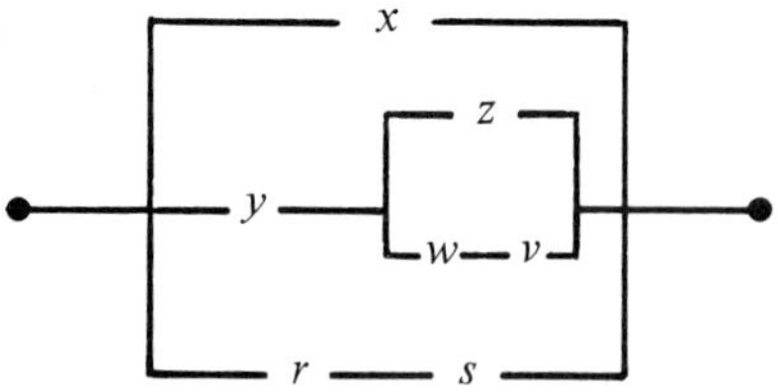

FIGURE A–33

(b)

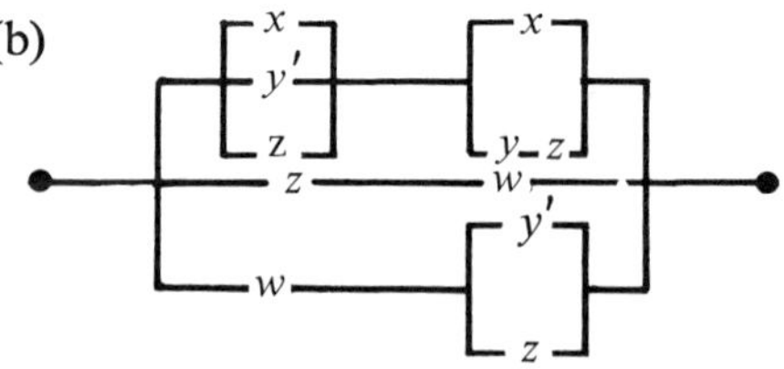

FIGURE A–34

4. (a) $(x + y)[zwv + t(r + s)]$
   (b) $z + v'[w' + x(v + z')]$

5. (a)

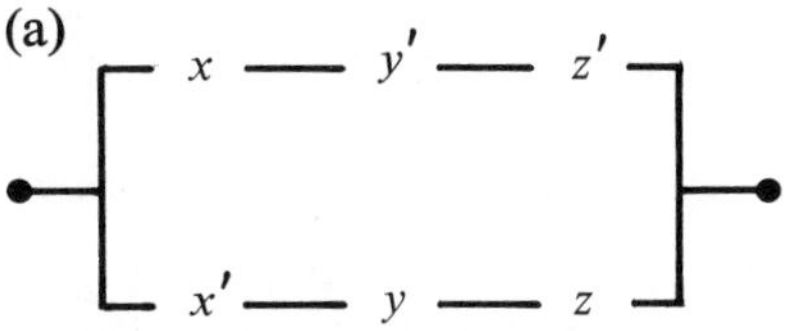

FIGURE A–35

(b)

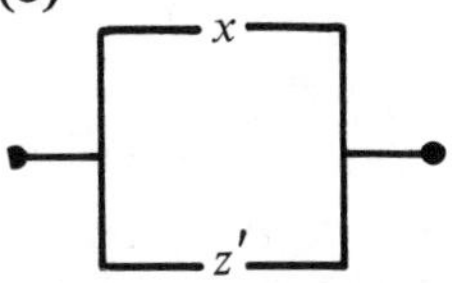

FIGURE A–36

6. (a) $x + yz$ (b) xz

Exercise Set 5.1

1. (a) Integers (c) Rationals (e) Rationals (g) Rationals
2. (a) Rational (c) Rational (e) Irrational
3. (a) -4 (c) 12 (e) $\sqrt{3}$ or $-\sqrt{3}$
4. (a) 0.6 (c) $0.\overline{222}$
5. Integers
7. Irrational because no sequence of digits repeats

Exercise Set 5.2

1. All these propositions are true.
3. (a) $\{\ldots, -3, -2, -1\}$ is not a set of natural numbers.
   (b) $\{1, \frac{1}{2}, \frac{1}{4}, \frac{1}{8}, \ldots\}$ is not a set of natural numbers.
   (c) $\{x$ such that x is a natural number and $x \times 1 \neq x\}$ is empty.

Exercise Set 5.3

1. (a) 6 divides 18 (e) 3 divides 21
   (c) 4 does not divide 26 (g) 8 does not divide 10
3. $96 = 4 \times 24$ $96 = 16 \times 6$
   $96 = 6 \times 16$ $96 = 32 \times 3$
5. 32, 16, 8, 4, 2, 1

6. (a) $q = 2$ and $r = 4$ (c) $q = 1$ and $r = 38$

7. (a) 0, 1, 2, 3, 4 (c) 0, 1, 2, 3, . . ., 22

Exercise Set 5.4

1. In addition to the primes found in Example 5.12, this set contains 101, 103, 107, 109, 113, 127, 131, 137, 139, 149, 151, 157, 163, 167, 173, 179, 181, 191, 193, 197, 199.

3. $14 = 11 + 3$, $16 = 11 + 5$, etc.

5. One possible set is {3, 7, 11, 19, 23, 31, 43, 47}.

Exercise Set 5.5

1. (a) $2 \times 3 \times 5 \times 11$ (e) $2^2 \times 3^2 \times 5^2$
   (c) $2^2 \times 5 \times 7$ (g) $2^2 \times 3^3 \times 5 \times 11$

2. (a) 3 (c) 15 (e) 1

3. (a) $3^3 \times 5 = 135$ (e) $2 \times 3 \times 5 \times 7 \times 17 \times 29 = 103{,}530$
   (c) $3^2 \times 5^2 = 225$

Exercise Set 5.6

1. (a) 4 (c) 0 (e) 4 (g) 3 (i) 0 (k) 1 (m) 4

2.

·	0	1	2	3	4	5
0	0	0	0	0	0	0
1	0	1	2	3	4	5
2	0	2	4	0	2	4
3	0	3	0	3	0	3
4	0	4	2	0	4	2
5	0	5	4	3	2	1

The identity is 1.
Only 1 and 5 have inverses.

3. (a) $3 + 4 + 5 + 6 = 18$ and $18 = 9 \times 2$; therefore 3456 is divisible by 9.
   (c) $3 + 5 + 7 + 3 + 1 = 19$ and 19 is not divisible by 9; therefore 35,731 is not divisible by 9.

REVIEW EXERCISES FOR CHAPTER 5

1. (a) Integers (d) Irrationals
   (b) Integers (e) Rationals
   (c) Rationals (f) Rationals
2. (a) and (c) through (e) are true.
3. (a) to (c) and (f) to (j) are true.
4. $2^2 \times 3 \times 5 \times 17$
5. (a) 2 (b) 2 (c) 10 (d) 2 (e) 4 (f) 2

Exercise Set 6.1

1. $\mathbf{a} = (2, 4, 0, 0, 0)$
3. None of the vectors are equal.
4. (a) Impossible. A column vector cannot be added to a row vector.
   (c) [6, 5]
   (e) $\begin{bmatrix} \frac{7}{3} \\ 3 \end{bmatrix}$
   (g) Impossible. The row vector must be on the left of the column vector.
   (i) Not possible. $\mathbf{f} \cdot \mathbf{g}$ is undefined because both $\mathbf{f}$ and $\mathbf{g}$ are column vectors.
5. $$[3, 7, 2, 10, 4] \begin{bmatrix} 0.25 \\ 0.35 \\ 0.83 \\ 0.04 \\ 0.52 \end{bmatrix} = 7.34$$
6. (a) [5, 5] for each sum.
   (c) [6, 10] for each answer.
   (e) Each answer is 35.
7. (a) Yes (c) Yes

Exercise Set 6.2

1. A is 3×4 C is 5×5 E is 1×4
2. (a) $a_{12} = 3$ (e) $d_{51} = 5$
   (c) b_{43} does not exist since **B** has no third column. (g) $e_{13} = 7$

3. (a) $\begin{bmatrix} 15 & 10 \\ 5 & 0 \\ 20 & 5 \end{bmatrix}$ (c) $\begin{bmatrix} 5 & 3 & 4 \\ 1 & 6 & 2 \end{bmatrix}$ (e) $\begin{bmatrix} 3 & 6 & 8 \\ 6 & 2 & 1 \end{bmatrix}$

4.
$$\begin{array}{c} \\ 1 \\ 2 \\ 3 \\ 4 \end{array} \begin{array}{c} \begin{array}{cccc} 1 & 2 & 3 & 4 \end{array} \\ \begin{bmatrix} 0 & 1 & 0 & 0 \\ 0 & 0 & 1 & 0 \\ 1 & 0 & 0 & 1 \\ 1 & 1 & 0 & 0 \end{bmatrix} \end{array}$$

5. (a) to (c) are true.

6. (a) $\begin{bmatrix} -2 & -3 \\ -1 & -4 \end{bmatrix}$ (c) $\begin{bmatrix} -1 \\ -8 \\ -6 \end{bmatrix}$

7. (a) $\begin{bmatrix} -2 & -4 & -6 \\ -1 & -2 & 0 \end{bmatrix}$ (c) $\begin{bmatrix} 0 & -2 & -3 \\ -1 & 3 & -2 \\ -4 & -8 & 7 \end{bmatrix}$

8. (a) $\begin{bmatrix} 1 & 0 \\ 1 & 0 \end{bmatrix}$ (c) $[-1, 1, 3]$

Exercise Set 6.3

1. (a) $\begin{bmatrix} 2 & 3 \\ 8 & 7 \end{bmatrix}$ (c) $[16, 9]$ (e) $\begin{bmatrix} 3 & 2 \\ 1 & 5 \end{bmatrix}$ (g) 22

2. (a) $\begin{bmatrix} 26 \\ 12 \end{bmatrix}$

3. $\mathbf{A}^2 = \begin{bmatrix} 11 & 4 \\ 8 & 3 \end{bmatrix}$ $\mathbf{A}^3 = \begin{bmatrix} 41 & 15 \\ 30 & 11 \end{bmatrix}$ $\mathbf{A}^4 = \begin{bmatrix} 153 & 56 \\ 112 & 41 \end{bmatrix}$

4. 9 *A*'s, 14 *B*'s, and 13 *C*'s

5. 1; 2

7.
$$\begin{array}{c} \\ A \\ B \\ C \\ D \end{array}\begin{array}{c} \begin{array}{cccc} A & B & C & D \end{array} \\ \begin{bmatrix} 0 & 1 & 1 & 0 \\ 0 & 0 & 0 & 1 \\ 0 & 1 & 0 & 0 \\ 1 & 0 & 1 & 0 \end{bmatrix} \end{array} = \mathbf{D} \qquad \begin{bmatrix} 0 & 1 & 0 & 1 \\ 1 & 0 & 1 & 0 \\ 0 & 0 & 0 & 1 \\ 0 & 2 & 1 & 0 \end{bmatrix} = \mathbf{D}^2$$

$$\mathbf{D} + \mathbf{D}^2 = \begin{bmatrix} 0 & 2 & 0 & 1 \\ 1 & 0 & 1 & 1 \\ 0 & 1 & 0 & 1 \\ 1 & 2 & 2 & 0 \end{bmatrix}$$

Team	Power
A	3
B	3
C	2
D	5

Exercise Set 6.4

1. (a) Inverses (c) Inverses

2. (a) Replace row 2 with row 2 − 2 times row 1.
   (c) Replace row 1 by row 1 − 2 times row 2.

3. (a) Replace row 2 by row 2 − row 1.
   (c) 1. Replace row 1 by row 1 − row 3.
   2. Replace row 2 by row 2 − 2 times row 2.

4. (a) To form *B*: Replace row 1 of *A* by row 1 − row 2.
   To form *C*: Replace row 2 of *B* by row 2 − row 1.
   To form *D*: Replace row 2 of *C* by one-seventh of row 2.
   To form *E*: Replace row 1 of *D* by row 1 + 2 times row 2.

5. If $\mathbf{I} = \begin{bmatrix} 1 & 0 & 0 \\ 0 & 1 & 0 \\ 0 & 0 & 1 \end{bmatrix}$ then $\mathbf{I}\cdot\mathbf{I} = \mathbf{I}$. Therefore **I** is its own inverse.

7. Let **A** and **B** be square matrices such that $\mathbf{AB} = \mathbf{I}$. Then

$$\begin{aligned} & \mathbf{BA} = \mathbf{BA} \\ \text{So}\quad & \mathbf{(BA)B} = \mathbf{(BA)B} \\ \text{or}\quad & \mathbf{(BA)B} = \mathbf{B(AB)} \\ & \mathbf{(BA)B} = \mathbf{BI} \\ & \mathbf{(BA)B} = \mathbf{B} \end{aligned}$$

Then it must be true that **BA** is an identity matrix. However, the identity matrix **I** is unique. Therefore $\mathbf{BA} = \mathbf{I}$.

Exercise Set 6.5

1. $\begin{bmatrix} \frac{3}{10} & \frac{1}{10} \\ \frac{-2}{5} & \frac{1}{5} \end{bmatrix}$

3. $\begin{bmatrix} \frac{1}{8} & \frac{1}{4} \\ \frac{-1}{4} & \frac{1}{2} \end{bmatrix}$

5. $\begin{bmatrix} \frac{3}{4} & \frac{-3}{4} & -1 \\ 0 & 1 & 0 \\ \frac{-1}{2} & \frac{1}{2} & 1 \end{bmatrix}$

7. $\begin{bmatrix} -1 & -2 & 2 \\ 1 & 1 & -1 \\ -2 & 0 & 1 \end{bmatrix}$

9. $\begin{bmatrix} \frac{-1}{19} & \frac{4}{19} & \frac{4}{19} \\ \frac{5}{19} & \frac{-1}{19} & \frac{-1}{19} \\ \frac{-7}{19} & \frac{-10}{19} & \frac{9}{19} \end{bmatrix}$

Exercise Set 6.6

1. (a) $x = 3, y = 7$ (c) $x = -\frac{5}{2}, y = 10, z = 2$

2. (a) $\begin{bmatrix} 1 & 1 \\ 2 & -1 \end{bmatrix}$ $\begin{bmatrix} 3 \\ 4 \end{bmatrix}$

$\begin{bmatrix} 1 & 1 \\ 0 & -3 \end{bmatrix}$ $\begin{bmatrix} 3 \\ -2 \end{bmatrix}$

$\begin{bmatrix} 1 & 1 \\ 0 & 1 \end{bmatrix}$ $\begin{bmatrix} 3 \\ \frac{2}{3} \end{bmatrix}$ $x = \frac{7}{3}, y = \frac{2}{3}$

$\begin{bmatrix} 1 & 0 \\ 0 & 1 \end{bmatrix}$ $\begin{bmatrix} \frac{7}{3} \\ \frac{2}{3} \end{bmatrix}$

(c) $x = 2, y = -4, z = \frac{1}{2}$

3. x barrels of brand A
y barrels of brand B
z barrels of brand C

$$2x + y + 2z = 5$$
$$3x + 3y + 6z = 13$$
$$4x + 5y + z = 9$$
$$x = \frac{2}{3}, y = 1, z = \frac{4}{3}$$

REVIEW EXERCISES FOR CHAPTER 6

1. (a) 3×2 (c) 2×2 (e) This product not possible.
(b) This sum is not possible. (d) 3×3 (f) 1×2

2. (a) $x = 1$ (b) $y = \frac{7}{2}$ (c) $x = 1, y = 0, z = 0, w = 1$
(d) $x = 4, y = 10, z = 5, w = 8$
(e) Not possible. The product should be a row vector.

3. (a) $\begin{bmatrix} 5 \\ 3 \\ 2 \end{bmatrix}$ (b) $\begin{bmatrix} 3 & 6 \\ 9 & -3 \\ 12 & 0 \end{bmatrix}$ (c) [8, 3] (d) $\begin{bmatrix} 5 & 5 \\ 7 & 10 \end{bmatrix}$

(e) $\begin{bmatrix} 4 & 1 & 9 & 5 & 3 \\ 0 & 0 & 0 & 0 & 0 \\ 15 & 6 & 36 & 21 & 9 \\ 5 & 2 & 12 & 7 & 3 \\ 2 & 1 & 5 & 3 & 1 \end{bmatrix}$ (f) **0**

4. (a)

	1	2	3	4	5
1	0	1	0	0	0
2	1	0	1	0	0
3	1	0	0	1	1
4	1	0	0	0	1
5	1	1	1	1	0

(c) $\begin{bmatrix} 2 & 2 & 1 & 1 & 1 \\ 5 & 3 & 3 & 2 & 2 \\ 8 & 5 & 4 & 4 & 4 \\ 6 & 3 & 3 & 2 & 2 \\ 9 & 7 & 5 & 5 & 4 \end{bmatrix}$

(b) $\begin{bmatrix} 1 & 0 & 1 & 0 & 0 \\ 1 & 1 & 0 & 1 & 1 \\ 2 & 2 & 1 & 1 & 1 \\ 1 & 2 & 1 & 1 & 0 \\ 3 & 1 & 1 & 1 & 2 \end{bmatrix}$ (d) 2

5. (a)

	June	July	Aug.
F	$45\frac{1}{4}$	$40\frac{1}{2}$	38
R	$27\frac{3}{4}$	26	$24\frac{1}{2}$
P	$39\frac{1}{2}$	38	36

(b) [100, 200, 80]

(c) [13235, 12290, 11580]

6. (a) Has no inverse (b) $\frac{1}{3}\begin{bmatrix} -1 & 2 & -9 \\ 2 & -1 & 3 \\ 0 & 0 & 3 \end{bmatrix}$

(c) $\frac{1}{5}\begin{bmatrix} 1 & 1 & 1 & -3 \\ 0 & 0 & 0 & 5 \\ -3 & 2 & 3 & -1 \\ 6 & -4 & -1 & 2 \end{bmatrix}$

7. x trucks of type 1 $\quad x + y + 2z = 6$
 y trucks of type 2 $\quad x + 2y + z = 8$
 z trucks of type 3 $\quad 2x + y = 5$
 $x = 1, y = 3, z = 1$

Exercise Set 7.1

1. (a) $\frac{5}{36}$ (c) $\frac{2}{9}$ (e) 0
2. (a) $\frac{1}{2}$ (c) $\frac{2}{13}$ (e) $\frac{1}{26}$ (g) $\frac{1}{4}$
3. $\{(H, H, H), (H, H, T), (H, T, H), (H, T, T), (T, H, H), (T, H, T), (T, T, H), (T, T, T)\}$
4. (a) $\frac{1}{8}$ (c) $\frac{7}{8}$
5. $\{(B, B, B, B), (B, B, B, G), (B, B, G, B), (B, B, G, G), (B, G, B, B), (B, G, B, G), (B, G, G, B), (B, G, G, G), (G, B, B, B), (G, B, B, G), (G, B, G, B), (G, B, G, G), (G, G, B, B), (G, G, B, G), (G, G, G, B), (G, G, G, G)\}$
6. (a) $\frac{1}{4}$ (c) $\frac{1}{4}$
7. $\frac{1}{2}$

Exercise Set 7.2

1. (b), (d), and (e) are mutually exclusive.
2. (a) $\frac{5}{18}$ (c) $\frac{4}{9}$ (e) $\frac{23}{36}$ (g) $\frac{1}{2}$
3. (a) $\frac{3}{13}$ (c) $\frac{1}{13}$ (e) $\frac{19}{22}$ (g) 1
5. (a) $\frac{3}{4}$ (c) 1
6. (a) $\frac{1}{365}$ (c) $\frac{1}{7}$
7. (a) $\frac{2}{3}$ (c) 0 (e) 1
8. (a) $\frac{11}{25}$ (c) $\frac{21}{25}$

Exercise Set 7.3

1. (a) $\frac{2}{11}$ (c) $\frac{1}{3}$ (e) $\frac{2}{5}$
2. (a) $\frac{3}{10}$ (c) $\frac{-2}{10}$
3. (a) $\frac{3}{10}$ (c) $\frac{2}{5}$
4. (a) $\frac{1}{5}$ (c) $\frac{1}{6}$
5. (a) $\frac{1}{16}$ (c) $\frac{1}{16}$

Exercise Set 7.4

1. (a) $\frac{1}{7}$ (c) $\frac{2}{7}$ (e) $\frac{3}{7}$
2. (a) $\frac{9}{49}$ (c) $\frac{12}{49}$ (e) $\frac{25}{49}$
3. (a) $P(A \cap \text{red}) + P(B \cap \text{red}) + P(C \cap \text{red}) = (\frac{1}{3})(\frac{3}{4}) + (\frac{1}{3})(\frac{1}{2}) + (\frac{1}{3})(\frac{1}{4}) = \frac{1}{2}$
4. (a) $\frac{1}{49}$ (c) $\frac{12}{49}$
5. $\frac{2}{5}$
6. (a) 0.195112 (c) 0.074088
7. $\frac{71}{360}$
8. (a) $\frac{7}{102}$ (c) $\frac{5}{51}$

Exercise Set 7.5

1. (a) $\frac{1}{7}$ (c) $\frac{1}{12}$
2. (a) $\frac{11}{1}$ (c) $\frac{11}{2}$
3. $\frac{1}{34}$
5. About \$4.67
7. \$300
9. (98% of \$6) − \$3.80 = \$2.08

REVIEW EXERCISES FOR CHAPTER 7

1. (1, 1), (1, 2), (1, 3), (1, 4), (2, 1), (2, 2), (2, 3), (2, 4), (3, 1), (3, 2), (3, 3), (3, 4), (4, 1), (4, 2), (4, 3), (4, 4)

   (a) $\frac{1}{16}$ (b) $\frac{1}{4}$ (c) $\frac{3}{8}$ (d) $\frac{7}{8}$
2. $(1, r)$, $(1, w)$, $(1, b)$, $(2, r)$, $(2, w)$, $(2, b)$, $(3, r)$, $(3, w)$, $(3, b)$, $(4, r)$, $(4, w)$, $(4, b)$, $(5, r)$, $(5, w)$, $(5, b)$, $(6, r)$, $(6, w)$, $(6, b)$

   (a) $\frac{1}{18}$ (b) $\frac{1}{6}$ (c) $\frac{4}{9}$ (d) $\frac{2}{3}$ (e) $\frac{2}{9}$
3. $\frac{9}{10}$
4. $\frac{1}{150}$; $\frac{149}{150}$
5. $\frac{1}{2}$
6. (a) $\frac{5}{12}$ (b) $\frac{2}{3}$ (c) $\frac{3}{4}$ (d) $\frac{3}{7}$

7. (a) $\frac{1}{3}$ (b) $\frac{1}{2}$ (c) $\frac{2}{3}$ (d) 1 (e) 0 (f) $\frac{1}{2}$ (g) $\frac{1}{6}$ (h) $\frac{2}{3}$

8. $\frac{1}{6}$

9. $\frac{1}{8}$

10. $\frac{1}{4}$; $\frac{1}{2}$; $\frac{1}{4}$; \$3

11.

Subject	Expected Value
Mathematics	\$0.40 loss
Chemistry	0.00
History	1.90 gain
English	0.00
P.E.	3.80 gain
Total	\$5.30

12.

Model		Expected Value
I	15,000(0.90) − 10,000 =	\$3,500
II	24,000(0.80) − 18,000 =	1,200
III	30,000(0.75) − 20,000 =	2,500
IV	60,000(0.65) − 30,000 =	9,000

The corporation should build type IV aircraft.

Exercise Set 8.1

1. (a) Mode is 8, median is 8, and mean is 7.5. Any one of these is a good representative.

   (c) Mode is 8, median is 8, and mean is 5.2. Either the median or the mode should be used to represent the sample.

2. (a)

Data point	Frequency
9	2
8	4
7	2
6	1
5	1

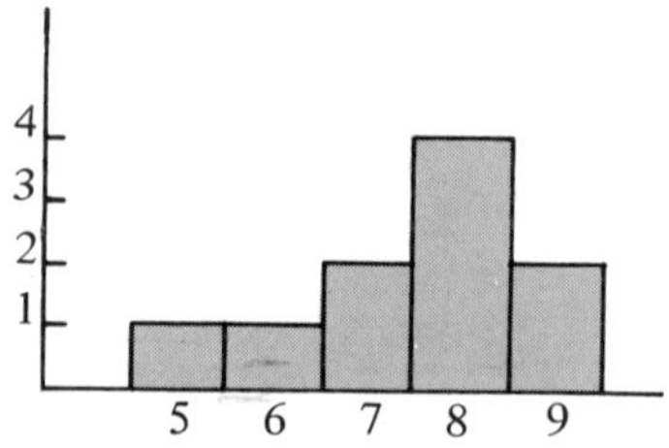

FIGURE A–37

(c)

Data point	Frequency
10	1
9	2
8	5
7	1
6	1
5	0
4	1
3	0
2	1
1	1

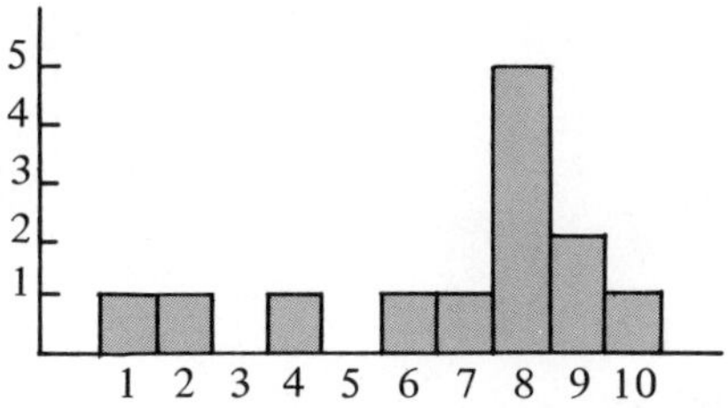

FIGURE A–38

3. (a) 22 (c) 79 (e) 484

5. (a) 81 (c) 81 (e) 60%

7. 1500

Exercise Set 8.2

1. (a) $\sqrt{1.55}$ or about 1.2 (c) $\sqrt{25.88}$ or about 5.1

3. $\bar{x} = 86.1$, $s = \sqrt{46.05}$ or about 6.8

5. $\bar{x} = 152.5$ hours, $s = \sqrt{5169}$ or about 72 hours

Exercise Set 8.3

1. (a) to (c) are binomial experiments.

3. (a)

Number of Heads x	Number of Ways x Can Occur	$P(x)$
0	1	$\frac{1}{16}$
1	4	$\frac{4}{16}$
2	6	$\frac{6}{16}$
3	4	$\frac{4}{16}$
4	1	$\frac{1}{16}$

(b)

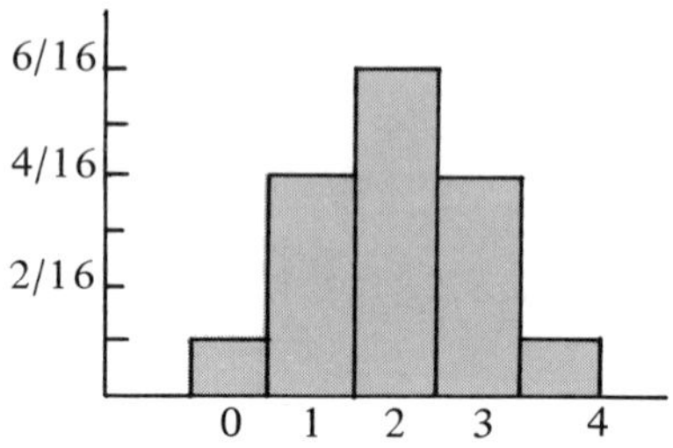

FIGURE A–39

(c) $\frac{5}{16}$
(d) $\frac{10}{16}$

5. $\frac{15}{32}$

7. About 0.253

9. (a) $\mu = 50 \quad \sigma = 5$
   (c) $\mu = 120 \quad \sigma = 6.9$

Exercise Set 8.4

1. (a) 15.5% (c) 47.9%
2. (a) 49.1% (c) 63.2% (e) 0%
3. 1220 pairs
5. (a) 0.309 (c) 0.045
7. 0.006

Exercise Set 8.5

1. More than 15
3. Reject the null hypothesis. She could guess as well only 0.6% of the time.
5. 10 or more

Exercise Set 8.6

1. (a) Positive (c) Positive (e) Positive

2. (a)

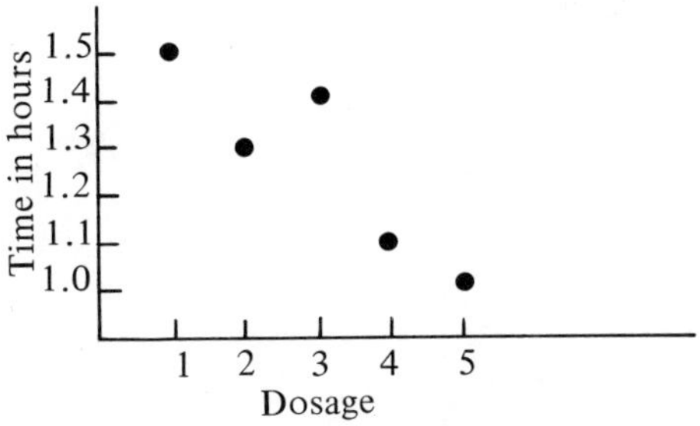

FIGURE A–40

3. (a)

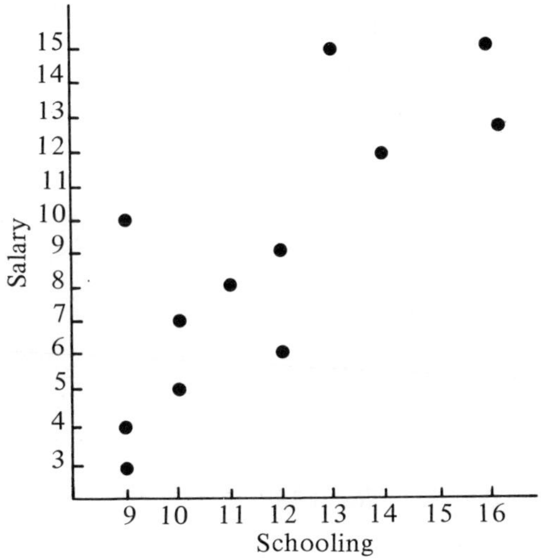

FIGURE A–41

(b) $r = 0.86$

5. (a) About 68 inches (b) About 69 inches

REVIEW EXERCISES FOR CHAPTER 8

1.

Speed	Frequency	Speed	Frequency
85	1	55	0
80	2	50	2
75	3	45	0
70	3	40	0
65	6	35	1
60	2		

Mean = 66.25
Median = 65
Mode = 65

2. Mean = 71.57, median = 71, mode = 71

3.

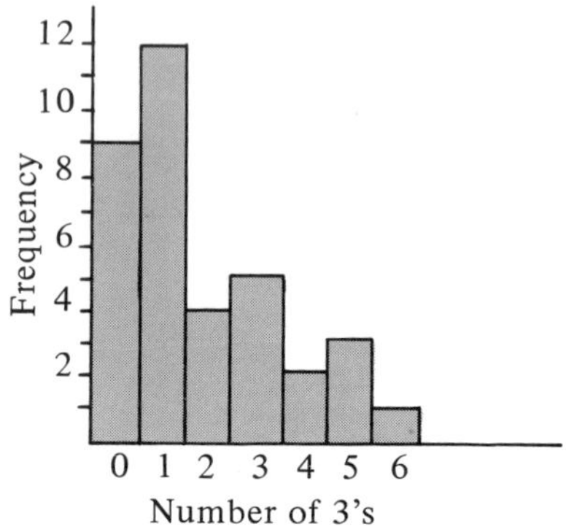

FIGURE A–42

4. $\bar{x} = 85.6$; $s = 5.8$; 12

5. (b)

6. (a) $\frac{1}{125}$ (b) $\frac{64}{125}$ (c) $\frac{61}{125}$

7. 0.3087

8. (a) 0.933 (b) 0.067 (c) 0.866

9. About 2355

10. $\mu = 12.5$ $\sigma = 3$
    (a) 0.212 (b) 0.006 (c) 0.788 (d) 0.242

11. $\mu = 120$, $\sigma = 9.2$; almost 0

12. If only 40 percent of the students smoke, one can expect to find 218 or more smokers in a sample of 500 students 6.7 percent of the time. Therefore the hypothesis of more than 40% smokers should be rejected at the 5 percent level.

13. If the probability that a right-handed person puts his left shoe on first is only $\frac{1}{2}$, then as many as 16 out of 20 right-handed people will put on their left shoe first in 1.8 percent of the samples. Therefore we may reject the hypothesis of chance ($p = \frac{1}{2}$) at any significance level greater than 1.8 percent.

14. (a)

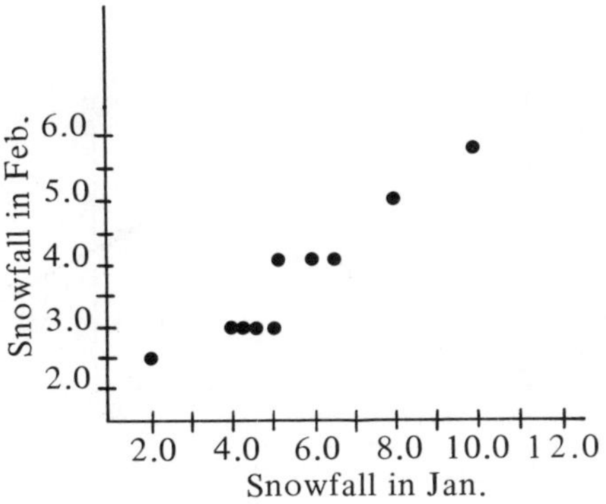

FIGURE A–43

(b) $r = 0.99$

Exercise Set 9.1

1. x = pounds of A
   y = pounds of B
   $x \geq 0, y \geq 0, 5x + 6y \leq 50, 3x + 2y \leq 30$
   Maximize $150x + 200y$

3. x = number of pounds of A
   y = number of pounds of B
   $x \geq 0, y \geq 0, 3x + y \geq 10, 6x + 5y \geq 13$
   Minimize $0.80x + 0.60y$

5. x = number of refrigerators
   y = number of stoves
   $x \geq 0, 0 \leq y \leq 50, x + y \leq 100, 200x + 100y \leq 12{,}000$
   Minimize $35x + 20y$

Exercise Set 9.2

1.

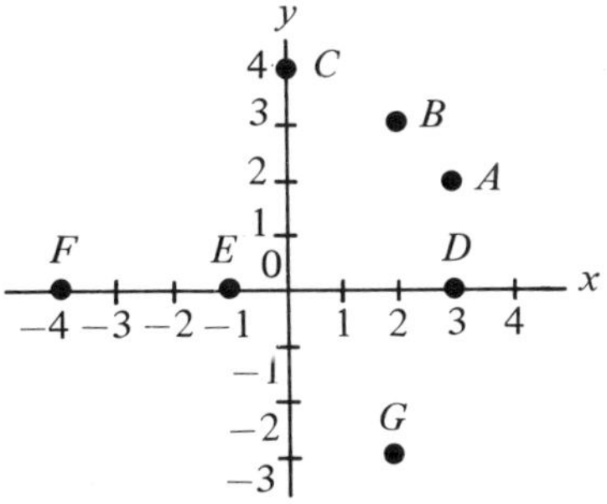

FIGURE A–44

2. (a)

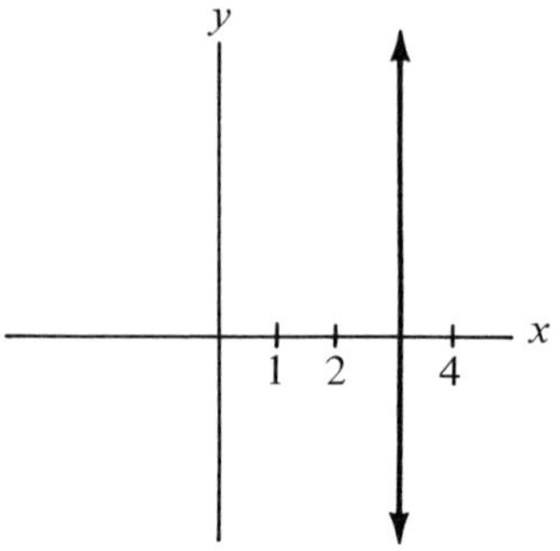

FIGURE A–45

(c)

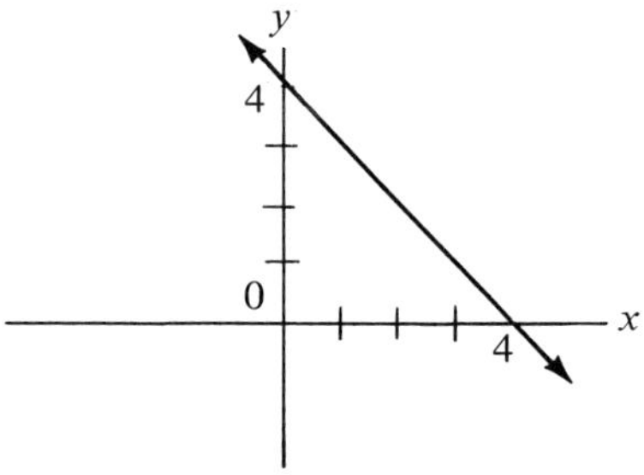

FIGURE A–46

(e)

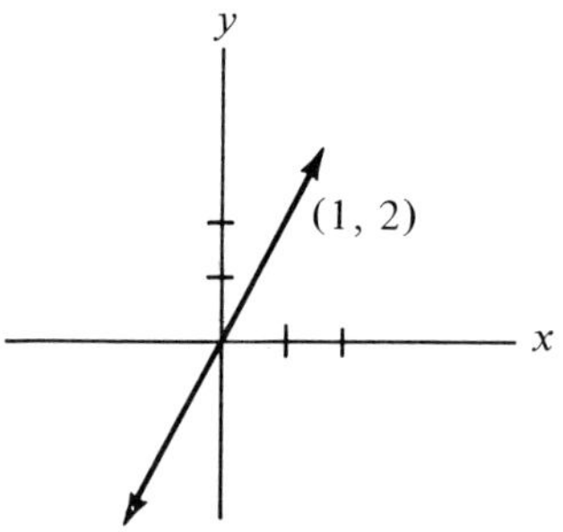

FIGURE A–47

(g)

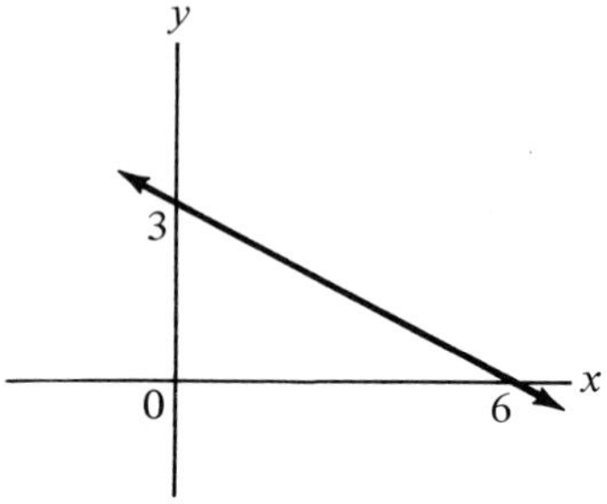

FIGURE A–48

(i)

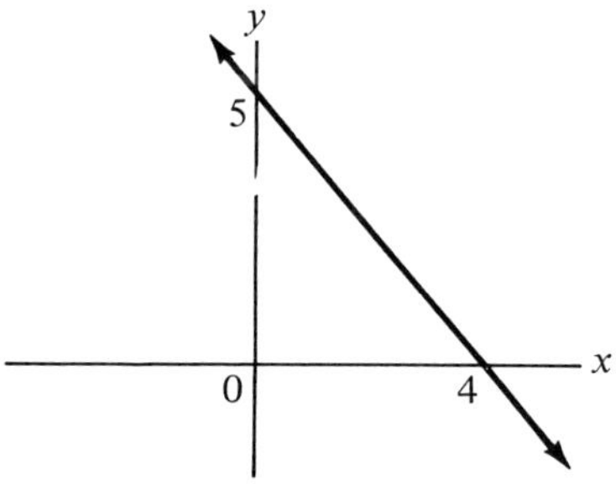

FIGURE A–49

3. (a) (3, 4) (b) (2, 2) (c) (1, 1)

Exercise Set 9.3

1.

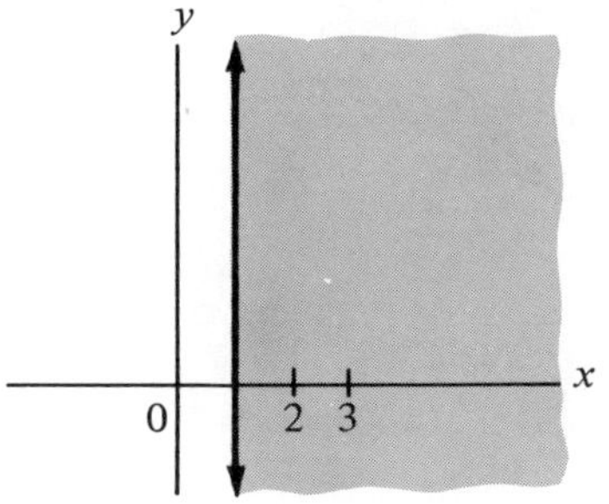

FIGURE A–50

3.

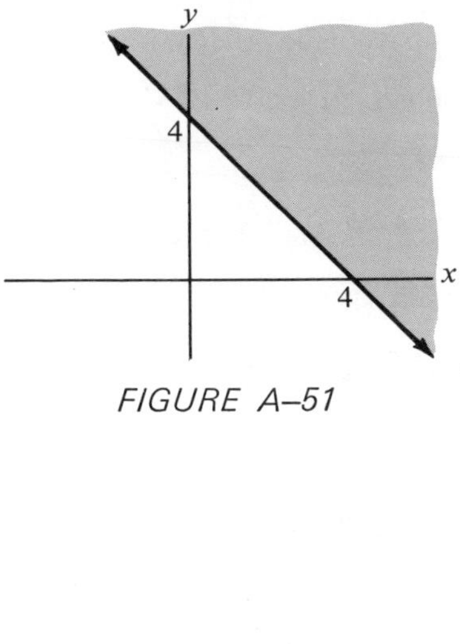

FIGURE A–51

5.

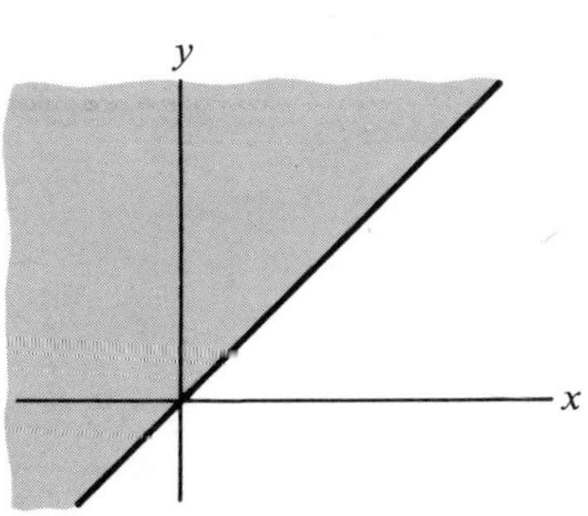

FIGURE A–52

7.

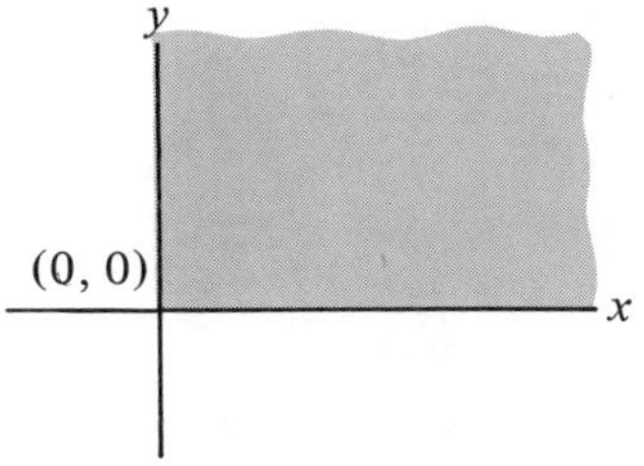

FIGURE A–53

9.

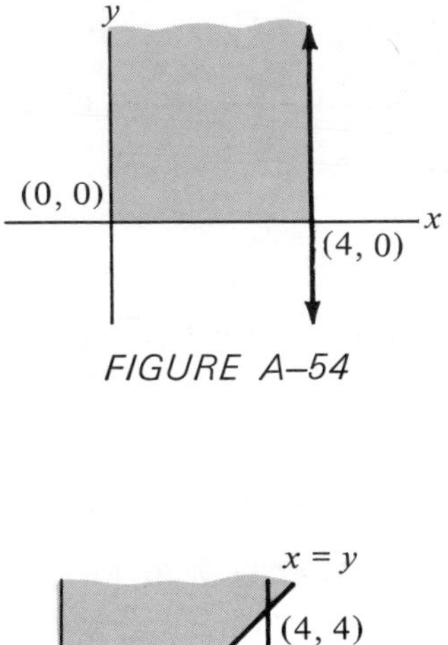

FIGURE A–54

11.

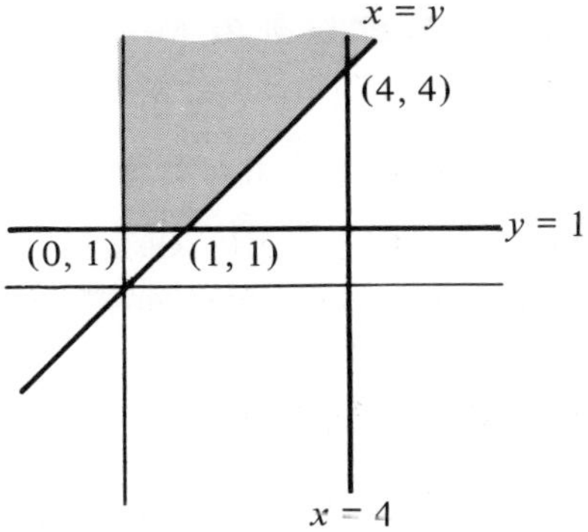

FIGURE A–55

13.

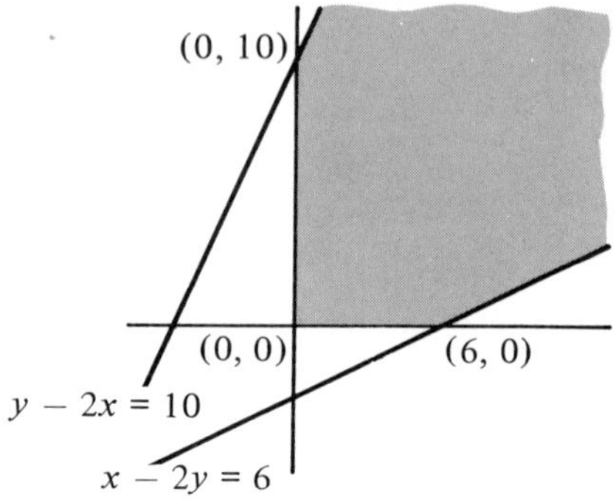

FIGURE A–56

Exercise Set 9.4

1. (b) (0, 1), (1, 1), (4, 1) (c) 7 at (4, 1) (d) −2 at (0, 1)
3. (b) (0, 1), (4, 1), (4, 2), (0, 3) (c) 140 at (4, 2) (d) 30 at (0, 1)
5. (b) (0, 4), (4, 0), (4.5, 0), (3, 3) (c) 1.65 at (3, 3) (d) 1 at (4, 0)
7. $8\frac{1}{3}$ pounds of vitamin B. Profit is \$1666.67.
9. $3\frac{1}{3}$ pounds of *A*. Cost is \$2.67.
11. 35 refrigerators and 50 stoves. Profit is \$2225.

REVIEW EXERCISES FOR CHAPTER 9

1. 18 at (2, 6)
2. 19 at any point on the line segment joining (5, 3) to (9.5, 0).
3. Maximum of 18 at (0, 6). Minimum of −48 at (−12, −14).
4. Maximum of 27 at (12, 0). Minimum of 11 at (3, 2).
5. Two type *A*'s and 6 type *B*'s. 1800 gallons of fuel.
6. Program *A* should be shown 2 times, and program *B* should be shown 4 times. Maximum number of viewers is 100,000.

Exercise Set 10.1

1. *R* play row 1, *C* play column 2; $v = 1$
3. *R* play row 1, *C* play column 2; $v = 3$

5. R play row 3, C play column 2; $v = 0$

7.
$$\begin{array}{c} \\ 1 \\ 2 \\ 3 \end{array}\begin{array}{c} \begin{array}{ccc} 1 & 2 & 3 \end{array} \\ \begin{bmatrix} 0 & -1 & -2 \\ 1 & 0 & -1 \\ 2 & 1 & 0 \end{bmatrix} \end{array}$$

Both players extend three fingers; $v = 0$

Exercise Set 10.2

1. (a) Strictly determined; $\mathbf{p} = [0, 1]$, $\mathbf{q} = \begin{bmatrix} 1 \\ 0 \end{bmatrix}$, $v = 2$

   (c) Not strictly determined.

   (e) Strictly determined; $\mathbf{p} = [1, 0]$, $\mathbf{q} = \begin{bmatrix} 1 \\ 0 \end{bmatrix}$, $v = 0$

   (g) Strictly determined; $\mathbf{p} = [0, 1, 0]$, $\mathbf{q} = \begin{bmatrix} 0 \\ 1 \\ 0 \end{bmatrix}$, $v = 3$.

2. (a) $[2, 2]$ (c) $[11, 11]$ (e) $[\frac{1}{2}, \frac{1}{2}]$ (g) $\frac{1}{2}$

3. (a) $\geq$ (c) $\leq$ (e) $\geq$

Exercise Set 10.3

1. $\mathbf{p} = [\frac{7}{9}, \frac{2}{9}]$, $\mathbf{q} = \begin{bmatrix} \frac{4}{9} \\ \frac{5}{9} \end{bmatrix}$, $v = \frac{37}{9}$

3. $\mathbf{p} = [0, 1]$, $\mathbf{q} = \begin{bmatrix} 1 \\ 0 \end{bmatrix}$, $v = 3$

5. $\mathbf{p} = [\frac{2}{5}, \frac{3}{5}]$, $\mathbf{q} = \begin{bmatrix} \frac{4}{5} \\ \frac{1}{5} \end{bmatrix}$, $v = \frac{12}{5}$

7. $\mathbf{p} = [0, 1]$, $\mathbf{q} = \begin{bmatrix} 1 \\ 0 \end{bmatrix}$, $v = 9$

9. $\mathbf{p} = [\frac{1}{2}, \frac{1}{2}]$, $\mathbf{q} = \begin{bmatrix} \frac{1}{2} \\ \frac{1}{2} \end{bmatrix}$, $v = \frac{1}{2}$

11.
$$\begin{array}{c} \\ 1 \\ 2 \end{array}\begin{array}{c} \begin{array}{cc} 1 & 2 \end{array} \\ \begin{bmatrix} 2 & -3 \\ -3 & 4 \end{bmatrix} \end{array} \qquad \mathbf{p} = [\tfrac{7}{12}, \tfrac{5}{12}], \quad \mathbf{q} = \begin{bmatrix} \frac{7}{12} \\ \frac{5}{12} \end{bmatrix}, \quad v = -\tfrac{1}{12}$$

13. (a) Toss three coins. Play row 1 if all three are heads; play row 2 otherwise.
 (c) Roll one die. Play row 1 if it lands 1 or 2; play row 3 if it lands 3, 4, 5, or 6.
 (e) Toss two coins. If they land both heads, play row 1. If they land one head and one tail, play row 2. If they land both tails, play row 3.

Exercise Set 10.4

1. $\mathbf{p} = [0, \tfrac{1}{2}, 0, \tfrac{1}{2}]$, $\mathbf{q} = \begin{bmatrix} \frac{3}{4} \\ \frac{1}{4} \end{bmatrix}$, $v = \tfrac{3}{2}$

3. $\mathbf{p} = [\tfrac{4}{5}, \tfrac{1}{5}]$, $\mathbf{q} = \begin{bmatrix} \frac{2}{5} \\ 0 \\ \frac{3}{5} \end{bmatrix}$, $v = \tfrac{17}{5}$

5. $\mathbf{p} = [1, 0]$, $\mathbf{q} = \begin{bmatrix} 0 \\ 0 \\ 1 \\ 0 \end{bmatrix}$, $v = 4$

7. $\mathbf{p} = [0, 0, 0, 1]$, $\mathbf{q} = \begin{bmatrix} 1 \\ 0 \end{bmatrix}$, $v = 10$

9. $\mathbf{p} = [\tfrac{1}{4}, \tfrac{3}{4}]$, $\mathbf{q} = \begin{bmatrix} 0 \\ \frac{1}{2} \\ \frac{1}{2} \\ 0 \\ 0 \end{bmatrix}$, $v = \tfrac{3}{2}$

11. $\mathbf{p} = [\tfrac{1}{5}, \tfrac{4}{5}]$, $\mathbf{q} = \begin{bmatrix} \frac{2}{5} \\ \frac{3}{5} \\ 0 \\ 0 \end{bmatrix}$, $v = \tfrac{18}{5}$

13.

$$\begin{array}{r} \\ \\ \text{Black 7} \\ \text{Black 5} \\ \text{Red 3} \end{array}\begin{array}{c} \begin{array}{cc} \text{Black} & \text{Red} \\ 5 & 7 \end{array} \\ \begin{bmatrix} 2 & 0 \\ 0 & -2 \\ -2 & 4 \end{bmatrix} \end{array} \qquad \mathbf{p} = [\tfrac{3}{4}, 0, \tfrac{1}{4}], \quad \mathbf{q} = \begin{bmatrix} \frac{1}{2} \\ \frac{1}{2} \end{bmatrix}, \quad v = 1$$

15.

$$\begin{array}{r} \\ 1 \\ 2 \\ 3 \end{array}\begin{array}{c} \begin{array}{cc} \text{Odd} & \text{Even} \end{array} \\ \begin{bmatrix} -1 & 1 \\ 2 & -2 \\ -3 & 3 \end{bmatrix} \end{array} \qquad \mathbf{p} = [0, \tfrac{3}{5}, \tfrac{2}{5}], \quad \mathbf{q} = \begin{bmatrix} \frac{1}{2} \\ \frac{1}{2} \end{bmatrix}, \quad v = 0$$

Exercise Set 10.5

1.

(a) $\mathbf{p} = [\tfrac{3}{8}, \tfrac{5}{8}], \quad \mathbf{q} = \begin{bmatrix} \frac{1}{2} \\ 0 \\ 0 \\ \frac{1}{2} \end{bmatrix}, \quad v = \tfrac{7}{2}$

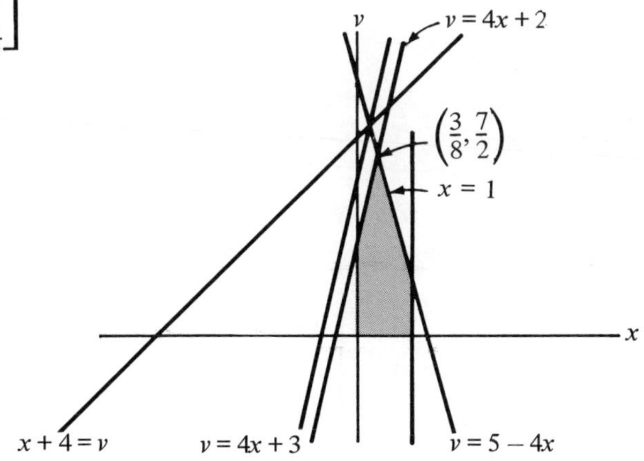

FIGURE A–57

(c) $\mathbf{p} = [\tfrac{1}{2}, \tfrac{1}{2}], \quad \mathbf{q} = \begin{bmatrix} 0 \\ 1 \\ 0 \end{bmatrix}, \quad v = 2$

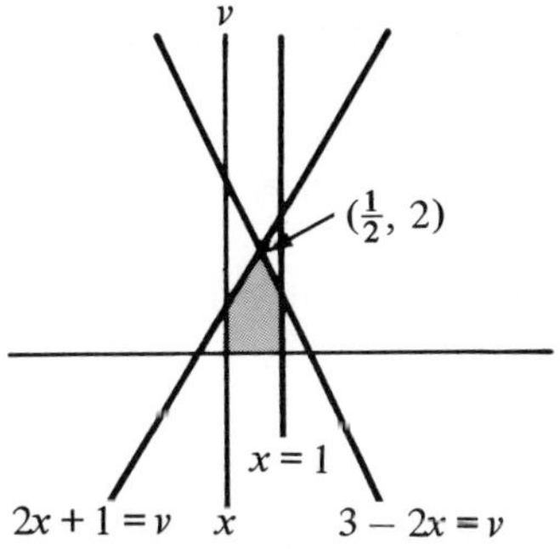

FIGURE A–58

2. (a) $\mathbf{p} = [\frac{4}{5}, \frac{1}{5}, 0]$, $\mathbf{q} = \begin{bmatrix} 0 \\ \frac{1}{5} \\ \frac{4}{5} \end{bmatrix}$, $v = \frac{9}{5}$

(c) $\mathbf{p} = [0, 0, 1]$, $\mathbf{q} = \begin{bmatrix} 0 \\ 0 \\ 1 \\ 0 \end{bmatrix}$, $v = 3$

(e) $\begin{array}{c|cc} & \frac{1}{2} & \frac{1}{4} \\ \hline \frac{1}{2} & \frac{1}{2} & 0 \\ \frac{1}{4} & 0 & \frac{1}{4} \end{array}$ $\mathbf{p} = [\frac{1}{3}, \frac{2}{3}]$, $\mathbf{q} = \begin{bmatrix} \frac{1}{3} \\ \frac{2}{3} \end{bmatrix}$, $v = \$\frac{1}{6}$

(g) $\begin{array}{c|ccc} & 1 & 2 & 3 \\ \hline 1 & -1 & -\frac{3}{2} & -3 \\ 2 & -1 & -2 & -\frac{5}{2} \end{array}$ $\mathbf{p} = [0, 1, 0]$, $\mathbf{q} = \begin{bmatrix} 0 \\ 0 \\ 1 \end{bmatrix}$, $v = -\frac{5}{2}$

REVIEW EXERCISES FOR CHAPTER 10

1. $\mathbf{p} = [1, 0]$, $\mathbf{q} = \begin{bmatrix} 0 \\ 1 \end{bmatrix}$, $v = 3$

2. $\mathbf{p} = [0, 1, 0]$, $\mathbf{q} = \begin{bmatrix} 0 \\ 1 \\ 0 \end{bmatrix}$, $v = 5$

3. $\mathbf{p} = [0, 0, 1, 0]$, $\mathbf{q} = \begin{bmatrix} 1 \\ 0 \\ 0 \\ 0 \end{bmatrix}$, $v = 1$

4. $\mathbf{p} = [1, 0]$, $\mathbf{q} = \begin{bmatrix} 0 \\ 1 \\ 0 \end{bmatrix}$, $v = -2$

5. $\mathbf{p} = [\frac{3}{4}, \frac{1}{4}]$, $\mathbf{q} = \begin{bmatrix} \frac{3}{4} \\ \frac{1}{4} \\ 0 \end{bmatrix}$, $v = \frac{3}{4}$

6. $\mathbf{p} = [1, 0], \quad \mathbf{q} = \begin{bmatrix} 0 \\ 1 \\ 0 \\ 0 \\ 0 \end{bmatrix}, \quad v = 0$

7. $\mathbf{p} = [\frac{1}{2}, \frac{1}{2}, 0], \quad \mathbf{q} = \begin{bmatrix} \frac{3}{4} \\ \frac{1}{4} \end{bmatrix}, \quad v = \frac{7}{2}$

8. $\mathbf{p} = [\frac{1}{3}, \frac{2}{3}, 0], \quad \mathbf{q} = \begin{bmatrix} \frac{1}{2} \\ 0 \\ \frac{1}{2} \end{bmatrix}, \quad v = 2$

9.

	A-H	A-C	2-S
A-S	−1	0	2
A-D	0	1	−1
2-H	−2	−2	2

$\mathbf{p} = [\frac{1}{4}, \frac{3}{4}, 0], \quad \mathbf{q} = \begin{bmatrix} \frac{3}{4} \\ 0 \\ \frac{1}{4} \end{bmatrix}, \quad v = -\frac{1}{4}$

10. Ex-Girl

Al	J	F
J	0	7
F	5	0

$\mathbf{p} = [\frac{5}{12}, \frac{7}{12}], \quad \mathbf{q} = \begin{bmatrix} \frac{7}{12} \\ \frac{5}{12} \end{bmatrix}, \quad v = \frac{35}{12}$

11. Small Company

Large Company	A	B	C	D
A	60	48	56	64
B	72	60	64	68
C	64	56	60	72
D	56	52	48	60

Both should build at town B. The large company will get 60 percent of the business.

12.

	Black 8	Red 4	Red 3
Black 9	−1	5	6
Black 5	−3	1	1
Red 7	1	−3	−4
Red 3	5	−1	0

$\mathbf{p} = (\frac{1}{2}, 0, 0, \frac{1}{2})$,

$\mathbf{q} = \begin{bmatrix} \frac{1}{2} \\ 0 \\ \frac{1}{2} \end{bmatrix}, \quad v = 2$

Index

1 2 3 4 5 6 7 8 9 10